Quantum Gravity

Mathematical Models and Experimental Bounds

Bertfried Fauser
Jürgen Tolksdorf
Eberhard Zeidler
Editors

Birkhäuser Verlag
Basel · Boston · Berlin

Editors:

Bertfried Fauser
Jürgen Tolksdorf
Eberhard Zeidler
Max-Planck-Institut für Mathematik in den
Naturwissenschaften
Inselstrasse 22–26
D-04103 Leipzig
Germany
e-mail: fauser@mis.mpg.de
        Juergen.Tolksdorf@mis.mpg.de
        EZeidler@mis.mpg.de

2000 Mathematical Subject Classification: primary 81-02; 83-02; secondary: 83C45;
81T75; 83D05; 83E30; 83F05

Library of Congress Control Number: 2006937467

Bibliographic information published by Die Deutsche Bibliothek
Die Deutsche Bibliothek lists this publication in the Deutsche Nationalbibliografie;
detailed bibliographic data is available in the Internet at <http://dnb.ddb.de>.

ISBN 978-3-7643-7977-3 Birkhäuser Verlag, Basel – Boston – Berlin

© 2007 Birkhäuser Verlag, P.O. Box 133, CH-4010 Basel, Switzerland
Part of Springer Science+Business Media
Printed on acid-free paper produced from chlorine-free pulp. TCF ∞
Cover design: Heinz Hiltbrunner, Basel
Printed in Germany
ISBN-10: 3-7643-7977-4   e-ISBN-10: 3-7643-7978-2
ISBN-13: 978-3-7643-7977-3   e-ISBN-13: 978-3-7643-7978-0

9 8 7 6 5 4 3 2 1   www.birkhauser.ch

CONTENTS

Preface

This Edited Volume is based on a workshop on "Mathematical and Physical Aspects of Quantum Gravity" held at the *Heinrich-Fabri* Institute in Blaubeuren (Germany) from July 28th to August 1st, 2005. This workshop was the successor of a similar workshop held at the same place in September 2003 on the issue of "Mathematical and Physical Aspects of Quantum Field Theories". Both workshops were intended to bring together mathematicians and physicists to discuss profound questions within the non-empty intersection of mathematics and physics. The basic idea of this series of workshops is to cover a broad range of different approaches (both mathematical and physical) to a specific subject in mathematical physics. The series of workshops is intended, in particular, to discuss the basic conceptual ideas behind different mathematical and physical approaches to the subject matter concerned.

The workshop on which this volume is based was devoted to what is commonly regarded as the biggest challenge in mathematical physics: the "quantization of gravity". The gravitational interaction is known to be very different from the known interactions like, for instance, the electroweak or strong interaction of elementary particles. First of all, to our knowledge, any kind of energy has a gravitational coupling. Second, since Einstein it is widely accepted that gravity is intimately related to the structure of space-time. Both facts have far reaching consequences for any attempt to develop a quantum theory of gravity. For instance, the first fact questions our understanding of "quantization" as it has been developed in elementary particle physics. In fact, this understanding is very much related to the "quantum of energy" encountered in the concept of photons in the quantum theory of electromagnetism. However, in Einstein's theory of gravity the gravitational field does not carry (local) energy. While general relativity is a local theory, the notion of gravitational energy is still a "global" issue which, however, is not yet well-defined even within the context of classical gravity. The second fact seems to clearly indicate that a quantum theory of gravity will radically chance our ideas about the structure of space-time. It is thus supposed that a quantum version of gravity is deeply related to our two basic concepts: "quantum" and "space-time" which have been developed so successfully over the last one-hundred years.

The idea of the second workshop was to provide a forum to discuss different approaches to a possible theory of quantum gravity. Besides the two major accepted roads provided by String Theory and Loop Quantum Gravity, also other ideas were discussed, like those, for instance, based on A. Connes' non-commutative geometry. Also, possible experimental evidence of a quantum structure of gravity was discussed. However, it was not intended to cover the latest technical results but instead to summarize some of the basic features of the existing ansätze to formulate a quantum theory of gravity. The present volume provides an appropriate cross-section of the discussion. The refereed articles are written with the intention to bring together experts working in different fields in mathematics and physics and who are interested in the subject of quantum gravity. The volume provides the reader with some overview about most of the accepted approaches to develop a quantum gravity theory. The articles are purposely written in a less technical style than usual and are mainly intended to discuss the major questions related to the subject of the workshop.

Since this volume covers rather different perspectives, the editors thought it might be helpful to start the volume by providing a brief summary of each of the various articles. Obviously, such a summary will necessarily reflect the editors' understanding of the subject matter.

The volume starts with an overview, presented by **Claus Kiefer**, on the main roads towards a quantum theory of relativity. The chapter is nontechnical and comprehensibly written. The author starts his article with a brief motivation why there is a need to consider a quantum theory of gravity. Next, he discusses several aspects of the different approaches presented in this volume. For instance, he contrasts background independent approaches with background dependent theories. The chapter closes with a brief summary of some of the main results obtained so far to achieve a quantum version of gravity.

In the second article **Claus Lämmerzahl** reports on the experimental status of quantum gravity effects. On the one hand, gravity is assumed to be "universal". On the other hand, quantum theory is regarded as being "fundamental". As a consequence, one should expect that a quantum theory of gravity will yield corrections to any physical process. Recent experiments, however, confirm with high precision the theory of relativity and quantum theory. Nonetheless, the author indicates how astrophysical as well as laboratory and satelite experiments may be improved in accuracy so that possible quantum gravity effects could be observed not too far in the future. Lämmerzahl describes at which scales and parameter ranges it could be more promising to push forward experimental efforts. The article closes with a discussion on recent proposals to increase experimental accuracy. A wealth of information is presented in a very readable style.

In their contribution the authors **Alfredo Macias** and **Hernando Quevedo** review in a very precise and compelling way the role of time in the process of (canonical) quantization. They discuss different approaches to solve the so-called "time paradox". The different conclusions drawn from their analysis imply that, after 70 years of attempts to quantize gravity, the fundamental "problem of time" is still an unresolved and fascinating issue.

In the search for quantum gravity the approach proposed by **Louis Kauffman** assumes non-commutative variables. In his contribution the author considers a non-commutative description of the world from an operational point of view. He introduces in a fascinating and straightforward approach a differential calculus that permits rephrasing some of the most basic notions known from classical differential geometry without the use of smooth manifolds. This includes, especially, the notion of Riemannian curvature, which is less simple to algebraically rephrase than the notion of Yang-Mills curvature of ordinary gauge theories. The discussion presented in this article should be contrasted with the contributions by Majid and Paschke as well as with the ideas presented by Grosse and Wulkenhaar concerning a non-commutative quantum field theory.

In contrast to the approach by Kauffman, the contribution to this volume by **Shahn Majid** uses the framework of Hopf algebras and (bi)covariant calculi on the former to address the problem of quantum gravity within an elaborated algebraic framework. The author starts by presenting the basic mathematical material in order to afterwards discuss a whole series of examples. These examples provide the reader with an introduction to a possible quantum theory of gravity on finite sets. The outlook of the article addresses further generalizations toward a purely functorial setting and a statement about the author's viewpoint on why this is needed to put quantum theory and gravity into a single framework. The author concludes his contribution with a number of remarks concerning some links to several other contributions of this volume.

Group field theory is an elaborated extension of Penrose's spin networks and spin foams. **Daniele Oriti** critically describes the challenges and achievements of group field theory. It seems possible that this way of generalized loop quantum gravity provides a richer framework that permits to handle problems like the Hamiltonian constraint. The author puts emphasis on the viewpoint that group field theory should be regarded as being a theoretical framework of its own.

Alain Connes' non-commutative geometry may provide an alternative approach to a quantum theory of gravity. In his contribution **Mario Paschke** gives an overview on the present status of this approach. The author focusses his attention on the role of the so-called "spectral action". In particular, he critically discusses the need to extend non-commutative geometry to Lorentzian signature and to study globally hyperbolic spectral triples. This viewpoint may provide a Lorentzian covariant and hence a more physically convincing approach to a non-commutative generalization of gravity. In this respect the article is closely related to the ideas concerning a covariant description of a perturbative quantum field theory as it is proposed by Brunetti and Fredenhagen in the next contribution.

Romeo Brunetti and **Klaus Fredenhagen** propose a certain background independent axiomatic formulation of perturbative quantum gravity. This formulation is based on a functorial mapping from the category of globally hyperbolic manifolds to the category of ∗-algebras. As explained in some details, the axioms for this functor are physically well motivated and permit to consider a quantum field as a fundamental local observable.

A major question of interest is the study of representations of the diffeomorphism group for any diffeomorphism invariant theory, like general relativity. In the case of globally hyperbolic space-times, these representations are related to the topology of spatially closed orientable 3-manifolds. One expects that a quantum theory of gravity would yield a super-selection structure that is induced by the topology of the classical (limiting) space. In his contribution to this volume, **Domenico Giulini** provides a well written introduction to this fascinating topic. He puts emphasis on a geometrical understanding of the mapping-class group of 3-manifolds and includes many illuminating pictures for illustration.

Next, **Christian Fleischhack** studies uniqueness theorems in loop quantum gravity analogous to the famous Stone – von Neumann theorem of ordinary quantum mechanics that guarantees the unitary equivalence of all the irreducible representations of the Heisenberg algebra. Due to the tremendously more complicated configuration space of loop quantum gravity, it is of utmost importance to know whether different quantization schemes may give rise to different physical predictions. Fleischhack's discussion of two uniqueness theorems proves that, under certain technical assumptions, an almost unique quantization procedure can be obtained for Ashtekar's formulation of a quantum gravity.

String theory is known to naturally include a spin-two field. When quantized, this field is commonly interpreted as graviton analogous to the photon in quantum electrodynamics. A basic object in any string theoretical formulation of a quantum theory of gravity plays the partition function defined in terms of appropriate functional integrals. A perturbative evaluation of the partition functions yields topological invariants of the background manifolds under consideration. **Kishore Marathe** discusses several aspects of the interplay between topological quantum field theory and quantum gravity. For instance, he discusses the Jones polynomial and other related knot invariants of low-dimensional smooth manifolds.

A rather different route to quantum gravity is proposed by **Felix Finster**. His "principle of the fermionic projector" summarizes the idea to start the formulation of a quantum theory of gravity from a set of points, a certain set of projectors related to these points and a discrete variational principle. The author summarizes the basic ideas how gravity and the gauge theory may be formulated within the framework presented in his contribution. Contrary to the common belief, the author considers locality and causality as fundamental notions only in the continuum limit of "quantum space-time".

Black holes are models of actual astrophysical effects involving strong gravitational fields. Hence, black holes are optimally suited as a theoretical laboratory for quantum gravity. In his article, **Thomas Mohaupt** deals with a string theoretical approach to black hole physics. The usual black hole quantum theory is heuristic and needs to be supported by a microscopic (statistical) theory. Formal arguments from string theory permit possible scenarios to construct densities of states that give rise to a statistical definition of entropy. Furthermore, Mohaupt's article demonstrates how effects even next to the leading order can be given a satisfactory explanation by the identification of different statistical ensembles.

Quantum mechanics originated in the attempt to understand experimental results which were in sharp contrast with Maxwell's electrodynamics. In this respect, one of the most crucial experimental effects was what is called today the "photon electric effect". Together with the black body radiation, the photon electric effect may be considered as the birth of the idea of the "quantum of (electromagnetic) energy". This idea, in turn, is known to have been fundamental for the development of the quantum theory of Maxwell's electrodynamics. The authors **Tekin Dereli** and **Robin W. Tucker** start out from the question of whether there is a similar effect related to the "quantum of gravitational energy". Analogous to electrodynamics one may look for a Hamiltonian that incorporates the energy of the classical gravitational field. In Einstein's theory of gravity this is known to be a non-trivial task. For this the authors introduce a different Lagrangian density which includes additional degrees of freedom and from which they derive an energy momentum tensor of the gravitational field. Moreover, the authors discuss specific gravitational (plane) wave like solutions of their generalized gravitational field equations which may be regarded as being similar to the electromagnetic plane waves in ordinary electromagnetism.

General relativity is known to be a perturbatively non-renormalizable theory. Such a theory needs the introduction of infinitely many free parameters ("counter terms"), which seems to spoil any predictive power of the corresponding quantum theory. Using the renormalization group **Oliver Lauscher** and **Martin Reuter** discuss the existence of non-Gaussian fixed points of the renormalization group flow such that the number of counter terms can be restricted to a finite number. Such a scenario can be obtained from numerical techniques called "asymptotic safety". Employing techniques from random walks one can show that a scale dependent effective theory which probes the nature of space-time on that particular scale can be obtained. The renormalization group trajectories permit discussing space-time properties on changing scales. While for large scales a smooth four dimensional manifold occurs, at small scales a fractal space-time of dimension two is obtained. Similar results are obtained using the idea of numerical dynamical triangulation which has been introduced by other groups.

The idea to change the structure of space-time at small distances in order to cure divergence problems in ordinary quantum field theory was introduced by Heisenberg and Schrödinger and was firstly published by Snyder. Recent developments concerning D-branes in string theory also support such a scenario. Yet another approach to a not point-like structure of space-time has been introduced by Dopplicher, Fredenhagen and Roberts in the 1990's. It is based on the idea to also obtain an uncertainty principle for the configuration space similar to the known phase space uncertainty of ordinary quantum mechanics. **Harald Grosse** and **Raimar Wulkenhaar** close this volume by a summary of their pioneering work on the construction of a specific model of a renormalizable quantum field theory on such a so-called "$\theta-$deformed" space-time. They also describe the relation of their model to multi-scale matrix models.

Acknowledgements

It is a great pleasure for the editors to thank all of the participants of the workshop for their contributions, which have made the workshop so successful. We would like to express our gratitude to the staff of the Max Planck Institute, especially to Regine Lübke, who managed the administrative work excellently. The editors would like to thank the German Science Foundation (DFG) and the Max Planck Institute for Mathematics in the Sciences in Leipzig (Germany) for their generous financial support. Furthermore, they would also like to thank Thomas Hempfling from Birkhäuser for the excellent cooperation.

Bertfried Fauser, Jürgen Tolksdorf and Eberhard Zeidler
Leipzig, October 1, 2006

Quantum Gravity
B. Fauser, J. Tolksdorf and E. Zeidler, Eds., 1–13

Quantum Gravity — A Short Overview

Claus Kiefer

Abstract. The main problems in constructing a quantum theory of gravity as well as some recent developments are briefly discussed on a non-technical level.

Mathematics Subject Classification (2000). Primary 83-02; Secondary 83C45; 83C47; 83F05; 83E05; 83E30.

Keywords. Quantum gravity, string theory, quantum geometrodynamics, loop quantum gravity, black holes, quantum cosmology, experimental tests.

1. Why do we need quantum gravity?

The task to formulate a consistent quantum theory of gravity has occupied physicists since the first attempts by Léon Rosenfeld in 1930. Despite much work it is fair to say that this goal has not yet been reached. In this short contribution I shall attempt to give a concise summary of the situation in 2005 from my point of view. The questions to be addressed are: Why is this problem of interest? What are the main difficulties? And where do we stand? A comprehensive technical treatment can be found in my monograph [1] as well as in the Proceedings volumes [2] and [3]. A more detailed review with focus on recent developments is presented in [4]. The reader is referred to these sources for details and references.

Why should one be interested in developing a quantum theory of the gravitational field? The main reasons are conceptual. The famous singularity theorems show that the classical theory of general relativity is incomplete: Under very general conditions, singularities are unavoidable. Such singularities can be rather mild, that is, of a purely topological nature, but they can also consist of diverging curvatures and energy densities. In fact, the latter situation seems to be realized in two important physical cases: The Big Bang and black holes. The presence of the cosmic microwave background (CMB) radiation indicates that a Big Bang has happened in the past. Curvature singularities seem to lurk inside the event horizon of black holes. One thus needs a more comprehensive theory to understand these

I thank Bertfried Fauser and Jürgen Tolksdorf for inviting me to this stimulating workshop.

situations. The general expectation is that a *quantum* theory of gravity is needed, in analogy to quantum mechanics in which the classical instability of atoms has disappeared. The origin of our universe cannot be described without such a theory, so cosmology remains incomplete without quantum gravity.

Because of its geometric nature, gravity interacts with all forms of energy. Since all non-gravitational degrees of freedom are successfully described by quantum fields, it would seem natural that gravity itself is described by a quantum theory. It is hard to understand how one should construct a unified theory of all interactions in a hybrid classical–quantum manner. In fact, all attempts to do so have up to now failed.

A theory of gravity is also a theory of spacetime. Quantum gravity should thus make definite statements about the behaviour of spacetime at the smallest scales. For this reason it has been speculated long ago that the inclusion of gravity can avoid the divergences that plague ordinary quantum field theories. These divergences arise from the highest momenta and thus from the smallest scales. This speculation is well motivated. Non-gravitational field theories are given on a fixed background spacetime, that is, on a non-dynamical structure. Quantum gravity addresses the background itself. One thus has to seek for a background-independent formulation. If the usual divergences have really to do with the smallest scales of the background spacetime, they should disappear together with the background.

A particular aspect of background independence is the 'problem of time', which should arise in any approach to quantum gravity, [1, 2, 6]. On the one hand, time is external in ordinary quantum theory; the parameter t in the Schrödinger equation,

$$i\hbar \frac{\partial \psi}{\partial t} = H\psi \, , \tag{1}$$

is identical to Newton's absolute time — it is *not* turned into an operator and is presumed to be prescribed from the outside. This is true also in special relativity where the absolute time t is replaced by Minkowski spacetime, which is again an absolute structure. On the other hand, time (as part of spacetime) is dynamical in general relativity. The implementation of gravity into the quantum framework should thus entail the disappearance of absolute time.

A major obstacle in the search for quantum gravity is the lack of experimental data. This is connected with the smallness of the scales that are connected with such a theory: Combining the gravitational constant, G, the speed of light, c, and the quantum of action, $\hbar$, into units of length, time, and mass, respectively, one arrives at the famous Planck units,

$$l_{\mathrm{P}} = \sqrt{\frac{\hbar G}{c^3}} \approx 1.62 \times 10^{-33} \, \mathrm{cm} \, , \tag{2}$$

$$t_{\mathrm{P}} = \sqrt{\frac{\hbar G}{c^5}} \approx 5.40 \times 10^{-44} \, \mathrm{s} \, , \tag{3}$$

$$m_{\mathrm{P}} = \sqrt{\frac{\hbar c}{G}} \approx 2.17 \times 10^{-5} \, \mathrm{g} \approx 1.22 \times 10^{19} \, \mathrm{GeV} \, . \tag{4}$$

To probe, for example, the Planck length with contemporary accelerators one would have to built a machine of the size of our Milky Way. Direct observations are thus expected to come mainly from the astrophysical side — the early universe and black holes. In addition there may of course occur low-energy effects such as a small violation of the equivalence principle [5, 3]. The search for such effects is often called 'quantum gravity phenomenology'. The irrelevance of quantum gravity in usual astrophysical investigations can be traced back to the huge discrepancy between the Planck mass and the proton mass: It is the small constant

$$\alpha_{\rm g} = \frac{Gm_{\rm pr}^2}{\hbar c} = \left(\frac{m_{\rm pr}}{m_{\rm P}}\right)^2 \approx 5.91 \times 10^{-39} \, , \tag{5}$$

where $m_{\rm pr}$ denotes the proton mass, that enters astrophysical quantities of interest such as stellar masses and stellar lifetimes.

On the road towards quantum gravity it is important to study all levels of interaction between quantum systems and the gravitational field. The first level, where experiments exist, concerns the level of quantum mechanical systems interacting with Newtonian gravity [7]. The next level is quantum field theory on a given (usually curved) background spacetime. Here one has a specific prediction: Black holes radiate with a *temperature* proportional to $\hbar$ ('Hawking radiation'),

$$T_{\rm BH} = \frac{\hbar \kappa}{2\pi k_{\rm B} c} \, , \tag{6}$$

where κ is the surface gravity. For a Schwarzschild black hole this temperature reads

$$T_{\rm BH} = \frac{\hbar c^3}{8\pi k_{\rm B} GM} \approx 6.17 \times 10^{-8} \left(\frac{M_\odot}{M}\right) \, {\rm K} \, . \tag{7}$$

The black hole shrinks due to Hawking radiation and possesses a finite lifetime. The final phase, where γ-radiation is being emitted, could be observable. The temperature (7) is unobservably small for black holes that result from stellar collapse. One would need primordial black holes produced in the early universe because they could possess a sufficiently low mass, cf. the review by Carr in [3]. For example, black holes with an initial mass of 5×10^{14} g would evaporate at the present stage of the universe. In spite of several attempts, no experimental hint for black-hole evaporation has been found. Primordial black holes can result from density fluctuations produced during an inflationary epoch. However, they can only be produced in sufficient numbers if the scale invariance of the power spectrum is broken at some scale, cf. [8] and references therein.

Since black holes radiate thermally, they also possess an *entropy*, the 'Bekenstein–Hawking entropy', which is given by the expression

$$S_{\rm BH} = \frac{k_{\rm B} c^3 A}{4G\hbar} = k_{\rm B} \frac{A}{4l_{\rm P}^2} \, , \tag{8}$$

where A is the surface area of the event horizon. For a Schwarzschild black hole with mass M, this reads

$$S_{\mathrm{BH}} \approx 1.07 \times 10^{77} k_{\mathrm{B}} \left(\frac{M}{M_\odot} \right)^2 . \tag{9}$$

Since the Sun has an entropy of about $10^{57} k_{\mathrm{B}}$, this means that a black hole resulting from the collapse of a star with a few solar masses would experience an increase in entropy by twenty orders of magnitude during its collapse.

It is one of the challenges of any approach to quantum gravity to provide a microscopic explanation for this entropy, that is, to derive (8) from a counting of microscopic quantum gravitational states according to the formula

$$S = -k_{\mathrm{B}} \mathrm{Tr}(\rho \ln \rho) , \tag{10}$$

where ρ is the density matrix corresponding to the relevant degrees of freedom.

An effect analogous to (6) already exists in flat spacetime: An observer who is accelerated through the Minkowski vacuum experiences a thermal spectrum of particles, the 'Davies–Unruh effect'. This effect may be measurable in the not-too-distant future.

Concerning now the attempt to construct a full quantum theory of gravity, the question arises: What are the main approaches? In brief, one can distinguish between

- *Quantum general relativity*: The most straightforward attempt, both conceptually and historically, is the application of 'quantization rules' to classical general relativity. This approach can be divided further into
    - *Covariant approaches*: These are approaches that employ four-dimensional covariance at some stage of the formalism. Examples include perturbation theory, effective field theories, renormalization-group approaches, and path integral methods.
    - *Canonical approaches*: Here one makes use of a Hamiltonian formalism and identifies appropriate canonical variables and conjugate momenta. Examples include quantum geometrodynamics and loop quantum gravity.
- *String theory (M-theory)*: This is the main approach to construct a unifying quantum framework of all interactions. The quantum aspect of the gravitational field only emerges in a certain limit in which the different interactions can be distinguished.
- Other fundamental approaches such as a direct quantization of topology, or the theory of causal sets.

I shall now briefly review the main approaches and some of their applications.

2. Quantum general relativity

2.1. Covariant approaches

The first attempt to construct a theory of quantum theory was, of course, the attempt to develop a perturbation theory. After all, major theories such as QED are only understood perturbatively. However, for gravity there is a distinctive feature: The perturbation theory is non-renormalizable, that is, infinitely many parameters must be introduced (and experimentally determined) in order to absorb the ensuing divergences. This seems to render the perturbation theory meaningless. Still, even a non-renormalizable theory may be able to yield definite predictions at low energies. This occurs on the level of effective field theories. One example is the quantum gravitational correction to the Newtonian potential calculated in [9]. Another example concerns the application of renormalization group methods [10]: The theory may be asymptotically safe, that is, possess a non-vanishing ultraviolet fixed point at the non-perturbative level. This would lead to running gravitational and cosmological 'constants' that could in principle provide an explanation for the dark matter and the dark energy in the universe.

A direct 'covariant quantization' of general relativity would be the formulation of a path-integral framework. This has been tried in both a Euclidean and a Lorentzian formalism: Whereas in the former one rotates the time parameter onto the imaginary axis, the latter leaves the Lorentzian signature of spacetime untouched. Besides their use in the search for boundary conditions in quantum cosmology (see below), much work has been devoted to a numerical analysis via 'Regge calculus' or 'dynamical triangulation'. The latter approach has recently provided some interesting results concerning the structure of space and time: The Hausdorff dimension of the resulting space is consistent with the (expected) number three, a positive cosmological constant is needed, and the universe behaves semiclassically at large scales, cf. [11] for a recent introductory review.

2.2. Canonical approaches

The major alternative to covariant quantization is to use a Hamiltonian framework — the framework of 'canonical quantization'. Under the assumption of global hyperbolicity, the classical spacetime is foliated into three-dimensional spaces each of which is isomorphic to a given three-manifold Σ. One then chooses an appropriate canonical variable and its momentum and turns their Poisson-bracket algebra into a commutator algebra. According to the choice of variables one can distinguish various subclasses of canonical quantization. The oldest is quantum geometrodynamics in which the canonical variable is the three-dimensional metric on Σ, and the momentum is linearly related to the extrinsic curvature. More recent approaches employ either a connection variable or its integral along a loop in Σ. The latter leads to what is now called 'loop quantum gravity'.

The central equations in all these approaches are the *quantum constraints*. The invariance of the classical theory under coordinate transformation leads to four (local) constraints: the Hamiltonian constraint,

$$\mathcal{H}_\perp \Psi = 0 \, , \tag{11}$$

and the three diffeomorphism (or momentum) constraints,

$$\mathcal{H}_a \Psi = 0 \, . \tag{12}$$

The total gravitational Hamiltonian reads (apart from boundary terms)

$$H = \int \mathrm{d}^3 x \ (N\mathcal{H}_\perp + N^a \mathcal{H}_a) \, , \tag{13}$$

where N ('lapse function') and N^a ('shift vector') are Lagrange multipliers. In the connection and loop approaches, three additional (local) constraints emerge because of the freedom to choose the local triads upon which the formulation is based.

2.2.1. Quantum geometrodynamics. Let me first address quantum geometrodynamics where the configuration variable is the three-metric. In this case one usually refers to (11) (or the full equation $H\Psi = 0$) as the 'Wheeler–DeWitt equation'. Here are some characteristics of quantum geometrodynamics:

- The wave functional Ψ depends on the *three*-dimensional metric, but because of (12) it is invariant under coordinate transformations on three-space.
- No external time parameter is present anymore — in this sense the theory is 'timeless'. This also holds for the connection and loop approaches. The 'problem of time' is thus 'solved' in an unexpected manner by avoiding any external time parameter at all.
- Such constraints result from any theory that is classically reparametrization invariant, that is, a theory without background structure.
- The Wheeler–DeWitt equation is (locally) hyperbolic, and one can thereby define a local 'intrinsic time' distinguished by the sign in this wave equation.
- This approach is a candidate for a non-perturbative quantum theory of gravity. But even if it is superseded by a more fundamental theory at the Planck scale (such as superstring theory, see below), it should approximately be valid away from the Planck scale. The reason is that general relativity is then approximately valid, and the quantum theory from which it emerges in the WKB limit is quantum geometrodynamics.

Quantum geometrodynamics has the main disadvantage of not (yet?) allowing a precise mathematical formulation on the general level. It, however, allows to tackle directly central physical questions on the level of models. One can thereby apply this framework to quantum cosmology and to black holes. Also the semiclassical approximation is, on the formal level, straightforward [1]: The external time emerges in an approximate way through some Born–Oppenheimer type of approximation scheme. An important ingredient is also the process of decoherence — the irreversible emergence of classical properties through the interaction with irrelevant degrees of freedom [12]. In this way major physical features of standard quantum theory find their applications in quantum gravity.

2.2.2. Connection and loop variables. Instead of the metric formulation of the last subsection one can use a different set of variables, leading to the connection or the loop formulation. Detailed expositions can be found, for example, in [13], see also [1]. Starting point are the 'new variables' introduced by Abhay Ashtekar in 1986,

$$E_i^a(x) = \sqrt{h}\, e_i^a(x) \,, \tag{14}$$

$$GA_a^i(x) = \Gamma_a^i(x) + \beta K_a^i(x) \,. \tag{15}$$

Here, $e_i^a(x)$ is the local triad (with i being the internal index), h is the determinant of the three-metric, $\Gamma_a^i(x)$ the spin connection, and $K_a^i(x)$ the second fundamental form. The parameter $\beta \in \mathbb{C}\backslash\{0\}$ denotes a quantization ambiguity similar to the θ-parameter ambiguity in QCD and is called the 'Barbero–Immirzi parameter'. The canonical pair of variables are the densitized triad $E_i^a(x)$ (this is the new momentum) and the connection $A_a^i(x)$ (the new configuration variable). They obey the Poisson-bracket relation

$$\{A_a^i(x), E_j^b(y)\} = 8\pi\beta\delta_j^i\delta_a^b\delta(x,y) \,. \tag{16}$$

The use of this pair leads to what is called 'connection dynamics'. One can rewrite the above constraints in terms of these variables and subject them to quantization. In addition, one has to treat the 'Gauss constraint' arising from the freedom to perform arbitrary rotations in the local triads (SO(3)- or SU(2)-invariance).

The *loop variables*, on the other hand, are constructed from a non-local version of the connection. The fundamental loop variable is the holonomy $U[A,\alpha]$ defined as the path-ordered product

$$U[A,\alpha] = \mathcal{P}\exp\left(G\int_\alpha A\right) \,, \tag{17}$$

where α denotes an oriented loop in Σ. The conjugate variable is the 'flux' of E_i^a through a two-dimensional surface $\mathcal{S}$ in Σ.

The original motivation for the introduction of these 'new variables' was the hope to simplify the Hamiltonian constraint (11). However, it was not yet possible to fulfil this hope, at least not in its original form. Instead, attention has focused on geometric operators in loop quantum gravity. Introducing a kinematic Hilbert space (that is, on the level before all constraints are implemented), a *discrete* structure for these operators appears. For example, one can define an 'area operator' $\hat{A}(\mathcal{S})$ (whose classical analogue is the area of the two-dimensional surface $\mathcal{S}$) and find the following spectrum,

$$\hat{A}(\mathcal{S})\Psi_S[A] = 8\pi\beta l_{\mathrm{P}}^2 \sum_{P\in S\cap\mathcal{S}} \sqrt{j_P(j_P+1)}\Psi_S[A] \,, \tag{18}$$

where the points P denote the intersections of the so-called 'spin network' with the surface, cf. [13] for details. It has to be emphasized, however, that the area operator (as well as the other geometric operators) does *not* commute with the constraints, that is, cannot be considered as an observable. It remains an open problem to construct a version that *does* commute.

In spite of these encouraging results, some important open problems remain, cf. [14]. These include:

- The presence of quantum anomalies in the constraint algebra could make the formalism inconsistent. How can anomalies be avoided?
- How does one have to treat the constraints? The Gauss constraint is easy to implement, but there are various subtleties and ambiguities with the diffeomorphism and Hamiltonian constraints.
- Can a semiclassical limit be obtained? Since the constraints assume now a form different from the one in geometrodynamics, a Born–Oppenheimer approach cannot be straightforwardly applied. An important role may be played by generalized coherent states. For large enough scales, the semiclassical approximations of loop quantum gravity and quantum geometrodynamics should of course coincide.
- Is a Hilbert-space structure really needed? After all, the introduction of a Hilbert space as a central ingredient in quantum mechanics is motivated by the probability interpretation and the conservation of probability with respect to external time ('unitarity'). But since the external time has disappeared in quantum gravity, a Hilbert space may not be needed.

3. String theory

The major alternative to quantum general relativity is string theory. Its starting point is completely different: Whereas quantum general relativity focuses on a separate quantum description for the gravitational field, string theory seems to implement the idea that the problem of quantum gravity can be dealt with only within a unified quantum theory of *all* interactions. Since string theory is covered at length in other contributions to this volume, I shall be short, cf. also [15] for a concise technical introduction. The main characteristics of string theory are:

- The appearance of gravity is inevitable. The graviton is an excitation of closed strings, and it appears via virtual closed strings in open-string amplitudes.
- String theory has as important ingredients the concepts of gauge invariance, supersymmetry (SUSY), and the presence of higher dimensions.
- It provides a unification of all interactions.
- String perturbation theory seems to be finite at each order, but the sum diverges.
- Only *three* fundamental constants are present: $\hbar, c$ and the string length l_s; all other physical parameters (masses and coupling constants, including the Planck mass) should in principle be derivable from them.
- A net of dualities connects various string theories and indicates the presence of an underlying non-perturbative theory (M-theory) from which the various string theories can be recovered in appropriate limits.

In spite of much progress, there are still many problems. They include

- Lack of experimental evidence.

- There are many ways to compactify the additional spacetime dimensions. Moreover, these additional dimensions may be non-compact as indicated by the existence of various 'brane models'. Without a solution to this problem, no definite relation to low-energy coupling constants and masses can be made.
- Background independence is not yet fully implemented into string theory, as can be recognized from the prominent role of the embedding space. The AdS/CFT theories discussed in recent years may come close to background independence in some respect.
- The Standard Model of particle physics, which is experimentally extremely well tested, has not yet been recovered from string theory.
- What is M-theory and what is the role of the 11th dimension which has emerged in this context?
- Quantum cosmology has not yet been implemented into the full theory, only at the level of the effective action ('string cosmology').

4. Loops versus strings – a few points

Both string theory and loop quantum gravity exhibit aspects of non-commutative geometry, cf. the contributions on this topic in this volume. This could be relevant for understanding space at the smallest scale. In loop quantum gravity, the three-geometry is non-commutative in the sense that area operators of intersecting surfaces do not commute. In string theory, for n coincident D-branes, the fields X^μ — the embeddings — and A_a — the gauge fields — become non-commuting $n \times n$ matrices. It has also been speculated that time could emerge from a timeless framework if space were non-commutative [16]. This would be an approach very different from the recovery of time in quantum geometrodynamics discussed above. Quite generally, one envisages a highly non-trivial structure of space(time) in string theory [17].

As mentioned above, the main areas of application for any theory of quantum gravity are black holes and quantum cosmology. What can the main approaches say about these issues [1, 4]? As for the black holes, there are three main questions: How does the final evaporation of the black hole proceed? Is information being lost in this process? Can one derive the Bekenstein–Hawking entropy (8) from quantum gravitational statistical theory? Whereas not much progress has been made on the first question, some results have been obtained for the remaining two. As for the problem of information loss, the opinion now seems to prevail that the full evolution is unitary, that is, there is no information loss. Unitarity here refers to the semiclassical time of the outside universe. This is consistent with the thermal nature of Hawking radiation, which can be understood to emerge from the interaction with irrelevant field degrees of freedom, that is, from decoherence [18]. Moreover, the black hole by itself cannot evolve unitarily because it is rendered an 'open quantum system' by the decohering influence of the irrelevant fields (which include Hawking radiation); unitarity holds only for the fully entangled quantum state of the hole plus the other fields [19].

As for the microscopic derivation of (8), the situation can be briefly summarized as follows [4],

- *Loop quantum gravity:* The microscopic degrees of freedom are given by the spin-network states. An appropriate counting procedure for the number of the relevant horizon states leads to an entropy that is proportional to the Barbero–Immirzi parameter β. The demand for the result to be equal to (8) then *fixes* β. Until 2004, it was believed that the result is

$$\beta = \frac{\ln 2}{\pi\sqrt{3}} \quad \left(\frac{\ln 3}{\pi\sqrt{2}}\right) , \qquad (19)$$

where the value in parentheses would refer for the relevant group to the choice of SO(3) instead of SU(2). The SO(3)-value would have exhibited an interesting connection with the quasi-normal modes for the black hole. More recently, it was found that the original estimate for the number of states was too small. A new calculation yields the following numerical estimate for β [20]: $\beta = 0.237532\ldots$ An interpretation of this value at a more fundamental level has not yet been given.

- *String theory:* The microscopic degrees of freedom are here the *D-branes*, for which one can count the quantum states in the weak-coupling limit (where no black hole is present). Increasing the string coupling, one reaches a regime where no D-branes are present, but instead one has black holes. For black holes that are extremal in the relevant string-theory charges (for extremality the total charge is equal to the mass), the number of states is preserved ('BPS black holes'), so the result for the black-hole entropy is the same as in the D-brane regime. In fact, the result is just (8), as it must be if the theory is consistent. This remains true for non-extremal black holes, but no result has been obtained for a generic black hole (say, an ordinary Schwarzschild black hole). More recently, an interesting connection has been found between the partition function, Z_{BH}, for a BPS black hole and a topological string amplitude, Z_{top}, [21],

$$Z_{\mathrm{BH}} = |Z_{\mathrm{top}}|^2 . \qquad (20)$$

Moreover, it has been speculated that the partition function can be identified with (the absolute square of) the Hartle–Hawking wave function of the universe [22]. This could give an interesting connection between quantum cosmology and string theory.

5. Quantum cosmology

Let me turn to quantum cosmology. If one assumes the universal validity of quantum theory (as is highly suggested by the experimental situation [23]), one needs a quantum theory for the universe as a whole: All subsystems are entangled with their environment, with the universe being the only strictly closed quantum system. Since gravity dominates on the largest scales, a theory of quantum gravity is needed for quantum cosmology. The main questions to be addressed are the following: [24]

- How does one impose boundary conditions in quantum cosmology?
- Is the classical singularity being avoided?
- How does the appearance of our classical universe follow from quantum cosmology?
- Can the arrow of time be understood from quantum cosmology?
- How does the origin of structure proceed?
- Is there a high probability for an inflationary phase? Can inflation itself be understood from quantum cosmology?
- Can quantum cosmological results be justified from full quantum gravity?
- Has quantum cosmology relevance for the measurement problem in quantum theory?
- Can quantum cosmology be experimentally tested?

There has been much progress on each of these questions, but final answers can probably only obtained after the correct quantum theory of gravity is available. For the status of each of these questions I refer to [24, 25, 1, 26]. Loop quantum gravity has recently been applied to cosmology by using spectra such as (18) for the size of the universe ('loop quantum cosmology' [27]). The Wheeler–DeWitt equation then assumes the form of a difference equation. It seems that one can get singularity avoidance from it. Moreover, the quantum modifications to the Friedmann equations arising from the loop approach seem to favour an inflationary scenario and could potentially be observed in the CMB anisotropy spectrum.

6. Some central questions about quantum gravity

I conclude this brief review with some central questions taken from [4] concerning the development of quantum gravity:

- Is unification needed to understand quantum gravity?
- Into which approaches is background independence implemented? (In particular, is string theory background independent?)
- In which approaches do ultraviolet divergences vanish?
- Is there a continuum limit for path integrals?
- Is a Hilbert-space structure needed for the full theory? (This has an important bearing on the interpretation of quantum states.)
- Is Einstein gravity non-perturbatively renormalizable? (Can the cosmological constant be calculated from the infrared behaviour of renormalization-group equations?)
- What is the role of non-commutative geometry?
- Is there an information loss for black holes?
- Are there decisive experimental tests?

References

[1] C. Kiefer, *Quantum Gravity* (Clarendon Press, Oxford, 2004).

[2] *Conceptual Problems in Quantum Gravity*, edited by A. Ashtekar and J. Stachel. Einstein Studies, Vol. 2 (Birkhäuser, Boston, 1991).

[3] *Quantum Gravity: From Theory to Experimental Search*, edited by D. Giulini, C. Kiefer, and C. Lämmerzahl. Lecture Notes in Physics, Vol. 631 (Springer, Berlin, 2003).

[4] C. Kiefer, Quantum gravity: general introduction and recent developments, Ann. Phys. (Leipzig) **15**, No. 1-2, 2005: 129–148.

[5] C. Lämmerzahl, contribution to this volume.

[6] A. Macías and H. Quevedo, contribution to this volume.

[7] C. Kiefer and C. Weber, On the interaction of mesoscopic quantum systems with gravity, Ann. Phys. (Leipzig) **14**, No 4, 2005: 253–278.

[8] D. Blais, T. Bringmann, C. Kiefer, and D. Polarski, Accurate results for primordial black holes from spectra with a distinguished scale, Phys. Rev. D **67**, 2003: 024024–024035.

[9] N. E. J. Bjerrum-Bohr, J. F. Donoghue, and B. R. Holstein, Quantum corrections to the Schwarzschild and Kerr metrics, Phys. Rev. D **68**, 2003: 084005–084021.

[10] O. Lauscher and M. Reuter, contribution to this volume.

[11] R. Loll, J. Ambjørn, and J. Jurkiewicz, The Universe from Scratch, Contemp. Phys. 47, 2006: 103–117, hep-th/0509010.

[12] E. Joos, H. D. Zeh, C. Kiefer, D. Giulini, J. Kupsch, and I.-O. Stamatescu, *Decoherence and the Appearance of a Classical World in Quantum Theory*, second edition (Springer, Berlin, 2003). See also www.decoherence.de.

[13] C. Rovelli, *Quantum Gravity* (Cambridge University Press, Cambridge, 2004).

[14] H. Nicolai, K. Peeters, and M. Zamaklar, Loop quantum gravity: an outside view, Class. Quantum Grav. **22**, 2005: R193–R247.

[15] B. Zwiebach, *A First Course in String Theory* (Cambridge University Press, Cambridge, 2004).

[16] S. Majid, Noncommutative model with spontaneous time generation and Planckian bound, J. Math. Phys. **46**, 2005: 103520–103520; See also his contribution to this volume.

[17] G. T. Horowitz, Spacetime in String Theory, New. J. Phys. **7**, 2005: 201 13pp electronic.

[18] C. Kiefer, Hawking radiation from decoherence, Class. Quantum Grav. **18**, 2001: L151–L154.

[19] H. D. Zeh, Where has all the information gone?, Phys. Lett. A **347**, 2005: 1–7.

[20] M. Domagala and J. Lewandowski, Black-hole entropy from quantum geometry, Class. Quantum Grav. **21**, 2004: 5233-5243;
K. Meissner, Black-hole entropy in loop quantum gravity, Class. Quantum Grav. **21**, 2004: 5245–5251.

[21] H. Ooguri, A. Strominger, and C. Vafa, Black hole attractors and the topological string, *Phys. Rev. D* **70**, 2004: 106007–106020.

[22] H. Ooguri, C. Vafa, and E. Verlinde, Hartle-Hawking Wave-Function for Flux Compactifications, Lett.Math.Phys. **74**, 2005: 311–342, hep-th/0502211.

[23] M. Schlosshauer, Experimental motivation and empirical consistency in minimal no-collapse quantum mechanics, Ann. Phys. (N.Y.) **321**, 2006: 112-149, quant-ph/0506199.

[24] C. Kiefer, Quantum cosmology: Expectations and results, Ann. Phys. (Leipzig) **15**, no. 4-5, 2006:316–325.

[25] D. H. Coule, Quantum cosmological models, Class. Quantum Grav. **22**, 2005: R125–R166.

[26] H. D. Zeh, *The physical basis of the direction of time*, fourth edition (Springer, Berlin, 2001). See also www.time-direction.de.

[27] M. Bojowald, Universe scenarios from loop quantum cosmology, Ann. Phys. (Leipzig) **15**, 2006: 326–341, astro-ph/0511557.

Claus Kiefer
Institut für Theoretische Physik
Universität zu Köln
Zülpicher Str. 77
50937 Köln
Germany
e-mail: Kiefer@thp.uni-koeln.de

Quantum Gravity
B. Fauser, J. Tolksdorf and E. Zeidler, Eds., 15–39

The Search for Quantum Gravity Effects

Claus Lämmerzahl

Abstract. The status of experimental tests of the foundations of present day standard physics is given. These foundations mainly consists of the Universality of Free Fall, the Universality of the Gravitational Redshift, the local validity of Lorentz invariance. This essentially determines the Riemannian structure of the gravitational interaction as well as essential features of the standard model. More is needed for fixing the Einstein field equations and the structure of quantum mechanical equations. After the extended review on these foundations, the magnitude of possible Quantum Gravity effects is discussed and a strategy for a dedicated search for such effects is presented. An outline of what can be expected from future experiments concludes this contribution.

Mathematics Subject Classification (2000). Primary 83B05; Secondary 83D05; 83C45; 83F05.

Keywords. Quantum gravity effects, experimental tests, equivalence principle, constant speed of light, universality of free fall, local Lorentz invariance, universality of gravitational redshift, Pioneer anomaly, fly-by anomaly.

1. Introduction

Finding a viable theory combining General Relativity (GR) and quantum theory is a major task of present day physics (see many other contributions in this volume). The search for such a theory can be supported by experiments which may give restrictions to a possible domain of theories or admissible ranges of parameters. At the end experiments also also have to confirm predictions made from Quantum Gravity (QG) theories.

There are two questions which have to be addressed: (i) where should we look for QG effects and (ii) how should we look for these effects? Since GR and quantum theory are applicable to *all* kinds of matter and since these two theories should be replaced by a new theory in principle *all* effects should show modifications

This work has been supported by the German Space Agency DLR.

compared with the standard results. Therefore, one trivial answer to the above questions is that *any* experiment is also an experiment searching for QG effects. However, from general considerations one may extract some kind of strategy of where it is preferable to look for such effects, that is, in which situations QG effects are expected to show up more pronounced than in other situations. This is one issue addressed in this contribution.

Another aspect is the present status of experimental test of the validity of the standard theories. The validity of a theory relies on the experimental verification of the basic principles or foundations of the theory under consideration. This experimental exploration defines the "landscape" of the experimentally explored regime which then is covered by standard theories. QG effects can be present only outside this explored domains.

These remarks define the outline of this paper: At first we recall the basic principles present day standard physics relies on. This is the Einstein Equivalence Principle which essentially determines the structure of all equations of motion. This is followed by a compact review of most of the experiments aimed at testing the Einstein Equivalence Principle. Then we come to open problems ("dark clouds") of present physics which found no explanation so far. This leads us to the issue of how we could search for QG effects. We discuss the expected magnitude of these effects and also where we might expect QG effects and how we should search for these effects. That is, we discuss some strategy for the search for QG effects. We close this contribution by outlining which experimental progress we can we expect in the future.

2. The basic principles of standard physics

The basic principle of standard physics is the Einstein Equivalence Principle which consists of three parts

The Universality of Free Fall (UFF): This principle states that in a gravitational field all kinds of structureless matter fall in the same way. In principle, this is an amazing fact; gravity is the only interaction with this property. In the frame of elementary particle theory this just means that all elementary particles fall in the gravitational field in the same way. Therefore one carries through experiments with various macroscopic species of materials with different proton to neutron ratio.

The Universality of the Gravitational Redshift (UGR): This universality means that all clocks based on non-gravitational physics behave in the same way when moved together in gravitational fields. This again is an amazing fact and means that in addition to all particles also all (non-gravitational) interactions (also represented by particles) couple to the gravitational field in the same way. Again, one has to test this for all kinds of clocks and to analyze the results in terms of the coupling of elementary particles to gravity.

The local validity of Lorentz invariance: This means that the outcome of all local, small-scale, experiments are independent of the orientation and the state of motion of the laboratory: it is not possible to single out a particular reference system from local experiments. In particular the (2–way) velocity of light should be constant and all limiting velocities of elementary particles are given by the velocity of light. This again is an amazing fact: Where do all the particles know from properties of the other particles? Since LLI is a property which applies to all physics, one has to perform tests with all physical systems. The Michelson-Morley, Kennedy-Thorndike, and Ives-Stilwell tests are the most well known tests of this type which have to completed with tests with electrons, protons, etc. which are given by, e.g., the Hughes-Drever experiments.

On the theoretical side some of these formulations may be still not satisfactorily defined. However, since from the experimental side it is clear and well defined what to do and since until now all experiments led to unique results and statements, any theoretical deficiency is of academic interest only and will not lead to modifications of the conclusions.

3. Experimental tests

Since GR as well as quantum theory cannot be correct in a strict sense, there should be some deviations from it. These deviations may appear as violations of local Lorentz invariance, the UFF, the UGR, or of the superposition principle, the unitary time development, etc., in quantum theory. Any deviations from standard theories have to be described by some modified dynamical equations and/or by the presence of additional fields or interactions. In many cases, the additional fields survive in some low–energy limit as, e.g., scalar fields leading to dilaton scenarios, for example.

3.1. Tests of the universality of free fall

The UFF holds for neutral point–like particles only. The corresponding tests are described in terms of the acceleration of these particles in the reference frame of the gravitating body: The Eötvös factor compares the (weighted) acceleration of two bodies $\eta = \frac{a_1 - a_2}{\frac{1}{2}(a_1 + a_2)}$. In the frame of Newton's theory this can be expressed as $\eta = \frac{\mu_1 - \mu_2}{\frac{1}{2}(\mu_1 + \mu_2)}$ with $\mu = m_g/m_i$ where m_g is the gravitational and m_i the inertial mass. Though there are no point particles it is possible experimentally to manufacture macroscopic bodies such that their higher gravitational multipoles either are very small or very well under control. This is used in the various tests of the UFF. The most precise tests carried through until now yields a verification of the UFF at the order $5 \cdot 10^{-13}$ [1]. Free fall tests led, due to the short time span of free fall, to slightly less accurate results [2, 3, 4]. UFF tests have also been carried through with quantum systems, that is, with neutrons [5, 6, 7] and atoms [8, 9].

In the case the particle is charged or possesses a spin, then standard GR shows that these particles couple to curvature. For a charged particle we have an interaction term of the form $D_u u = \frac{e}{m} R(\cdot, u)$ [10], where D_u is the Christoffel covariant derivative along u and R is the Ricci tensor. A particle with spin experiences an acceleration $D_u u = \lambda_C R(u, S) u$, where S is the spin vector of the particle and $R(\cdot, \cdot)$ the curvature operator. Therefore, in principle UFF is always violated for charged particles and particles with spin. However, nevertheless it makes sense to look for an anomalous coupling of spin and charge to the gravitational field. For a charged particle one may look, on the Newtonian level, for extra coupling terms of the form $\kappa q U$ where q is the charge, U the Newtonian potential and κ some parameter of the dimension 1/(specific charge). This necessarily leads to a charge dependent gravitational mass and, thus, to a charge dependent Eötvös coefficient [11]. Until now only one test of the UFF for charged particles has been carried out [12] with a precision of approx. 10% only.

Motivations for anomalous spin couplings came from the search for the axion, a candidate for the dark matter in the universe [13] and from general schemes of violation of Lorentz invariance, e.g. [14]. In these models spin may couple to the gradient of the gravitational potential or to gravitational fields generated by the spin of the gravitating body. The first case can easily be tested by weighting polarized bodies what gave that for polarized matter the UFF is valid up to the order 10^{-8} [15].

Until now, no tests of the UFF with antimatter has been carried through. However, corresponding experiments are in preparation [16] with an anticipated accuracy 10^{-3} on ground and possibly 10^{-5} in space.

Physical system	Experiment	Method	Accuracy
neutral bulk matter	Adelberger et al [1]	torsion balance	$5 \cdot 10^{-13}$
polarized matter	Ritter et al [15]	weighting	10^{-8}
charged particles	Witteborn & Fairbank [12]	time–of–flight	10^{-1}
Quantum system	Chu & Peters [8]	atom interferometry	10^{-9}
antimatter	not yet carried through		

3.2. Tests of the universality of the gravitational redshift

For a test of this principle the run of clocks based on different physical principles has to be compared during their common transport through a gravitational potential. Clocks that have been used are:

- Light clocks which frequency is defined by standing electromagnetic waves between two mirrors separated by a length L.
- Atomic clocks based on electronic hyperfine transitions which are characterized by $g(m_e/m_p)\alpha^2 f(\alpha)$ where g and f are some functions, and m_e and m_p are the electron and proton mass, respectively, and where α is the fine structure constant.
- Atomic clocks based on electronic fine structure transitions characterized by α^2.

- Atomic clocks based on electronic principal transitions given by some $f(\alpha)$.
- Molecular clocks based on rotational transitions characterized by m_e/m_p
- Molecular clocks based on vibrational transition given by $\sqrt{m_e/m_p}$.
- Gravitational clocks based on revolution of planets or binary systems.
- Pulsar clocks based on the spin of stars (or, equivalently but less precise, the rotation of the Earth).
- Clocks based on the decay time of elementary particles. (Since this is not periodic, this effects has not been used to define a time standard. However, it has been used to test time dilation effects.)

On a phenomenological level one describes the possibility that the gravitational redshift of clocks may depend on the used clock by an additional clock–dependent parameter α_{clock}:

$$\nu(x_1) = \left(1 - (1 + \alpha_{\text{clock}}) \frac{U(x_1) - U(x_0)}{c^2}\right) \nu(x_0). \tag{3.1}$$

Then the comparison of two collocated clocks is given by

$$\frac{\nu_1(x_1)}{\nu_2(x_1)} \approx \left(1 - (\alpha_{\text{clock2}} - \alpha_{\text{clock1}}) \frac{U(x_1) - U(x_0)}{c^2}\right) \frac{\nu_1(x_0)}{\nu_2(x_0)}. \tag{3.2}$$

If this frequency ratio does not depend on the value of the gravitational potential then the gravitational redshift is universal. This again is a null–test of the tested quantity $\alpha_{\text{clock2}} - \alpha_{\text{clock1}}$. It is obviously very preferable to have available a large difference in the gravitational potential. Below is a list of various clock comparison experiments. In all these experiments the variation of the gravitational field was induced by the motion of the Earth around the Sun what implies that the used clocks should have a very good long term stability.

Comparison	Result	Experiment
Cs – Resonator	$|\Delta\alpha| \leq 2 \cdot 10^{-2}$	Turneaure & Stein 1987 [17]
Mg – Cs (fine structure)	$|\Delta\alpha| \leq 7 \cdot 10^{-4}$	Godone et al 1995 [18]
Resonator – I_2 (electronic)	$|\Delta\alpha| \leq 4 \cdot 10^{-2}$	Braxmaier et al 2002 [19]
Cs – H-Maser (hf)	$|\Delta\alpha| \leq 2.5 \cdot 10^{-5}$	Bauch & Weyers 2002 [20]

3.3. Tests of local Lorentz invariance

Lorentz invariance can be based on two postulates only. These two postulates are

Postulate 1: The speed of light c is constant.
Postulate 2: The relativity principle.

The first postulate may be replaced by the statement that light is a unique phenomenon, that is, between an event and a worldline there are two and only two light rays. The light ray, in particular, does not depend on the trajectory the emission event lies on. Otherwise there will be more than two light rays. The second postulate then makes sure that the measured velocity of light does not depend on its direction and on the velocity of the observer.

The two postulates have some immediate consequences which all can be tested in experiments:

- The velocity of light, c, does not depend on
    - the velocity of the source (what is a statement of the uniqueness of the phenomenon)
    - the velocity of the observer,
    - the direction of propagation,
    - the polarization or frequency of the light ray.
- The relativity principle implies that
    - the limiting velocity of all particles is the speed of light

$$c = c_+ = c_- = c_\nu = v_p^{\max} = v_e^{\max} = v_{\mathrm{grav}}$$

    (If these velocities are not equal then this defines a preferred frame by the condition that in this frame both limiting velocities are isotropic. This is the SR aspect of the $TH\epsilon\mu$–formalism (see, e.g., [21, 22] and references therein. This is in contradiction to the second postulate, with the consequence that c is universal and, thus, can be interpreted as *geometry*.)
    - that *all* physics is the same in *all* inertial systems, that is, experimental results do not depend on the
        * orientation of the laboratory and
        * on the velocity of the laboratory.

Accordingly, we have the following classes of experiments:

3.3.1. Constancy of c. The independence of the speed of light from the velocity of the source has been examined by astrophysical observations as well as in laboratory experiments. We mention just two of them: (i) the observation by Brecher [23] who analyzed the time of arrival of X–rays emitted from distant a bright star orbiting a dark, heavy central star. We model a hypothetical dependence of the speed of light c from the velocity of the source by $c' = c + \kappa v$, where v is the velocity of the source and κ some parameter which is 0 in SR and 1 in Galilean kinematics. If the star is moving away from the Earth, the emitted light is slower than the light emitted when the star is moving toward the Earth. Therefore, light emitted toward the Earth may overtake the light emitted earlier. The images of the star seen on Earth may show an achronological order. Since this has never been observed one can derive $\kappa \leq 10^{-10}$. A laboratory version of this has been carried through at CERN [24]. Protons hitting a Beryllium target created π^0 mesons possessing a velocity of $v = 0.99975c$. These π^0 mesons decay within 10^{-16} s into photons which velocity has been measured and compared with the velocity of photons emitted from π^0 mesons at rest. No difference in the speed of the photons has been found leading to $\kappa \leq 10^{-6}$. Though this is not as good as the Brecher result, it shows the result for a velocity of the source being almost the speed of light. The independence of the speed of light from the velocity of the source cannot be demonstrated more convincingly than by this experiment.

3.3.2. Universality of c. The next step is to show the equality of all limiting velocities. The equality of the maximum speed of electrons, photons in various velocity ranges, neutrinos and muons has been tested by a variety of experiments, see e.g. [25, 26, 27, 28], which all result in a relativity equality at the 10^{-6} level. Astrophysical observations of radiation from the supernova SN1987A yield for the comparison of photons and neutrinos an estimate which is two orders of magnitude better. Furthermore, from astrophysical observations one can conclude that the vacuum shows no birefringence up to an order $|(c_+ - c_-)/c_+| \leq 2 \cdot 10^{-32}$ [29].

3.3.3. Isotropy of c. Due to newly emerged technologies many new tests of the principles of SR, in particular of the three classical tests, have recently been carried through with much improved accuracy. The relative difference of the speed of light in different directions is now smaller than $\Delta_\vartheta c/c \leq 10^{-16}$ [30, 31, 32] and is approaching the 10^{-17} level.

3.3.4. Independence of c from the velocity of the laboratory. The difference of the speed of light in differently moving inertial systems is now smaller than $\Delta_v c/c \leq 10^{-16}$ [33]. These model independent estimates can be converted into estimates of parameters in certain tests theories. Within the kinematical Robertson–Mansouri–Sexl test theory the two–way velocity of light which is involved only in these experiments is expressed as $c(\vartheta, v)/c = 1 + A\frac{v^2}{c^2} + B\frac{v^2}{c^2} \cos^2 \vartheta$, where v is the velocity of the laboratory with respect to a preferred frame which usually is taken to be the cosmological frame. In this case the above estimates translate into $|A| \leq 3 \cdot 10^{-7}$ and $|B| \leq 2 \cdot 10^{-10}$ [31]. The experimental results can also be interpreted within a dynamical test theory for the electromagnetic field like the Standard Model Extension [34] or even more general models [35]. Within a dynamical test theory for the electromagnetic field like the Standard Model Extension [34] or even more general models [35].

3.3.5. Time dilation. Also the time dilation factor has been confirmed with much better accuracy using ions in an storage ring moving with a velocity of $v = 0.064\,c$. With the time dilation factor γ in the parametrization $\gamma(v) = 1 + \left(\frac{1}{2} + \alpha\right)\frac{v^2}{c^2} + \dots$ (for SR we have $\alpha = 0$) the most recent experiment gave $|\alpha| \leq 2.2 \cdot 10^{-7}$ [36]. Time dilatation has also be verified by the decay of moving elementary particles, see [37] for a recent version of these experiments where time dilation has been verified at the 10^{-3} level. Though these experiments are not as precise as the spectroscopic ones they prove the time dilatation for a different physical process and, thus, its universality. Another class of time dilation experiments are rotor experiments [38, 39, 40, 41]. The achieved accuracy was $|\alpha| \leq 10^{-5}$. It has been claimed [42] that an accuracy of $|\alpha| \leq 10^{-7}$ has been obtained, but no paper appeared. This type of experiment has also been interpreted as the proof of the isotropy of the one–way velocity of light since in these experiments light only runs from the emitter at the centre of rotation to the absorber mounted on the rotors. However, during the rotation a constant distance from the axis of rotation to the rotor has been taken as granted. Since this this constancy depends on the synchronization procedure, also these experiments finally also test only the two–way velocity of light.

3.3.6. Isotropy in the matter sector. Other aspects of violations of SR are an anomalous inertial mass tensor [43] in the Schrödinger equation or orientation–dependent spin effects for massive particles, and effects related to higher order derivatives in the Maxwell or Dirac equations. Anomalous mass tensors are looked for in the famous Hughes–Drever experiments [44, 45] and modern versions of it [46, 47, 48] constraining these effects by an order of 10^{-30}. Also an anomalous coupling of the spin to some given cosmological vector or tensor fields destroys the Lorentz invariance. Recent tests gave no evidence for an anomalous spin coupling neither to the neutron [49, 50], the proton [51], nor the electron [52, 53] which are all absent to the order of 10^{-31} GeV (for a recent review on the experimental search for anomalous spin couplings, see [54]).

Spectroscopy of anti–hydrogen which may yield information about the validity of the *PCT* symmetry is in a planning status.

More complete reviews of the experimental status of the foundations of SR can be found in [22, 55, 56]. For a description of technological applications of SR one may contact [57].

3.4. Implications for the equations of motion

The EEP not only determines the metrical structure of gravity but also fixes the structure of the equations of motion of the matter content in the universe, that is, of the

- Maxwell equations,
- Dirac equation (which in the non–relativistic limit leads to the Schrödinger equation),
- Structure of the Standard Model.

Since gravity is what can be explored by the dynamics of tests matter like point particles, matter fields, etc., it is clear that any restriction in that dynamics also restricts the degrees of freedom of the gravitational field.

3.4.1. Implication for point particles and light rays. As a particular well worked out example of that scheme is the Ehlers–Pirani–Schild (EPS) constructive axiomatic approach to the gravitational field [58, 59]. This approach is based on the most simple physical object one can think of: structureless point particles and light rays. Assuming (i) that there are only two light rays connecting one space–time point with a (nearby) trajectory introduces a conformal metrical structure, that is, a Riemannian metric up to a position–dependent conformal factor. This establishes pointwise a Lorentzian structure given by the metric $g_{\mu\nu}$. Assuming furthermore (ii) that the trajectory of a structureless uncharged point particle is completely determined by stating its position and velocity – what is equivalent to the UFF – gives a path structure (the set of all paths up to a reparametrization) on the manifold. Requiring (iii) that the path structure is compatible with the conformal structure in the sense that for each direction inside the light cone there is a particle trajectory leads to a Weylian structure (otherwise it would be possible to single out a preferred frame which violates LLI). The last requirement, namely that there is no second clock effect which amounts to requiring the universality of clocks in gravitational fields and, thus, the validity of the UGR, finally gives a Riemannian structure, that is

$$D_v v = \alpha v \tag{3.3}$$

as the equation of motion for a point particle where α is an arbitrary function. Here D is the covariant derivative based on the the Christoffel connection calculated from the Riemannian metric $g_{\mu\nu}$.

3.4.2. Implication for spin-$\frac{1}{2}$ particles.
Another example of a physical system are spin-$\frac{1}{2}$ particles described by a multicomponent complex vector field. Though the operational definition of such a field has not yet been worked out in a fully satisfactory way, the consequences of the requirement of the EEP on the dynamics of such a matter field are quite clear.

The first step is to set up a general dynamics for such a matter field by requiring the existence of a mapping $\psi_t = U(t, t_0)\psi_{t_0}$ where t is related to a foliation of space-time. Now we require this dynamics to be linear which implies U to be a linear operator. The requirement that there are no solutions propagating with an infinite velocity has the consequences that the dynamics is local (that is, the general dynamical equation is equivalent to a linear system of partial differential equations) and that the system of partial differential equations has to be of first order [60]. In a next step LLI requires that there is only one characteristic cone for this system. This defines a Riemannian metric. Furthermore, the requirement of LLI also implies that there are no other tensor fields which can couple to the matter field. Therefore, we end up with the Dirac equation coupled to a Riemannian space–time metric as the only gravitational field [60]

$$0 = i\gamma^\mu D_\mu \psi - m\psi \tag{3.4}$$

where γ^μ are the Dirac matrices fulfilling the Clifford algebra $\gamma^\mu\gamma^\nu + \gamma^\nu\gamma^\mu = 2g^{\mu\nu}$ and $D_\mu\psi = \partial_\mu\psi + \Gamma_\mu\psi$ is the covariant spinorial derivative with $\Gamma_\mu = -\frac{1}{2}(D_\mu h_a^\nu) h_b^\rho g_{\nu\rho} G^{ab}$. The $G^{ab} = \frac{1}{4}[\gamma^a, \gamma^b]$ are the spinorial generators of the Lorentz transformations, and h_a^μ are the tetrads defined by $h_a^\mu h_b^\nu g_{\mu\nu} = \eta_{ab}$ where η_{ab} is the Minkowski metric. The requirement of UFF or, equivalently, UGR implies that the mass m has to be a constant [60].

3.4.3. Implications for the Maxwell field.
A similar procedure can be carried through for the electromagnetic field. Starting from some general Maxwell equations which are of first order in the electromagnetic field strength tensor and first order in the derivative, the requirement of LLI again amounts to the requirement that the characteristic cones are not allowed to split. From that consideration a coupling to a space–time metric follows [35, 61]. LLI also requires that there are

no other fields the Maxwell equations can couple to. Therefore we arrive at the ordinary Maxwell equations minimally coupled to the space-time metric

$$D_\nu \left(g^{\mu\rho} g^{\nu\sigma} F_{\rho\sigma}\right) = 4\pi j^\mu, \qquad \partial_{[\mu} F_{\nu\rho]} = 0. \qquad (3.5)$$

One interesting feature is that, thus, LLI also is responsible for the validity of charge conservation. Also a coupling to a pseudoscalar field which gives Ni's axion [62] is forbidden since the derivative of a scalar field is a vector which defines a preferred direction and, thus, breaks Lorentz Invariance. Therefore, for the electromagnetic field, LLI alone is enough to show that only a space-time metric can couple to the electromagnetic field.

Another aspect has been discussed by Ni [62] who showed that when using generalized Maxwell equations in order to describe the behaviour of a neutral electromagnetically bound system made up of charged particles this neutral system will violate the UFF. Therefore, also the requirement of the UFF forces the Maxwell equations to be of a certain structure.

3.4.4. Summary. We found that the EEP implies the ordinary equations of motion for the physical system under consideration, that is

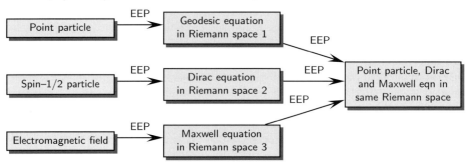

In each case, the gravitational field which is compatible with the EEP is a Riemannian metric. Until now it is possible that each physical system follows another metric. However, in a last step using LLI one requires that, e.g., the maximum velocities of all kinds of matter is the same. This implies that the metrics governing the motion of point particles, of the spin-$\frac{1}{2}$ particles and of the electromagnetic field, are all the same. Therefore we have a universality of the causal cones. As a overall consequence we have that the EEP implies that *gravity is a unique metrical theory*.

3.5. Implications for the gravitational field

Now we have to set up the equations from which the space–time metric can be determined. Until now there is no unique way to derive the Einstein field equation from simple requirements. The presently used scheme relates all possible physical sources of the gravitational field, that is, mass density, pressure, mass currents, etc. to various components of the metric in a combinatorial way which is motivated by the determination of the Newtonian gravitational potential from the mass density. This is the so–called PPN formalism, see below. This is a very general parametrization of all metrical gravitational theories. Within this parametrization one can calculate all the well–known measurable effects like Perihelion shift, red-shift, light bending etc. and also effects which are not present in Einstein's theory like the Nordtvedt effect, effects with a preferred frame, effects related to momentum non–conservation etc. A comparison with the conservation of all these effects then shows that the estimates for all these parameters are compatible with the set of parameters characterizing Einstein's GR. The compatibility in general is at the 10^{-4} level. As a consequence, the space–time metric is determined from the Einstein field equations

$$R_{\mu\nu} - \frac{1}{2} R g_{\mu\nu} = \frac{8\pi G}{c^2} T_{\mu\nu} \tag{3.6}$$

From the Einstein field equation we obtain $D_\nu T^{\mu\nu} = 0$ which are the equations of motion of the matter which creates the gravitational field. This gives back the equations of motion for the point particles, the electromagnetic field and of the Dirac equation.

3.6. Tests of predictions – determination of PPN parameters

All aspects of Einstein's GR are experimentally well tested and confirmed. The tests split into tests of the EEP which we have described above, and tests of the predictions of GR. No single test contradicts its foundation or its predictions. These predictions are

3.6.1. Solar system effects. If the EEP is proven to be valid, then gravity has to be described by a space–time metric $g_{\mu\nu}$. This metric has to be determined by the material content of the universe. From combinatorial considerations of the various contributions to a space–time metric, one obtains a metric in a parametrized from

$$g_{00} = -1 + 2\alpha \frac{U}{c^2} - 2\beta \frac{U^2}{c^4} \tag{3.7}$$

$$g_i := g_{0i} = 4\mu \frac{(\boldsymbol{J} \times \boldsymbol{r})_i}{c^3 r^3} \tag{3.8}$$

$$g_{ij} = (1 + 2\gamma) \frac{U}{c^2} \tag{3.9}$$

where U and $\boldsymbol{J}$ are the Newtonian potential and the angular momentum, respectively, of the gravitating body and the parameters β and γ are just the two most important ones out of 10 possible post–Newtonian parameters, see [21] and references therein. We added the parameter α in order to describe redshift experiments. In Einstein's GR $\alpha = \beta = \gamma = 1$.

Perihelion shift: Contrary to Newtonian celestial mechanics, the planetary el-lipses precess within the orbital plane by the angle

$$\delta\varphi = \frac{2(\alpha + \gamma) - \beta}{3} \frac{6\pi G M}{c^2 a^2 (1 - e^2)} \tag{3.10}$$

per turn what amounts to $45''$ per century for Mercury. This has been observed long before the raise of GR. Due to competing a cause by the Sun's quadrupole moment, the relativistic effect can be confirmed at the 10^{-3} level only.

Gravitational redshift: The speeding up of clocks when being brought to large heights has been best tested by the GP-A mission where a H-maser in a rocket has been compared with a H-maser on ground with the result $|\alpha_{\mathrm{H-maser}} - 1| \leq 7 \cdot 10^{-5}$ [63].

Deflection of light: This was the first prediction of Einstein's GR which has been confirmed by observation only four years after setting up the complete theory. Within the PPN description light is deflected by an angle

$$\delta = \frac{\alpha + \gamma}{2} \frac{4GM_\odot}{c^2 b} , \qquad (3.11)$$

where b is the impact parameter and $M_\odot$ is the mass of the Sun. Today's observations use VLBI what leads to a confirmation of Einstein's theory at the 10^{-4} level [64].

Gravitational time delay: Electromagnetic signals move slower in stronger gravitational fields. Therefore, light or radio signals need a longer time of propagation when the Sun comes nearer to the trajectory of the electromagnetic signals

$$\delta t = 2(\alpha + \gamma) \frac{GM_\odot}{c^3} \ln \frac{4 x_{\mathrm{sat}} x_{\mathrm{Earth}}}{d^2} , \qquad (3.12)$$

where x_{sat} and x_{Earth} are the distance between the Sun and the satellite and the Earth, respectively, and d is the closest approach of the radio signal to the Sun. The best test of this phenomenon has been carried through recently by the Cassini mission where the effect was approx. 10^{-4} s. Due to an new technique using multiband signalling the change of the frequency of the signals due to the change of the impact parameter has been measured. This led to a confirmation of Einstein up to $|\gamma - 1| \leq 10^{-5}$ [65].

Lense–Thirring effect: The rotation of a gravitating body results in a genuine post–Newtonian gravitomagnetic field which, of course, influences the equation of motion of bodies and also the rates of clocks. The influence of this field on the trajectory of satellites results in a motion of the nodes, which has been measured by observing the LAGEOS satellites via laser ranging. Together with new data of the Earth's gravitational field obtained from the CHAMP and GRACE satellites the confirmation recently reached the 10 % level [66].

Schiff effect: The gravitational field of a rotating gravitating body also influences the rotation of gyroscopes. This effect is right now under exploration by the GP-B mission. The science mode and the post–mission calibration mode has been completed, and data analysis will be completed in spring 2007. The accuracy of the measurement of the Schiff effect should be better than 1%.

Test	Experiment	Parameter
Gravitational redshift [63]	GP-A	$\lvert \alpha - 1 \rvert \leq 1.4 \cdot 10^{-4}$
Perihelion shift [21]	Astrophys. observation	$\left\lvert \frac{2(\alpha+\gamma)-\beta}{3} - 1 \right\rvert \leq 10^{-4}$
Light deflection [64]	VLBI	$\lvert \gamma - 1 \rvert \leq 10^{-4}$
Gravitational time delay [65]	Cassini	$\lvert \gamma - 1 \rvert \leq 2 \cdot 10^{-5}$
Lense–Thirring effect [66]	LAGEOS	$\sim 10\%$
Schiff effect [67]	GP-B	$\sim 0.5\%$ (expected)

3.6.2. Strong gravity and gravitational waves. GR in the strong field regime has been proven to be valid to very high accuracy by means of the observation of binary systems, see e.g. [22]. The existence of gravitational wave as predicted by GR has been indirectly proven by the energy loss experienced by binary systems. A new realm of strong gravity tests has been opened up by the discovery of the first double pulsar PSR J0737 3039A and B [68].

4. Unsolved problems: first hints for new physics?

Furthermore, though all experiments and observations are well described by standard physics, there are some problems in gravitational physics, still lacking a convincing solution. These problems are (i) dark matter, (ii) dark energy, and (iii) the Pioneer anomaly. One may also include the recently observed effects of (iv) a secular increase of the Astronomical Unit, (vi) an unexplained quadrupole anomaly in the cosmic microwave background, and (vi) the fly–by anomaly. Like the Pioneer anomaly, the latter effects seem to be related to the physics of and/or in the Solar system only.

(i) Dark matter has been introduced in order to "explain" the gravitational field needed for the rotation curves and the gravitational lensing of galaxies [69]. It also appears in the spectral decomposition of the cosmic microwave background radiation [70]. Since no particle has been found which can be identified as constituents of dark matter, the notion "dark matter" is just a synonymous for the fact that the gravitational field as seen by stars and light rays is stronger than expected from the observed possible sources. As an alternative to dark matter it is also possible to modify the equations for the gravitational field or to modify the equations of motion governing the behaviour of particles in gravitational fields. As example for the modification of the gravitational field, a Yukawa potential has been discussed for the explanation of galactic rotation curves [71]. A modification of the equation of motion is MOND which very successfully can "explain" a large range of observed galactic rotation curves [72]. This MOND ansatz recently has been put into the form of the relativistic field theory [73].

(ii) Similarly, recent observations of the Lyman-alpha forest lines indicate, according to the pioneering work of Priester and others [74, 75, 76], that the expansion of the universe is accelerating and that 75% of the total energy density consists of a mysterious Dark Energy component with negative pressure

[77, 78]. This and data from type Ia supernovae, the WMAP measurements of the cosmic microwave background [79, 80], as well as of the galaxy power spectrum suggest – when compared with standard cosmological models – the real existence of the Dark Energy rather than a modification of the basic laws of gravitation [81]. However, also in this case there are attempts to give an explanation in terms of modified field equations, see, e.g., [82]. Recently it has been claimed that Dark Energy or, equivalently, the observed acceleration of the universe can be explained by inhomogeneous cosmological models, such as the spherically–symmetric Lemaitre–Tolman–Bondi model, see, e.g., [83, 84, 85].

(iii) And at last we have the anomalous acceleration observed for the Pioneer 10 and 11 spacecraft [86, 87]. Though it most probably seems to be a systematic effect, no explanation has been found. Therefore, if this anomalous acceleration is of true gravitational origin, then it is an important question whether this effect is related to the issue of dark matter and dark energy.

(iv) Furthermore, there is a recently reported observation of a secular increase of the Astronomical Unit [88, 89, 90]. This increase is too large in order to be induced by the cosmological expansion. These new results have already been used in order to analyze planetary data in the framework of a particular brane model [91, 92].

(v) Last year an anomaly of the microwave background radiation has been reported [93]: Beside the dipole "anomaly" which can be understood in terms of a relative motion of the Sun with respect to the cosmological frame, an quadrupole and also an octopole anomaly has been found which geometrical orientation seems to be related to the orientation of the Solar system. This might perhaps indicate another still unknown physical phenomenon related to the physics of the Solar system.

(vi) Though it is not clear whether this has something to do with the above phenomena, we mention a fly–by anomaly experienced by satellite navigators during the last years [94]. This fly–by anomaly consists of an unexplained velocity increase during the fly–by of satellites at the Earth. This velocity increase is several mm/s which is by many orders above the measurement accuracy. An increase in the velocities may indicate a slightly stronger gravitational field than given by Newtonian theory. Unfortunately, no systematic study of this effect has been carried through until now.

There are other approaches which try to explain the effects usually ascribed to dark matter and dark energy. These approaches invoke a modification of the laws of gravitation or, more generally, a modification of the standard model. If no particle associated to dark matter or dark energy can be detected in the future (until now all attempts in that direction failed) these two schemes of explanation are on the same footing. In any case, dark matter, dark energy and the Pioneer anomaly, if it proves to be not a systematic error, very probably are hints to new areas of physics.

5. On the magnitude of quantum gravity effects

Since the energy scale of QG effects at the first sight is of the order of the Planck energy $E_{QG} \sim E_{Planck} \sim 10^{28}$ eV which laboratory energy scale are of the order of eV (for experiments in high energy laboratories the energy may reach GeV), the expected effects in the laboratory should be of the order $\sim 10^{-28}$. Though this is not really encouraging, there is some hope for circumventing this. First, $E_{QG} \sim E_{Planck}$ is presently just a hypothesis, there are many examples that QG effects might be amplified, and, astonishingly, there are devices which, in principle, may reach the accuracy of 10^{-28}.

At first, the QG energy scale E_{QG} might be smaller than the Planck energy. Arguments for that are:

- If QG lives in "large extra dimensions", then additional scales like compactification scales can come in which may enhance some effects. As an example, we mention the deviation from Newton's law at small distances which can be very large, even larger than Newtonian gravity.
- In string–theory–motivated "dilaton scenarios" it seems to be possible that the UFF may be violated at the 10^{-13} level [95, 96]. Furthermore, also the PPN parameter γ may differ from its GR value 1 by $|\gamma - 1| \sim 10^{-5}$.
- Low–energy data suggest that electroweak and strong interactions may unify at GUT energy scale of $\sim 10^{16}$ GeV, perhaps also gravity would be of the same strength as the other interactions at that scale.

Another way to "beat" the QG scale is to use some amplification by large factors given by the experimental setup. Examples for that are

- The neutral kaon system: QG may affect properties of the neutral–kaon system which are then amplified by peculiarly small mass difference between long–lived and the short–lived kaons $M_{L,S}/|M_L - M_S| \sim 10^{14}$ [97].
- It is generally expected that QG leads to fluctuations of space–time (in the metric, for example). These fluctuations should induce some fundamental and universal noise in physical systems like photons, electrons, atoms etc. For a wide class of fluctuations this should be observable as some $1/f$–noise which offers the opportunity that the effect increases for long time scales. Experimentally, such noise has been searched for in high precision long–term stable optical resonators [98].
- Sensitivity of some clock–comparison experiments is amplified by $m_p c^2/(h\nu) \sim 10^{18}$ where ν is the clock frequency and m_p the proton mass.

As last point we may mention that there are a few exceptional experimental devices which are in principle able to reach the accuracy of 10^{-28}. One such device is the next generation LIGO gravitational wave interferometer. With a measurement time of abut one year this device might be sensitive to QG induced modifications of the dispersion relation [99].

As a consequence, laboratory experiments may very well be capable to search for possible QG effects. Furthermore, also compared with the search for QG effects through astrophysical observation where one has access to very high energy cosmic rays there are some advantages of low energy laboratory experiments.

While astrophysical observations one has access to ultra high energies of more than 10^{21} eV these observations are plagued with the disadvantage that there is no systematic repeatability of these observations and that there is no unique interpretation of the results.

Accordingly the advantages of a controlled laboratory search consists in the repeatability of the experiment and the possibility of a systematic variation of initial and boundary conditions. This can be used for a unique identification of the cause of the effect as well as for an improvement of the precision of the result. Another advantage is that certain regimes can be accessed in th laboratory only. For example, ultra–low temperatures or ultrastable devices like optical resonators can be build and maintained in the laboratory only.

Due to stability and repeatability of experiments, laboratory searches for QG effects may be as promising as astrophysical observation

6. How to search for quantum gravity effects

Since QG replaces GR and/or quantum theory which are both universally valid QG should affect *all* physical phenomena. However, not all phenomena are equally sensitive to the expected QG modifications. Therefore one needs some kind of strategy for the search for QG effects.

Standard physics is supported by all present experimental data. These data have been obtained within some standard domain of experimental accessibility, that is, for some energy range, some velocity, distance, temperature range, etc. Therefore, one first attempt to search for QG effect is to explore new regimes, that is, to go to higher energies, to lower temperatures, to longer distances, to longer and shorter time scales, etc.: A search for new effects needs the exploration of new non–standard experimental situations. These situations are, for example,

Extreme high energies: This regime is well suited for the search for deviations from the standard dispersion relation for elementary particles.

Extreme low energies: With low temperatures one may search for fundamental noise arising from space–time fluctuations, which may lead to a fundamental decoherence of quantum systems at temperatures lower than 500 fK. Such temperatures may be achieved in BECs.

Extreme large distances: Gravity at long distances became the subject of discussion very recently since the unexplained phenomena dark energy, dark matter, and the Pioneer anomaly are related to large distances. Consequently QG induced modifications of gravity has bee proposed, see e.g. [100]. Furthermore, the detection of ultra–low frequency gravitational waves which give information about the very early universe where QG effects are surely more pronounced, also need very long distances.

Extreme small distances: Here, 'small distance' means small in relation to the explored interaction. For gravity, the sub-mm domain is a very small distance. Consequently, following suggestions form higher dimensional theories, violation of Newton's law at sub–mm distances have been looked for.

Extreme long timescales: Long timescales are in favour for a search of fundamental noises and also in the search for the time–dependence of fundamental constants.

Extreme short timescales: On the one hand short time scales are related to high energies but also in a better resolution of the dynamics of various physical systems.

7. Outlook

We present some of the ongoing projects related to improvements of fundamental physics tests:

Isotropy of c: New and improved laboratory set–ups may lead to an improvement of tests of the isotropy of the velocity of light by one to two orders thus approaching the 10^{-18}. In space this may be improved by further 2 orders of magnitude thus reaching the 10^{-20} level.

UFF: Since Earth bound experiments testing the UFF reached some principle threshold, there are attempts to improve this accuracy by performing such tests in space. The French MICROSCOPE project which is scheduled to be launched in 2009 should improve the limit to the 10^{-15} level [101], and there are plans to improve that further by an ESA–NASA mission STEP to the 10^{-18} level [102]. Furthermore, the nowadays standardized production of anti Hydrogen stimulated plans to perform free fall tests with anti hydrogen [16] which then would be the first UFF test with anti matter.

PPN γ: Some future astronomy and fundamental physics space missions like Gaja [103], LATOR [104], and ASTROD [105] have the capability to measure the PPN parameter γ with a hugely improved accuracy of 10^{-9}. Having in mind the possibility that dilation scenarios predict a deviation of γ from unity at the 10^{-5} level [96] this would be a serious test of these kinds of QG scenarios.

PPN β: Also the PPN parameter β can be measured by ASTROD with 10^{-9} accuracy.

UGR: In the near and far future there will be various clock missions which considerably may improve time measurement and related fundamental tests. The CNES/ESA mission PHARAO/ACES will bring atomic clocks on board of the ISS in space which will have a 10^{-16} stability [106]. PHARAO is a clock based on laser–cooled atoms. In future optical clocks will have a stability of 10^{-18}. Proposed space missions like SPACETIME [107, 108] and OPTIS [109] with a set of different clocks may give improvements of tests of the UGR up to an accuracy of 10^{-10}. Very important are future anti–hydrogen clocks which will have same precision as H-clocks and which will yield the first test of UGR for "anti clocks".

BEC: The present world record for the lowest temperature is 500 fK; in the near future it is feasible that, owing to an undisturbed expansion in free fall, a temperature below 1 fK is possible. This opens a new temperature regime with perhaps new phenomena and also leads to an improvement for the preparation of atoms for atomic interferometry and atomic clocks.

Gravity at large distances: New information about gravity at large distances may be obtained from a new analysis of the complete set of data from the Pioneer 10 and 11 missions. Furthermore, a new mission, a Deep Space Gravity Explorer mission, is in a proposing stage [110].

Condensed matter: Much more precise tests of the renormalization group theory using fluid Helium can be performed in gravity–free environment since gravity destroys the symmetry of the systems [106]. Unfortunately, these tests which were planned to be performed on the ISS have been cancelled.

UHECR: New detectors for high energy cosmic rays in Argentina and other places will yield much more information about e.g. dispersion relations.

Atom interferometry: New refined methods in atomic interferometry have the capability to measure the fine structure constant α with an accuracy of 10^{-10} which is one order of improvement compared with the present accuracy [111]. This leads to new checks of QED.

In summary we can state that:

- All kinds of experimental tests have to be improved.
- Experimental progress is also important for daily life technology.
- For a search for QG effects there is no preference for astrophysical observations, laboratory experiments are equally well suited for that task, and finally:
- If we suppose the seminal paper [112] as a starting point, so far we have had *only 8 years of a dedicated search for QG effects.*

Acknowledgements

I like to thank H. Dittus and H. Müller for fruitful discussions. This work has been supported by the German Aerospace Agency DLR.

References

[1] S. Baeßler, B.R. Heckel, E.G. Adelberger, J.H. Gundlach, U. Schmidt, and H.E. Swanson. Improved tests of the equivalence principle for gravitational self–energy. *Phys. Rev. Lett.*, 83:3585, 1999.

[2] T.M. Niebauer, M.P. McHugh, and J.E. Faller. Galilean test for the fifth force. *Phys. Rev. Lett.*, 59:609, 1987.

[3] K. Kuroda and N. Mio. Test of composition–dependent force by a free–fall interferometer. *Phys. Rev. Lett.*, 62:1941, 1989.

[4] K. Kuroda and N. Mio. Limits on a possible composition–dependent force by a Galilean experiment. *Phys. Rev.*, 42:3903, 1990.

[5] L. Koester. Absolutmessung der kohärenten Streulängen von Wasserstoff, Kohlenstoff und Chlor sowie Bestimmung der Schwerebeschleunigung für freie Neutronen mit dem Schwerkraft–Refraktometer am FRM. *Z. Physik*, 198:187, 1967.

[6] L. Koester. Verification of the equivalence principle of gravitational and inertial mass for the neutron. *Phys. Rev.*, D 14:907, 1976.

[7] S.A. Werner and H. Kaiser. Neutron interferometry – macroscopic manifestation of quantum mechanics. In J. Audretsch and V. de Sabbata, editors, *Quantum Mechanics in Curved Space–Time*, volume NATO ASI Series B 230, page 1. Plenum Press, New York, 1990.

[8] A. Peters, K.Y. Chung, and S. Chu. Measurement of gravitational acceleration by dropping atoms. *Nature*, 400:849, 1999.

[9] A. Peters, K.Y. Chung, and S. Chu. High-precision gravity measurements using atom interferometry. *Metrologia*, 38:25, 2001.

[10] B.S. DeWitt and R.W. Brehme. Radiation Damping in a Gravitational Field. *Ann. Phys. (NY)*, 9:220, 1960.

[11] H. Dittus, C. Lämmerzahl, and H. Selig. Testing the universality of free fall for charged particles. *Gen. Rel. Grav.*, 36:571, 2004.

[12] F.C. Witteborn and W.M. Fairbank. Experimental comparison of the graviational force on freely falling electrons and metallic electrons. *Phys. Rev. Lett.*, 19:1049, 1967.

[13] J.E. Moody and F. Wilczek. New macroscopic forces? *Phys. Rev.*, D 30:130, 1984.

[14] R. Bluhm. Overview of the SME: Implications and phenomenology of Lorentz violation. In J. Ehlers and C. Lämmerzahl, editors, *Special Relativity*, volume LNP, page in press. Springer–Verlag, Berlin, 2006.

[15] C.-H. Hsieh, P.-Y. Jen, K.-L. Ko, K.-Y. Li, W.-T. Ni, S.-S. Pan, Y.-H. Shih, and R.-J. Tyan. The equivalence principle experiment for spin–polarized bodies. *Mod. Phys. Lett.*, 4:1597, 1989.

[16] J. Walz and T.W. Haensch. A proposal to measure antimatter gravity using ultracold antihydrogen atoms. *Gen. Rel. Grav.*, 36:561, 2004.

[17] J.P. Turneaure and S.R. Stein. Development of the superconducting cavity oscillartor. In J.D. Fairbank, B.S. Deaver, C.W.F. Everitt, and P.F. Michelson, editors, *Near Zero*, page 414. Freeman, New York, 1988.

[18] A. Godone, C. Novero, and P. Tavella. Null gravitational redshift experiments with nonidentical atomic clocks. *Phys. Rev.*, D 51:319, 1995.

[19] C. Braxmaier, H. Müller, O. Pradl, J. Mlynek, A. Peters, and S. Schiller. Test of relativity using a cryogenic optical resonator. *Phys. Rev. Lett.*, 88:010401, 2002.

[20] A. Bauch and S. Weyers. New experimental limit on the validity of local position invariance. *Phys. Rev.*, D 65:081101(R), 2002.

[21] C.M. Will. *Theory and Experiment in Gravitational Physics (Revised Edition)*. Cambridge University Press, Cambridge, 1993.

[22] C.M. Will. The confrontation between general relativity and experiment. *Living Rev. Relativity*, 2001, ww.livingreviews.org/lrr-2001-4.

[23] K. Brecher. Is the speed of light independent of the velocity of the source? *Phys. Rev. Lett.*, 39:1051, 1977.

[24] T. Alväger, F.J.M. Farley, J. Kjellmann, and I. Wallin. Test of the second postulate of Special Relativity in the GeV region. *Phys. Lett.*, 12:260, 1964.

[25] B.C. Brown, G.E. Masek, T. Maung, E.S. Miller, H. Ruderman, and W. Vernon. Experimental comparison of the velocities of eV (visible) and GeV electromagnetic radiation. *Phys. Rev.*, 30:763, 1973.

[26] Z.G.T. Guiragossian, G.B. Rothbart, M.R. Yearian, R.A. Gearhart, and J.J. Murray. Relative velocity measurement of electrons and gamma rays at 15 GeV. *Phys. Rev. Lett.*, 34:335, 1975.

[27] J. Alspector, G.R. Kalbfleisch, N. Baggett, E.C Fowler, B.C. Barish, A. Bodek, D. Buchholz, F.J. Sciulli, E.J. Siskind, L. Stutte, H.E. Fisk, G. Krafczyk, D.L. Nease, and O.D. Fackler. Experimental comparison of neutrino and muon velocities. *Phys. Rev. Lett.*, 36:837, 1976.

[28] G.R. Kalbfleisch, N. Baggett, E.C. Fowler, and J. Alspector. Experimental comparison of Neutrino, Antoneutrino, and Muon velocities. *Phys. Rev. Lett.*, 43:1361, 1979.

[29] A. Kostelecky and M. Mewes. Signals for Lorentz violation in electrodynamics. *Phys. Rev.*, D 66:056005, 2002.

[30] P. Antonini, M. Okhapkin, E. Göklü, and S. Schiller. Test of constancy of speed of light with rotating cryogenic optical resonators. *Phys. Rev.*, A 71:050101, 2005.

[31] P. Stanwix, M.E. Tobar, P. Wolf, C.R. Susli, M. and F. Locke, E.N. Ivanov, J. Winterflood, and F. van Kann. Test of Lorentz invariance in electrodynamics using rotating cryogenic sapphire microwave oscillators. *Phys. Rev. Lett.*, 95:040404, 2005.

[32] S. Herrmann and A. Peters. *Special Relativity: Will It Survive the Next 101 Years?*, In J. Ehlers and C. Lämmerzahl, editors, volume LNP, to appear. Springer–Verlag, Berlin, 2006.

[33] P. Wolf, S. Bize, A. Clairon, G. Santarelli, M.E. Tobar, and A.N. Luiten. Improved test of Lorentz invariance in electrodynamics. *Phys. Rev.*, D 70:051902, 2004.

[34] D. Colladay and V.A. Kostelecky. Lorentz–violating extension of the standard model. *Phys. Rev.*, D 58:116002, 1998.

[35] C. Lämmerzahl, A. Macias, and H. Müller. Lorentz invariance violation and charge (non–)conservation: A general theoretical frame for extensions of the Maxwell equations. *Phys. Rev.*, D 71: 025007, 2005.

[36] S. Saathoff, S. Karpuk, U. Eisenbarth, G. Huber, S. Krohn, R. Muñoz-Horta, S. Reinhardt, D. Schwalm, A. Wolf, and G. Gwinner. Improved test of time dilation in special relativity. *Phys. Rev. Lett.*, 91:190403, 2003.

[37] J. Bailey, K. Borer, F. Combley, H. Drumm, F. Krienen, F. Langa, E. Picasso, W. van Ruden, F.J.M. Farley, J.H. Field, W. Flegl, and P.M. Hattersley. Measurements of relativistic time dilatation for positive and negative muons in a circular orbit. *Nature*, 268:301, 1977.

[38] M. Ruderfer. First–order terrestrial ether drift experiment using Mössbauer radiation. *Phys. Rev. Lett.*, 5:191, 1960.

[39] M. Ruderfer. First–order terrestrial ether drift experiment using Mössbauer radiation (erratum). *Phys. Rev. Lett.*, 7:361, 1961.

[40] D.C. Champeney, G.R. Isaak, and A.M. Khan. An 'aether drift' experiment based on the Mössbauer effect. *Phys. Lett.*, 7:241, 1963.

[41] D.C. Champeney, G.R. Isaak, and A.M. Khan. A time dilation experiment based on the Mössbauer effect. *Proc. Phys. Soc.*, 85:583, 1965.

[42] G.R. Isaak. The Mössbauer effect: Application to relativity. *Phys. Bull.*, 21:255, 1970.

[43] M.P. Haugan. Energy conservation and the principle of equivalence. *Ann. Phys.*, 118:156, 1979.

[44] V.W. Hughes, H.G. Robinson, and V. Beltran-Lopez. Upper limit for the anisotropy of inertial mass from nuclear resonance experiments. *Phys. Rev. Lett.*, 4:342, 1960.

[45] R.W.P. Drever. A search for the anisotropy of inertial mass using a free precession technique. *Phil. Mag.*, 6:683, 1961.

[46] J.D. Prestage, J.J. Bollinger, W.M. Itano, and D.J. Wineland. Limits for spatial anisotropy by use of nuclear–spin–polarized ^{9}Be$^+$ ions. *Phys. Rev. Lett.*, 54:2387, 1985.

[47] S.K. Lamoreaux, J.P. Jacobs, B.R. Heckel, F.J. Raab, and E.N. Fortson. New limits on spatial anisotropy from optically pumped ^{201}Hg and ^{199}Hg. *Phys. Rev. Lett.*, 57:3125, 1986.

[48] T.E. Chupp, R.J. Hoara, R.A. Loveman, E.R. Oteiza, J.M. Richardson, and M.E. Wagshul. Results of a new test of local Lorentz invariance: A search for mass anisotropy in ^{21}Ne. *Phys. Rev. Lett.*, 63:1541, 1989.

[49] D. Bear, R.E. Stoner, R.L. Walsworth, V.A. Kostelecky, and C.D. Lane. Limit on Lorentz and *CPT* violation of the neutron using a two–species noble–gas maser. *Phys. Rev. Lett.*, 85:5038, 2000.

[50] D. Bear, R.E. Stoner, R.L. Walsworth, V.A. Kostelecký, and C.D. Lane. Erratum: Limit on Lorentz and *CPT* violation of the neutron using a two-species noble-gas maser. *Phys. Rev. Lett.*, 89:209902(E), 2002.

[51] M.A. Humphrey, D.F. Phillips, E.M. Mattison, and R.L. Walsworth. Testing *CPT* and Lorentz symmetry with hydrogen masers. *Phys. Rev.*, A 68:063807, 2003.

[52] L.-S. Hou, W.-T. Ni, and Y.-C. Li. Test of cosmic spatial isotropy for polarized electrons using rotatable torsion balance. *Phys. Rev. Lett.*, 90:201101, 2003.

[53] B.R. Heckel. Torsion balance test of Lorentz symmetry violation. In V.A. Kostelecky, editor, *CPT and Lorentz Symmetry III*, page 133. Word Scientific, Singapore, 2004.

[54] R.L. Walsworth. Tests of Lorentz–symmetry in the spin–coupling sector. In J. Ehlers and C. Lämmerzahl, editors, *Special Relativity*, volume LNP, page in press. Springer–Verlag, Berlin, 2006.

[55] G. Amelino-Camelia, C. Lämmerzahl, A. Macias, and H. Müller. The search for quantum gravity signals. In A. Macias, C. Lämmerzahl, and D. Nunez, editors, *Gravitation and Cosmology*, page 30. AIP Conference Proceedings 758, Melville, New York, 2005.

[56] D. Mattingly. Modern tests of Lorentz invariance. *Living Reviews* **8**, http://www.livingreviews.org/lrr-2005-5 (cited on April 15, 2006).

[57] C. Lämmerzahl. Relativity and techology. *Ann. Phys. (Leipzig)*, 15:5, 2006.

[58] J. Ehlers, F.A.E. Pirani, and A. Schild. The geometry of free fall and light propagation. In L. O'Raifeartaigh, editor, *General Relativity, Papers in Honour of J.L. Synge*, page 63. Clarendon Press, Oxford, 1972.

[59] J. Ehlers. Survey of general relativity theory. In W. Israel, editor, *Relativity, Astrophysics and Cosmology*, page 1. Reidel, Dordrecht, 1973.

[60] J. Audretsch and C. Lämmerzahl. A new constructive axiomatic scheme for the geometry of space-time. In Majer U. and Schmidt H.-J., editors, *Semantical Aspects of Space-Time Geometry*, page 21. BI Verlag, Mannheim, 1993.

[61] C. Lämmerzahl and F.W. Hehl. Riemannian light cone from vanishing birefringence in premetric vacuum electrodynamics. *Phys. Rev.*, D 70:105022, 2004.

[62] W.-T. Ni. Equivalence principles and electromagnetism. *Phys. Rev. Lett.*, 38:301, 1977.

[63] R.F.C. Vessot, M.W. Levine, E.M. Mattison, E.L. Blomberg, T.E. Hoffmann, G.U. Nystrom, B.F. Farrel, R. Decher, P.B. Eby, C.R. Baughter, J.W. Watts, D.L. Teuber, and F.D. Wills. Test of relativistic gravitation with a space–borne hydrogen maser. *Phys. Rev. Lett.*, 45:2081, 1980.

[64] S.S. Shapiro, J.L. Davis, D.E. Lebach, and J.S. Gregory. Measurement of the Solar gravitational deflection of radio waves using geodetic very–large baseline interferometry data 1979 – 1999. *Phys. Rev. Lett.*, 92:121101, 2004.

[65] B. Bertotti, L. Iess, and P. Tortora. A test of general relativity using radio links with the Cassini spacecraft. *Nature*, 425:374, 2003.

[66] I. Ciufolini. Frame-dragging and Lense-Thirring effect. *Gen. Rel. Grav.*, 36:2257, 2004.

[67] C.W.F. Everitt, S. Buchman, D.B. DeBra, G.M. Keiser, J.M. Lockhart, B. Muhlfelder, B.W. Parkinson, J.P. Turneaure, and other members of the Gravity Probe B team. Gravity Probe B: Countdown to launch. In C. Lämmerzahl, C.W.F. Everitt, and F.W. Hehl, editors, *Gyros, Clocks, and Interferometers: Testing Relativistic Gravity in Space*, page 52. Springer–Verlag, Berlin, 2001.

[68] M. Kramer, I.H. Stairs, R.N. Manchester, M.A. MacLaughlin, A.G. Lyre, R.D. Ferdman, M. Burgag, D.R. Lorimer, A Possenti, N. D'Amico, J. Sarkission, B.C. Joshi, P.C.C. Freire, and F. Camilo. Strong field tests of gravity with the double pulsar. *Ann. Phys. (Leipzig)*, 15:34, 2006.

[69] T. Sumner. Experimental searches for dark matter. *Living Rev. Relativity*, 5:http://www.livingreviews.org/lrr–2002–4 (cited on 20.11.2005), 2002.

[70] W. Hu. Mapping the dark matter through the cosmic microwave background damping tail. *Astrophys. J.*, 557:L79, 2001.

[71] R.H. Sanders. Anti–gravity and galaxy rotation curves. *Astron. Astrophys.*, 136:L21, 1984.

[72] R.H. Sanders and S.S. McGaugh. Modified Newtonian Dynamics as an Alternative to Dark Matter. *Ann.Rev.Astron.Astrophys.*, 40:263, 2002.

[73] J.D. Bekenstein. Relativistic gravitation theory for the modified Newtonian dynamics paradigm. *Phys. Rev.* D 70:083509, (2004).

[74] C. van de Bruck and W. Priester. The Cosmological Constant Λ, the Age of the Universe and Dark Matter: Clues from the Lyman-α–Forest. In H.V. Klapdor–Kleingrothaus, editor, *Dark Matter in Astrophysics and Particle Physics 1998:*

Proceedings of the Second International Conference on Dark Matter in Astrophysics and Particle, Inst. of Physics, London, 1998.

[75] W. Priester and J.O. Overduin. How dominant is the vacuum? *Naturwiss.* 88:229, 2001.

[76] M. Tegmark et al. Cosmological parameters from SDSS and WMAP. *Phys. Rev.*, D 69:103501, 2004.

[77] A.G. Riess, A.V. Filippenko, and P. Challis et al. Measurements of omega and lambda from 42 high–redshift supernovae. *Astron. J.*, 116:1009, 1998.

[78] S. Perlmutter, G. Aldering, and G. Goldhaber et al. Measurements of omega and lambda from 42 high–redshift supernovae. *Astroph. J.*, 517:565, 1999.

[79] C.L. Bennet et al. First-year Wilkinson Microwave Anisotropy Probe (WMAP) observations: Preliminary maps and basic results. *Astrophys. J. Suppl. Ser.*, 148:1, 2003.

[80] N.N. Spergel et al. First-year Wilkinson Microwave Anisotropy Probe (WMAP) observations: Determination of cosmological parameters. *Astrophys. J. Suppl. Ser.*, 148:175, 2003.

[81] P.J.E. Peebles and B. Ratra. The cosmological constant and dark energy. *Rev. Mod. Phys.*, 75:559, 2003.

[82] S. Nojiri and S.D. Odintsov. Introduction to Modified Gravity and Gravitational Alternative for Dark Energy. hep-th/0601213, 2006.

[83] M.N. Celerier. Do we really see a cosmological constant in the supernovae data? *Astron. Astroph.*, 353:63, 2000.

[84] R.A. Vanderveld, E.E. Flanagan, and I. Wasserman. Mimicking Dark Energy with Lemaitre-Tolman-Bondi Models: Weak Central Singularities and Critical Points. Phys. Rev. D 74: 023506, 2006, astro-ph/0602476.

[85] P.S. Apostolopoulos, N. Broudzakis, N. Tetradis, and E. Tsavara. Cosmological Acceleration and Gravitational Collapse. astro-ph/0603234, 2006.

[86] J.D Anderson, P.A. Laing, E.L. Lau, A.S. Liu, M.M. Nieto, and S.G. Turyshev. Indication, from Pioneer 10/11, Galileo, and Ulysses Data, of an Apparent Anomalous, Weak, Long–Range Acceleration. *Phys. Rev. Lett.*, 81:2858, 1998.

[87] J.D. Anderson, P.A. Laing, E.L. Lau, A.S. Liu, M.M. Nieto, and S.G. Turyshev. Study of the anomalous acceleration of Pioneer 10 and 11. *Phys. Rev.*, D 65:082004, 2002.

[88] G.A. Krasinsky and V.A. Brumberg. Secular increase of astronomical unit from analysis of the major planets motions, and its interpretation. *Celest. Mech. & Dyn. Astron.*, 90:267, 2004.

[89] E.V. Pitjeva. High-precision ephemerides of planets-EPM and determinations of some astronomical constants. *Sol. Sys. Res.*, 39:176, 2005.

[90] E.M. Standish. The Astronomical Unit now. In D.W. Kurtz, editor, *Transits of Venus: New Views of the Solar System and Galaxy, Proceedings IAU Colloquium No. 196*, page 163. Cambridge University Press, Cambridge, Cambridge, 2005.

[91] L. Iorio. On the effects of the Dvali-Gabadadze-Porrati braneworld gravity on the orbital motion of a test particle. *Class. Quant. Grav.* 22:5271–5281, 2005, gr-qc/0504053.

[92] L. Iorio. Secular increase of the Astronomical Unit and perihelion precessions as tests of the Dvali-Gabadadze-Porrati multi-dimensional braneworld scenario. gr-qc/0508047.

[93] D.J. Schwarz, G.D. Starkman, D. Huterer, and C.J. Copi. Is the low-ℓ microwave background cosmic? *Phys. Rev. Lett.*, 93:221301, 2004.

[94] T. Morley. Private communication. 2005.

[95] T. Damour, F. Piazza, and G. Veneziano. Runaway dilaton and equivalence principle violations. *Phys. Rev. Lett.*, 89:081601, 2002.

[96] T. Damour, F. Piazza, and G. Veneziano. Violations of the equivalence principle in a dilaton–runaway scenario. *Phys. Rev.*, D 66:046007, 2002.

[97] J. Ellis, J.L. Lopes, N.E. Mavromatos, and D.V. Nanopoulos. Precision test of CPT–symmetry and quantum mechanics in the neutral kaon system. *Phys. Rev.*, D 53:3846, 1996.

[98] S. Schiller, C. Lämmerzahl, H. Müller, C. Braxmaier, S. Herrmann, and A. Peters. Experimental limits for low-frequency space-time fluctuations from ultrastable optical resonators. *Phys. Rev.*, D 69:027504, 2004.

[99] G. Amelino-Camelia and C. Lämmerzahl. Quantum-gravity-motivated Lorentz-symmetry tests with laser interferometers. *Class. Quantum Grav.*, 21:899, 2004.

[100] D. Dvali, G. Gabadadze, and M. Porrati. 4D gravity on a brane in 5D Minkowski space. *Phys. Lett.*, B 485:208, 2000.

[101] P. Touboul. MICROSCOPE, testing the equivalence principle in space. *Comptes Rendus de l'Aced. Sci. Série IV: Physique Astrophysique*, 2:1271, 2001.

[102] N. Lockerbie, J.C. Mester, R. Torii, S. Vitale, and P.W. Worden. STEP: A status report. In C. Lämmerzahl, C.W.F. Everitt, and F.W. Hehl, editors, *Gyros, Clocks, and Interferometers: Testing Relativistic Gravity in Space*, page 213. Springer–Verlag, Berlin, 2001.

[103] S. Klioner. Testing relativity with space astrometry missions. In H. Dittus, C. Lämmerzahl, and S. Turyshev, editors, *Lasers, Clocks and Drag–Free:*, in press. Springer Verlag, Berlin, 2006.

[104] S. Turyshev and the LATOR science team. Fundamental physics with the Laser Astrometric Test Of Relativity. In *Trends in Space Science and Cosmic Vision 2030*. ESA, Noordwijk, 2005; gr-qc/0506104.

[105] T.-Y. Huang, C. Lämmerzahl, W.-T. Ni, A. Rüdiger, and Y.-Z. Zhang (Guest Eds.). ASTROD, proceedings of the 1st symposium. *Int. J. Mod. Phys.*, D 11 (7):947 – 1158, 2002.

[106] C. Lämmerzahl, G. Ahlers, N. Ashby, M. Barmatz, P.L. Biermann, H. Dittus, V. Dohm, R. Duncan, K. Gibble, J. Lipa, N.A. Lockerbie, N. Mulders, and C. Salomon. Experiments in Fundamental Physics scheduled and in development for the ISS. *Gen. Rel. Grav.*, 36:615, 2004.

[107] L. Maleki and J. Prestage. SpaceTime mission: Clock test of relativity at four solar radii. In C. Lämmerzahl, C.W.F. Everitt, and F.W. Hehl, editors, *Gyros, Clocks, and Interferometers: Testing Relativistic Gravity in Space*, page 369. Springer–Verlag, Berlin, 2001.

[108] L. Maleki. SPACETIME – a Midex proposal. JPL 2001.

[109] C. Lämmerzahl, C. Ciufolini, H. Dittus, L. Iorio, H. Müller, A. Peters, E. Samain, S. Scheithauer, and S. Schiller. An Einstein mission for improved tests of special and general relativity. *Gen. Rel. Grav.*, 36:2373, 2004.

[110] H. Dittus and the Pioneer Explorer Collaboration. A mission to explore the Pioneer anomaly. In *Trends in Space Science and Cosmic Vision 2030*. ESA, Noordwijk, 2005; gr-qc/0506139.

[111] H. Müller, 2005. private communication.

[112] G. Amelino-Camelia, J. Ellis, N.E. Mavromatos, D.V. Nanopoulos, and S. Sarkar. Tests of quantum gravity from observations of gamma–ray bursts. *Nature*, 393:763, 1998.

Claus Lämmerzahl
ZARM
University of Bremen
Am Fallturm
28359 Bremen
Germany
e-mail: `laemmerzahl@zarm.uni-bremen.de`

Quantum Gravity
B. Fauser, J. Tolksdorf and E. Zeidler, Eds., 41–60

Time Paradox in Quantum Gravity

Alfredo Macías and Hernando Quevedo

Abstract. The aim of this work is to review the concepts of time in quantum mechanics and general relativity to show their incompatibility. We show that the absolute character of Newtonian time is present in quantum mechanics and also partially in quantum field theories which consider the Minkowski metric as the background spacetime. We discuss the problems which this non-dynamical concept of time causes in general relativity that is characterized by a dynamical spacetime.

Mathematics Subject Classification (2000). Primary 46N50; Secondary 85A99.

Keywords. Quantum gravity, time problem, quantization methods.

1. Introduction

Our present description of the laws of physics may be characterized as obtained from two types of constituents. The first type of constituent are theoretical frameworks which apply to *all* physical phenomena at *any* instant. These "universal" or "frame" theories are Quantum Theory, i.e., all matter is of microscopic origin, Special and General Relativity, i.e., all kinds of matter locally have to obey the principles of Lorentz symmetry and behave in the same way in gravitational fields, and Statistical Mechanics which is a method to deal with all kinds of systems for a large number of degrees of freedom. The second type of constituent is non–universal and pertains to the description of the four presently–known interactions: the electromagnetic, the weak, the strong, and the gravitational. The first three interactions are all described within a single formalism, in terms of a gauge theory. So far only gravity has not been successfully included into that scheme. One reason for that might be that gravity appears on both sides: it is an interaction but it is at the same time also a universal theory. Universal theories like relativity and gravity are geometric in origin and do not rely on the particular physical system under consideration, whereas a description in terms of a particular interaction heavily makes use of the particular particle content. Therefore, gravity plays a distinguished role which may be the reason for the difficulty encountered in attempting to unify the other interactions with gravity and attempting to quantize gravity [1, 2].

The concepts of time in quantum mechanics and general relativity are drastically different from each other.

One one hand, time in quantum mechanics is a Newtonian time, i.e., an absolute time. In fact, the two main methods of quantization, namely, canonical quantization method due to Dirac and Feynman's path integral method are based on classical constraints which become operators annihilating the physical states, and on the sum over all possible classical trajectories, respectively. Therefore, both quantization methods rely on the Newton global and absolute time. The absolute character of time in quantum mechanics results crucial for its interpretation, i.e., matrix elements are evaluated at fixed time, and the internal product is unitary, i.e., conserved in time, and it implies conservation of the total probability. Therefore, time is part of the classical background, which is needed for the interpretation of measurements. Moreover, the introduction of a time operator in quantum mechanics is thus problematic. The time parameter appears explicitly in the Schrödinger equation, together with the imaginary unit. Since time is absolute it can be factorized, for instance, in the canonical quantization, reducing the quantization problem to the construction of a Hilbert space for stationary states.

The transition to (special) relativistic quantum field theories can be realized by replacing the unique absolute Newtonian time by a set of timelike parameters associated to the naturally distinguished family of relativistic inertial frames. In this manner, the time continues to be treated as a background parameter.

On the other hand, time in general relativity is dynamical and local. Hence, it is not an absolute time. The geometry of spacetime influences material clocks in order to allow them to display proper time, and the clocks react on the metric changing the geometry. Therefore, the metric itself results to be a clock, and the quantization of the metric can be understood as a quantization of time [3].

The above mentioned quantization methods, when applied to general relativity lead to the Wheeler–DeWitt equation [4]. It is well known that, as a second order functional differential equation, the Wheeler–DeWitt equation presents familiar problems when one tries to turn the space of its solutions into a Hilbert space [5].

In full, general relativity does not seem to possess a natural time variable, while quantum theory relies quite heavily on a preferred time. Since the nature of time in quantum gravity is not yet clear, the classical constraints of general relativity do not contain any time parameter, and one speaks of the *time paradox*.

The aim of the present work is to review the concepts of time in both quantum mechanics and general relativity. Our understanding of time is in the context of the canonical quantization approach to quantum mechanics and quantum field theory. This is why we review in section 2 the axioms of canonical quantization, emphasizing the role of time at each step. Then, in section 3, we discuss the role of the time parameter in general relativity and establish its dynamical character.

Sections 4 and 5 are devoted to brief descriptions of how time enters the problem of canonical quantization on minisuperspaces and midisuperspaces, respectively. Section 6 contains a discussion on the main approaches used to attack the problem of time. Finally, section 7 is devoted to the conclusions.

2. Time in canonical quantization

Quantum theory is based on a certain procedure of quantization of a classical system which consists of a series of axioms. The standard and most used procedure is canonical quantization, whose starting point is the Hamiltonian describtion of the classical system. It is interesting that, like any other physical theory, there is no proof for quantum theory. The only thing we know for sure is that the experimental observations of Nature do not contradict the predictions of quantum theory, at least within the range of measurements accessible to current experimental devices. In canonical quantization time plays a very important role in all the axioms which are postulated as the fundamentals of this method. First, the mere fact that one needs to know the Hamiltonian of the system implies that a certain time parameter has to be chosen in order to define the variables in phase space. To be more specific let us briefly recall the main axioms of canonical quantization.

In the case of quantum mechanics for a system with only *bosonic degrees of freedom* these axioms can be stated as follows:

I) There exists a Hilbert space **H** for the quantum system and the elements of **H** are the quantum states $|\psi\rangle$ of the system. The Hilbert space is supposed to be equipped with an inner product, i.e. a positive definite Hermitian norm on **H**. Often the inner product of two elements $|\phi\rangle$ and $|\psi\rangle$ of **H** is denoted as $\langle\phi|\psi\rangle$.

II) A classical observable A is replaced by a Hermitian operator $\hat{A}$ acting on elements of **H**. When the observable A is measured, the result must coincide with one of the eigenvalues of $\hat{A}$. It is also assumed that for any physical state $|\psi\rangle \in$ **H**, there exists an operator for which the state $|\psi\rangle$ is one of its eigenstates.

III) If q_i and p_j ($i, j = 1, 2, ...n =$ number of bosonic degrees of freedom of the system), are the variables in phase space $\mathbf{R}^{2n}$, the corresponding operators must obey the commutation relations at a fixed time t

$$[\hat{q}_i, \hat{q}_j] = 0, \quad [\hat{p}_i, \hat{p}_j] = 0, \quad [\hat{q}_i, \hat{p}_j] = i\delta_{ij} , \tag{2.1}$$

where we are using units with $\hbar = 1$. This axiom can be generalized to include the case of phase spaces other than $\mathbf{R}^{2n}$ (see, for instance, [6, 7, 8]).

IV) If $\hat{A}$ does not depend explicitly on time, its evolution in time is determined by Heisenberg's evolution equation:

$$\frac{d\hat{A}}{dt} = \frac{1}{i}[\hat{A}, \hat{H}] . \tag{2.2}$$

The formal solution of this equation $\hat{A}(t) = e^{i\hat{H}t}\hat{A}(0)e^{-i\hat{H}t}$ can be used to obtain the equivalent Schrödinger picture in which the operators $\hat{A}(0)$ are time–independent and instead the states become time–dependent through the unitary transformation $|\psi(t)\rangle = e^{-i\hat{H}t}|\psi\rangle$. Then, the evolution of a state of the physical system turns out to be determined by the Schrödinger equation

$$i\frac{d}{dt}|\psi(t)\rangle = \hat{H}|\psi(t)\rangle .$$

(2.3)

V) In general, the observation of A in a physical system at a fixed time t yields random results whose expectation value is given by

$$\langle A\rangle_t = \frac{\langle\psi|\hat{A}(t)|\psi\rangle}{\langle\psi|\psi\rangle} .$$

(2.4)

These are the axioms that lie on the basis of canonical quantization for classical systems with a finite number of degrees of freedom. The time parameter t plays a very important role in determining the phase space, i.e. the choice of canonical positions q_i and momenta p_i. Fortunately, the time used in classical mechanics is the absolute Newtonian time which is defined up to constant linear transformations. Thus, the conjugate momenta are determined up to a multiplicative constant which does not affect the main structure of the phase space. This absolute time is then used with no changes in the quantization scheme described in the above axioms.

Time enters explicitly in axioms III and V, since the commutation relations must be satisfied at a given moment in time and the results of any observation lead to expectation values which are well–defined only if time is fixed. This crucial role of time can be rephrased in terms of the wave function. Indeed, if we define the wave function $\psi(t, x)$ as $\psi(t, x) = \langle x|\psi(t)\rangle$, fixing its normalization, means that it must be normalized to one at a fixed time.

The equation of evolution (2.3) represents changing relations amongst the fundamental entities (operators) of this construction. This equation indicates which operator has to be used to describe the physical system at a given time. When time changes, Heisenberg's equation explains which operator in Hilbert space corresponds to the new state of the physical system.

These observations indicate that in canonical quantization time is an "external" parameter. It is not a fundamental element of the scheme, but it must be introduced from outside as an absolute parameter which coincides with the Newtonian time. Since there is no operator which could be associated with time, it is *not* an observable.

The transition to quantum field theory is performed in a straightforward manner, although many technical details have to be taken into account [9]. The main variables are now the value of the field $\varphi(\mathbf{x})$ at each spatial point and the conjugate momentum $\pi(\mathbf{x})$ for that particular value. The collection of all the values of the field, together with the values of the conjugate momenta, represents the variables of the new phase space. Axioms I – V are then postulated for the corresponding phase space variables. Some changes are necessary in order to consider the new "relativistic" time. In particular, the commutation relation

$$[\hat{\varphi}(x), \hat{\varphi}(y)] = 0 \tag{2.5}$$

is valid for any spacetime points x and y which are spacelike separated. The main difference in the treatment of fields is that the time parameter is that of special relativity. Instead of the absolute Newtonian time, we now have a different parameter associated to each member of the distinguished class of inertial frames. The two absolute concepts of Newtonian physics, i.e. space and time, are now replaced by the single concept of spacetime. Nevertheless, in special relativity spacetime retains much of the Newtonian scheme. Although it is not possible to find an absolute difference between space and time, spacetime is still an element of the quantum theory which does not interact with the field under consideration. That is to say, spacetime remains as a background entity on which one describes the classical (relativistic) and quantum behavior of the field.

In other words, one could say that an observer with the ability to "see" only the physical characteristics of spacetime cannot determine if he/she is "living" on a spacetime with a classical or a quantum field. Spacetime in quantum field theory is therefore an external entity like the absolute external time in quantum mechanics. The dynamics of the field does not affect the properties of spacetime which is therefore a nondynamical element of the theory.

3. Time in general relativity

To implement the canonical quantization procedure in general relativity one needs to find the classical Hamiltonian. As mentioned above, such a formulation requires an explicit choice of time or, equivalently, a slicing of spacetime into spatial hypersurfaces associated to the *preferred* chosen time. This is the Arnowitt–Deser–Misner (ADM) [10] approach which splits spacetime into space and time. The pseudo–Riemannian manifold describing the gravitational field is therefore topologically equivalent to $\mathbf{R} \times \Sigma_t$, where $\mathbf{R}$ represents the "time axis", and Σ_t are constant–time hypersurfaces, each equipped with a set of three coordinates $\{x^i\}$ and a non–degenerate 3–metric q_{ij}. The relationship between the local geometry on Σ_t and the 4–geometry can be recovered by choosing an arbitrary point on Σ_t with coordinates x^i and displacing it by an infinitesimal amount dt normal to Σ_t. The result of this infinitesimal displacement induces an infinitesimal change in proper time τ, which can be written as $d\tau = N dt$, where $N = N(x^\mu)$ is the lapse function, and an infinitesimal change in spatial coordinates, which can be written as $x^i(t + dt) = x^i(t) - N^i dt$, where $N^i = N^i(x^\mu)$ is the shift vector. Then the 4–dimensional interval connecting the starting x^i and ending $x^i + dx^i$ points of this infinitesimal displacement is given by the ADM–metric

$$ds^2 = -N^2 dt^2 + q_{ij}(dx^i + N^i dt)(dx^j + N^j dt) . \tag{3.1}$$

Notice that this splitting of spacetime explicitly depends on the choice of the time parameter t. Indeed, the tensorial quantities N, N^i, and q_{ij} can be given different values by means of a general diffeomorphism.

The Einstein–Hilbert action on a manifold M with vanishing cosmological constant reads

$$S_{EH} = \frac{1}{16\pi G} \int L d^4 x = \frac{1}{16\pi G} \int_M \sqrt{-g} R d^4 x \pm \frac{1}{8\pi G} \int_{\partial M} \sqrt{q} K d^3 x . \quad (3.2)$$

In terms of (3.1) the action becomes a function of the intrinsic metric q_{ij} and its derivatives of first order in time. The boundary term in (3.2) is necessary in the variation to cancel terms that arise after integrating by parts [11]. It is positive (negative) in case of spacelike (timelike) components of ∂M and vanishes when the manifold is spatially compact. The phase space is then constructed by means of the configuration space variables q_{ij} and their canonically conjugate momenta $\pi^{ij} = \partial L / \partial(\partial_t q_{ij})$ which are related to the extrinsic curvature of the 3–dimensional hypersurface Σ_t as embedded in the 4–dimensional spacetime. The resulting Hamiltonian turns out to be that of a constrained system, indicating that the phase space variables are not independent. A straightforward calculation shows that the Einstein–Hilbert action can be written as (dropping boundary terms)

$$S_{EH} = \int dt \int_{\Sigma} d^3 x \left(\pi^{ij} \partial_t q_{ij} - N H_{\perp} - N^i H_i \right) . \quad (3.3)$$

Since this action does not contain time derivatives of N and N^i, their variation leads to the Hamiltonian constraint (super–Hamiltonian constraint)

$$H_{\perp} := 16\pi G G_{ijkl} \pi^{ij} \pi^{kl} - \frac{1}{16\pi G} \sqrt{q} \, ^{(3)}R = 0, \quad (3.4)$$

and the constraint of spatial diffeomorphisms (super–momentum constraint)

$$H^i(\mathbf{x}) = -2 \, ^{(3)}\nabla_j \pi^{ij} = 0 . \quad (3.5)$$

Here q is the determinant and $^{(3)}R$ the curvature scalar of the 3–metric q_{ij}. The covariant derivative with respect to q_{ij} is denoted by $^{(3)}\nabla_j$. The DeWitt supermetric is defined as

$$G_{ijkl} := \frac{1}{2\sqrt{q}} (q_{ik} q_{jl} + q_{jk} q_{il} - q_{ij} q_{kl}) . \quad (3.6)$$

Einstein's field equations are now the standard Hamilton equations for the corresponding action with the Poisson brackets defined according to

$$\{q_{ij}(\mathbf{x}), \pi^{kl}(\mathbf{x}')\} = \delta^k_{(i} \delta^l_{j)} \delta(\mathbf{x}, \mathbf{x}') . \quad (3.7)$$

This special slicing, in which the structure of the spatial hypersurfaces Σ_t is determined as the $t = \text{const.}$ surfaces, leads to the first computational complication for the algebra of diffeomorphisms. In fact, the diffeomorphism invariance in the starting 4–dimensional spacetime is well defined in terms of the corresponding

Lie group. When this spacetime diffeomorphism invariance is projected along and normal to the spacelike hypersurfaces Σ_t, one obtains

$$\{H_i(\mathbf{x}), H_j(\mathbf{x}')\} = H_i(\mathbf{x}')\partial_j^{\mathbf{x}}\delta(\mathbf{x}, \mathbf{x}') - H_j(\mathbf{x})\partial_i^{\mathbf{x}'}\delta(\mathbf{x}, \mathbf{x}') \, , \tag{3.8}$$

$$\{H_i(\mathbf{x}), H_\perp(\mathbf{x}')\} = H_\perp(\mathbf{x})\partial_i^{\mathbf{x}}\delta(\mathbf{x}, \mathbf{x}') \, , \tag{3.9}$$

$$\{H_\perp(\mathbf{x}), H_\perp(\mathbf{x}')\} = q^{ij}(\mathbf{x})H_i(\mathbf{x})\partial_j^{\mathbf{x}'}\delta(\mathbf{x}, \mathbf{x}') - q^{ij}(\mathbf{x}')H_i(\mathbf{x}')\partial_j^{\mathbf{x}}\delta(\mathbf{x}, \mathbf{x}') \, . \tag{3.10}$$

The fact that the right–hand side of Eq.(3.10) contains the 3–metric explicitly implies that the projected algebra of constraints is not a Lie algebra. This is a consequence of the choice of time which leads to considerable computational complications for quantization [12]. One could try to choose a specific gauge in accordance to the invariance associated with the algebra (3.8)–(3.10), then solve the constraints (3.4) and (3.5), and finally quantize the resulting system with the "true" degrees of freedom. It turns out that in general the final equations are tractable only perturbatively, and lead to ultraviolet divergences (for further details see, for example, [13, 14]).

An alternative approach consists in applying the canonical quantization procedure to the complete collection of variables in phase space. The variables q_{ij} and π^{jk} are declared as operators $\hat{q}_{ij}$ and $\hat{\pi}^{jk}$ which are defined on the hypersurface Σ_t and satisfy the commutation relations

$$[\hat{q}_{ij}(\mathbf{x}), \hat{q}_{kl}(\mathbf{x}')] = 0 \, ,$$
$$[\hat{\pi}^{ij}(\mathbf{x}), \hat{\pi}^{kl}(\mathbf{x}')] = 0 \, ,$$
$$[\hat{q}_{ij}(\mathbf{x}), \hat{\pi}^{kl}(\mathbf{x}')] = i\delta_{(i}^k\delta_{j)}^l\delta(\mathbf{x}, \mathbf{x}') \, . \tag{3.11}$$

According to Dirac's quantization approach for constrained systems, the operator constraints must annihilate the physical state vectors, i.e.,

$$\hat{H}_\perp\Psi[q] = 0 \, , \tag{3.12}$$

$$\hat{H}_i\Psi[q] = 0 \, , \tag{3.13}$$

at all points in Σ_t. If the standard representation

$$\hat{q}_{ij}\Psi[q] := q_{ij}\Psi[q] \, , \qquad \hat{\pi}^{ij}\Psi[q] := -i\frac{\delta\Psi[q]}{\delta q_{ij}} \, , \tag{3.14}$$

is used, the constraint $\hat{H}_i\Psi[q] = 0$ requires that $\Psi[q]$ behaves as a constant under changes of the metric q_{ij} induced by infinitesimal diffeomorphisms of the 3-dimensional hypersurface Σ_t. In this specific representation the Hamiltonian constraint (3.12) becomes the Wheeler–DeWitt equation

$$-16\pi G\, G_{ijkl}\frac{\delta^2\Psi[g]}{\delta q_{ij}\delta q_{kl}} - \frac{1}{16\pi G}\,{}^{(3)}R\Psi[g] = 0 \, . \tag{3.15}$$

In canonical quantization this is considered as the main dynamical equation of the theory, since classically the function(al) $H_\perp$ is associated with the generator of displacements in time–like directions. That is to say, $H_\perp$ is the generator of the classical evolution in time. By analogy with quantum mechanics or quantum field theory one expects that the Wheeler–DeWitt equation (3.15) determines the evolution among quantum states. Unfortunately, Eq.(3.15) makes no reference to time, i.e., all the quantities entering it are defined on the 3–dimensional hypersurface Σ_t. This is one of the most obvious manifestations of the problem of time in general relativity. The situation could not be worse! We have a quantum theory in which the main dynamical equation can be solved without considering the evolution in time.

Some researchers interpreted this result as an indication of the necessity of a completely different *"timeless"* approach to quantum theory [15, 16, 17]. This approach is still under construction and although, in principle, some conceptual problems can be solved some other problems related to "time ordering" and "time arbitrariness" appear which are, at best, as difficult as the above described problems of time.

On the other hand, the most propagated interpretation of the problem of time of the Wheeler–DeWitt equation (3.15) is that time must be reintroduced into the quantum theory by means of an auxiliary physical entity whose values can be correlated with the values of other physical entities. This correlation allows in principle to analyze the evolution of physical quantities with respect to the *"auxiliary internal time"*. Since there is no clear definition of the auxiliary internal time, one can only use the imagination to choose a quantity as the time parameter. For instance, if we have a physical quantity which classically depends linearly on time, it could be a good candidate for an auxiliary internal time. Although the linearity seems to be a reasonable criterion, it is not a necessary condition. Examples of this type of auxiliary internal time are the very well analyzed minisuperspaces of quantum cosmology. In particular, one could select the auxiliary internal time as one of the scale factors of homogeneous cosmological models. The volume element which is a combination of scale factors would be also a good choice since in most cases it evolves in cosmic time and reproduces the main aspects of cosmological evolution. The volume element has also been used recently in loop quantum cosmology [18, 19]. Certain low energy limits in string theory contain a tachyonic field which linearly evolves in time and, consequently, could be used as auxiliary internal time for quantization [20]. We will consider these examples with some more details in section 6.

Nevertheless, it is not clear at all if the procedure of fixing an auxiliary internal time can be performed in an exact manner and, if it can be done, whether the results of choosing different auxiliary times can be compared and are somehow related. Finally, a most controversial point is whether such an auxiliary time can be used to relate the usual concepts of spacetime.

In the last section we mentioned that the canonical quantization procedure implies that the fields to be quantized are defined on a background spacetime. In quantum field theory, the Minkowski spacetime with its set of preferred inertial frames plays the role of background spacetime. In general relativity there is no place for a background metric. In fact, the entries of the metric are the physical entities we need to quantize. This rises a new problem. If we success in quantizing the spacetime metric, we will obtain quantum fluctuations of the metric which make impossible the definition of spacelike, null or timelike intervals. But the starting commutation relations require the existence of a well–defined spacetime interval. For instance, the first commutation relation of Eq. (3.11) is usually interpreted as reflecting the fact that the points $\mathbf{x}$ and $\mathbf{x}'$ are separated by a spacelike interval. However, there is no background metric to define this causal structure. Moreover, if we would choose an arbitrary background metric, the quantum fluctuations of that metric could completely change the causal character of the interval. So we are in a situation in which if we want to solve the original problem, we must violate one the most important postulates needed to find the solution. Obviously, this is not a good situation to begin with.

4. Canonical quantization in minisuperspace

The first attempt at minisuperspace quantization is due to DeWitt [21], although the concept of minisuperspace was introduced by Misner [22] some years later. At that time Wheeler [23] suggested the idea of superspace as the space of all three–geometries as the arena in which the geometrodynamics develops. A particular four–geometry being a trajectory in this space. Later, Misner applied the Hamiltonian formulation of general relativity to cosmological models, having in mind the quantization of these cosmological models. He introduced the concept of minisuperspace and minisuperspace quantization or quantum cosmology to describe the evolution of cosmological spacetimes as trajectories in the finite dimensional sector of the superspace related to the finite number of parameters, needed to describe the $t = const.$ slices of the models and the quantum version of such models, respectively.

In the early 70's the minisuperspace models and their quantum version were extensively studied, however, the interest in them decreased at the middle of this decade till Hartle and Hawking [24] revived the field in the early 80's emphasizing the path–integral approaches. This started a lively resurgence of interest in minisuperspace quantization.

In 1987 Macías, Obregón, and Ryan [25] introduced the supersymmetric quantum cosmology approach by applying ($N = 1$) supergravity to quantum minisuperspaces in order to obtain the square root of the Wheeler–DeWitt equation, which governs the evolution of the quantum cosmological models in the standard approach. In 1988 D'Eath and Hughes [26] constructed a locally supersymmetric 1–dimensional model for quantum cosmology, based on a particular case of the Friedmann–Robertson–Walker spacetime (see also [27]). Later on, these results where generalized to include Bianchi cosmological models, supersymmetric matter, and cosmological constant [28, 29, 30].

In 1994 Carrol, Freedman, Ortíz, and Page [33], showed that there is no–physical states in $N = 1$ supergravity, unless there exist an infinite set of gravitino modes. In 1998 Macías, Mielke, and Socorro [31] showed that there are non–physical states in supersymmetric quantum cosmology.

As stated in [34], one of the greatest difficulties with quantum cosmology has always been the seductive character of its results. It is obvious that taking the metric of a cosmological model, which is truncated by an enormous degree of imposed symmetry and simply plugging it into a quantization procedure cannot give an answer that can be in any way interpreted as a quantum gravity solution. What people do is to assume that one can represent the metric as a series expansion in space dependent modes, the cosmological model being the homogeneous mode, and that in some sense one can ignore the dependence of the state function on all inhomogeneous modes. This artificial freezing of modes before quantization represents an obvious violation of the uncertainty principle and cannot lead to an exact solution of the full theory. However, the results of applying this untenable quantization procedure have always seemed to predict such reasonable and internally consistent behavior of the universe that it has been difficult to believe that they have no physical content.

The minisuperspace is often known as the homogeneous cosmology sector, as mentioned above, infinitely many degrees of freedom are artificially frozen by symmetries. This reduction is so drastic that only an unphysical finite number of degrees of freedom is left. The requirement of homogeneity limits the allowed hypersurfaces to the leaves of a privileged foliation, which is labeled by a single "time" variable. One can parametrize such hypersurfaces of homogeneity by the standard Euler angle coordinates and characterize the spatial metric uniquely by three real parameters, Ω, $\beta_\pm$. The Ω is related to the volume of the hypersurface Σ as follows:

$$\Omega = \ln \int_\Sigma d^3x |q(x)|^{1/2} \,. \tag{4.1}$$

The β parameters describe the anisotropy of the hypersurface Σ. Due to the symmetry of the model, the super–momentum constraints are identically satisfied, while the super–Hamiltonian constraint reduces to:

$$H_\perp = -p_\Omega^2 + p_+^2 + p_-^2 + \exp(-\Omega)\left[V(\beta_+, \beta_-) - 1\right] = 0 \,. \tag{4.2}$$

The potential $[V(\beta_+, \beta_-) - 1]$ is a combination of exponential terms, it vanishes at the origin and it is positive outside of it [35]. The parameter Ω is usually considered as a kind of "auxiliary internal time" (see section 6). A systematic analysis of the global time problem for homogeneous cosmological models seems to lead, quite generally, to the lack of a global time function. Even the volume time Ω is not globally permitted, for instance in oscillating models, since the universe would attend a given value $\Omega < \Omega_{max}$ at least twice, once when expanding, and once when recontracting.

Additionally, the Wheeler–DeWitt equation based on one particular choice of time variable, like Ω in this case, may give a different quantum theory than the same equation based on another choice of the time variable. This is what Kuchař called *the multiple choice problem* [5].

It is dangerous to draw conclusions from minisuperspace models to full quantum gravity. Minisuperspace spacetimes possess a privileged foliation by leaves of homogeneity which does not exist in a generic spacetime. Kuchař and Ryan [34] showed that even in the simple case of a microsuperspace (a reduced minisuperspace) the result of canonical quantization is not related to the quantization of the seed minisuperspace. One should try to avoid common practice, which consists of solving a time problem for a model way down in the hierarchy, and jumping to the conclusion that the time problems of quantum gravity are removed by the same treatment.

5. Canonical quantization in midisuperspace

The simplest generalization of the homogeneous models are the Gowdy cosmological models, since they possesses two Killing vectors and therefore two ignorable coordinates, reducing the problem to time (as in standard quantum cosmology) and one spacial coordinate, which completely eliminates homogeneity and leads to a system with an infinite number of degrees of freedom, i.e. a true field theory on a midisuperspace. Gowdy cosmologies are widely studied midisuperspace models.

Moreover, the canonical quantization of $N = 1$ supergravity in the case of a midisuperspace described by Gowdy T^3 cosmological models has been already studied in [36]. The quantum constraints were analyzed and the wave function of the universe was derived explicitly. Unlike the minisuperspace case, it was shown that physical states in midisuperspace models do exist. The analysis of the wave function of the universe leads to the conclusion that the classical curvature singularity present in the evolution of Gowdy models is removed at the quantum level due to the presence of the Rarita–Schwinger field. Since this supegravity midisuperspace model shares the same problem as other midisuperspace models, which consists in the lacking of a well–defined time parameter, in this work a classical solution was used to drive the evolution in time.

The midisuperspace models provide a canonical description of Einstein spacetimes with a group of isometries. Symmetries remove infinitely many degrees of freedom of the gravitational field, but there remain still infinitely many degrees of freedom. In spite of this simplification, the midisuperspace constraints of general relativity are still complicated functionals of the canonical variables.

The study of midisuperspace models and covariant field systems, like string models, indicates that if there exists an auxiliary internal time which converts the old constraints of general relativity into a Schrödinger equation form, such a time variable is a non–local functional of the geometric variables.

The Gowdy T^3 cosmological models have been analyzed in the context of non–perturbative canonical quantization of gravity [37, 38, 39]. The arbitrariness in the selection of a time parameter is a problem that immediately appears in the process of quantization. For a specific choice of time it was shown that there does not exist an unitary operator that could be used to generate the corresponding quantum evolution. Therefore, even in the case of midisuperspace models there is no natural time parameter.

6. The problem of time

Quite a lot of different proposals have been made over the years on how to interpret time in quantum gravity, i.e., the time paradox. Kuchař [5] classified them in three basic approaches. It should be stressed that the boundaries of these interpretations are not clearly defined:

1. *Internal Time.* Time is hidden among the canonical variables and it should be identified prior to quantization. The basic equation upon the interpretation is based in a Schrödinger equation, not a Wheeler–DeWitt one. Nevertheless, this interpretation is susceptible to the multiple choice problem, i.e., the Schrödinger equation based on different time variables may give different quantizations:

   (a) *Matter clocks and reference fluids.* The standard of time is provided by a matter system coupled to geometry, instead of the geometry itself. The intrinsic geometry and extrinsic curvature of a spacelike hypersurface enters into the constraints of general relativity in a very complicated way. Nothing in the structure of the mentioned constraints tell us how to distinguish the true dynamical degrees of freedom from the quantities which determine the hypersurface. The founding fathers of general relativity suggested a conceptual devise which leads exactly to that, i.e., the reference fluid. The particles of the fluid identify space points and clocks carried by them identify instants of time. This fixes the reference frame and the time foliation. In this frame and on the foliation, the metric rather than the geometry becomes measurable. The concept of reference fluid goes back to Einstein [40], and to Hilbert [41] who formalized the idea that the coordinate system should be realized by a realistic fluid carrying clocks which keep a causal time. They imposed a set of inequalities ensuring that the worldlines of the reference fluid be timelike and the leaves of the time foliation be spacelike.

   The reference fluid is traditionally considered as a tenuous material system whose back reaction on the geometry can be neglected. There is just matter but not enough to disturb the geometry. Instead of deriving the motion of the fluid from its action, one encodes it in coordinate conditions. Those are statements on the metric which holds in the coordinate system of the fluid and are violated in any other coordinate system. Such standpoint makes difficult to consider the reference fluid as a physical object which in quantum gravity could assume the role of an apparatus for identifying spacetime events.

   In order to turn the reference fluid into a physical system, it is possible to picture the fluid as a realistic material medium and devise a Lagrangian which describes its properties. By adding this Lagrangian to the Einstein–Hilbert Lagrangian, the fluid is coupled to gravity. Another possibility is to impose the coordinate condition before variation by adjoining them to the action via Lagrange multipliers. The additional terms in the action are parameterized and interpreted as matter source.

(b) *Cosmological time.* In one special case, the reference fluid associated with a coordinate condition allows a geometrical interpretation. This is the unimodular coordinate condition, i.e., $|g_{\mu\nu}|^{1/2} = 1$, fixing the spacetime volume element. These unimodular coordinates were proposed by Einstein [42]. By imposing the unimodular condition before rather than after variation, a law of gravitation with unspecified cosmological constant is obtained [43], reducing the reference fluid to a cosmological term. The cosmological constant appears as a canonical conjugate momentum to a time coordinate, i.e., the cosmological time.

The path–integral version of this approach has been used by Sorkin [44, 45] to show that in a simple model of unimodular quantum cosmology the wave function remains regular as the radius of the universe approaches the classical singularity, but its evolution is non–unitary.

Moreover, it has been shown by Heneaux and Teilteboim [46] that the increment of the cosmological time equals the four–volume enclosed between the initial and the final hypersurfaces.

Unruh and Wald [47] suggested that any reasonable quantum theory should contain a parameter, called Heraclitian time, whose role is to set the conditions for measuring quantum variables and to provide the temporal order of such measurements. The problem with this suggestion is that the cosmological time is not in any obvious way related with the standard concept of time in relativity theory. The basic canonical variables, the metric q_{ij} and its conjugate momentum π^{ij} are always imposed to be measured on a single spacelike hypersurface rather than at a single cosmological time. In order to be able to introduce a particular hypersurface, one needs to specify functions of three coordinates, instead of a single real parameter, i.e., the absolute time of Newtonian mechanics. Consequently, it remains the question in what sense the cosmological time sets the conditions of quantum measurements.

The cosmological time does not fix the conditions for a measurement uniquely, since it it cannot differentiate between the infinitely many possible hypersurfaces of the equivalence class, in order to know in which one the geometrical variables are to be measured. In other words, the hypersurfaces parametrized with different values of the cosmological time are allowed to intersect and cannot be causal ordered as the Heraclitian time requires. Therefore, cosmological time (Heneaux–Teilteboim volume) is not a functional time. Relativity time is a collection of all spacelike hypersurfaces and no single parameter is able to label uniquely so many events.

(c) *Time and tachyons.* The specific form of the low energy action of the tachyon dynamics reads [20]:

$$S = -\int d^{p+1}x V(T)\sqrt{1 + \eta^{\mu\nu}\partial_\mu T \partial_\nu T}\,, \qquad (6.1)$$

where $p = 9$ for strings type IIA or IIB, and $p = 25$ for bosonic strings, $\eta^{\mu\nu} = \mathrm{diag}(-1, 1, 1, ..., 1)$, and $V(T)$ is the potential of the tachyon T. Sen [20] proposed that, at the classical level, solutions of the equations of motion of the field theory described by (6.1), at "*late time*" are in one to one correspondence with a configuration of non–rotating, non interacting dust. At "*late time*" the classical vacuum solutions of the equations of motion approach the $T = x^0 = $ time coordinate, making T a candidate for describing time at the classical level. On the other hand, at "*late time*" the quantum theory of the tachyon T coupled to gravity leads to a Wheeler–DeWitt equation independent of T, whereas for "*early time*" or "*finite time*" the resulting Wheeler–DeWitt equation has a non–trivial dependence on T in the considered region.

Nevertheless, it is well known that the classical tachyon dynamics, when quantized coupled to gravity, may not describe correctly the physics arising from the quantum string theory. Additionally, since the tachyon is identified with a configuration of non–rotating and incoherent dust, its role as time variable shares all the diseases, mentioned above, of reference fluids. Therefore, even in string theory the time paradox remains unsolved.

2. *Wheeler–DeWitt framework.* The constraints are imposed in the metric representation leading to a Wheeler–DeWitt equation. The dynamical interpretation asserts that the solutions would be insensitive to the time identification among the metric functions. This interpretation has to deal with the fact that the Wheeler–DeWitt equation presents familiar problems when one tries to turn the space of its solutions into a Hilbert space. Hence, the statistical interpretation of the theory is based on the inner product. Moreover, if there is a Killing vector, no energy operator commutes with the general relativity constraint $H_\perp$, and the construction of the Hilbert space fails. Even if there exists a timelike Killing vector, the positivity of the inner product requires that the potential in the super–Hamiltonian is non–negative.

The semiclassical interpretation hides the problem of time behind an approximation procedure. It claims that the Wheeler–DeWitt equation for a semiclassical state approximately reduces to the Schrödinger equation, and the Klein–Gordon norm reduces to the Schrödinger norm. Unfortunately, it achieves the positivity of the norm at an unacceptable price of suspending the superposition principle [49, 50]. When the semiclassical interpretation is applied to quantum gravity properly the problem of separating the classical modes defining time from the quantum modes arises. In other words, this means that quantum gravity would have a probabilistic interpretation only if it is classical.

3. *Quantum gravity without time.* This interpretation claims that one does not need time to interpret quantum gravity and quantum mechanics in general. Time may emerge in particular situations, but even if it does not, quantum states still allow a probabilistic interpretation [15, 16, 17].

Its difficulty stands on the fact how to explain quantum dynamics in terms of constants of motion. The existing proposals are ambiguous, since the replacement of the classical global time parameter by an operator is ambiguous and its consequences lead to the multiple choice problem and to the problem of how to construct a Hilbert space [5].

As it is well known, in the canonical formalism, gauge transformations are generated by constraints linear in the momenta, and they move a point in the phase space along, to what is usually called, an orbit of the gauge group. Moreover, two points on the same orbit are physically indistinguishable and represent two equivalent descriptions of same physical state. An observable should not depend on description of the chosen state, the state must be the same along the given orbit, i.e., its Poisson bracket with all the constraints must vanish.

On one hand, all the physical content of general relativity is contained in the constraints and the observables are those dynamical variables that have vanishing Poisson brackets with all constraints, in particular, due to the fact that the diffeormorphim constraint generates a gauge, i.e., the group of spatial coordinates diffeomorphisms. Therefore, any observable in general relativity must be invariant under diffeomorphisms.

On the other hand, the super–Hamiltonian constraint generates the dynamical change of the geometrical dynamical variables from one hyper-surface to another, i.e., any dynamical variable which commutes with the super–Hamiltonian must be the same on every hypersurface and it must be constant of motion. Nevertheless, in order to be able to maintain that the quantum observables are those which commute with all the constraints of general relativity seems to imply that our quantum universe can never change. The transformations generated by the super–Hamiltonian should not be interpreted as gauge transformations. Two points on the same orbit of the super–Hamiltonian transformations are two events in the dynamical evolution of the system which are distinguishable instead of been two descriptions of the same physical state.

Second quantization. There exists a belief that the second quantization solves the problem of time in quantum theory of a relativistic particle. The second quantization approach to quantum field theory is based on the construction of a Fock space, i.e., one takes a one–particle Hilbert space $F_{(1)}$. From the direct product of the one–particle states, the states which span the N–particle sector $F_{(N)}$ are constructed. The Fock space F is then the direct sum of all such sectors, i.e., $F = F_{(0)} \oplus F_{(1)} \oplus F_{(2)} \oplus \cdots$, where $F_{(0)}$ is spanned by the vacuum state. It is clear that the Fock space F can be defined only if the one–particle state $F_{(1)}$ is a Hilbert space. This brings us to the Hilbert space problem for a relativistic particle. The absence of a privileged one–particle Hilbert space structure is the source of ambiguities in constructing a unique quantum field theory on a dynamical background [5].

A closer look to the second quantization approach reveals that it does not really solve the problem of time evolution and its formalism resists an operational interpretation, like the problems presented by the indefinite inner product of the Klein–Gordon interpretation, which are faced by suggesting that the solutions of the Wheeler–DeWitt equation are to be turned to operator. This is analogous to subjecting the relativistic particle, whose state is described by the Klein–Gordon equation, to second quantization.

In full, the second quantization merely shifts the problem of time to a different level without really solving it.

7. Conclusions

Since the concepts of time in quantum mechanics and general relativity are drastically different from each other, generalizations of the usual quantum theory are required to deal with quantum spacetime and quantum cosmology [51]. That is due to the fact that the usual framework for quantum theory assumes a fixed background spacetime geometry. Physical states are defined on spacelike hypersurfaces in this geometry and evolve unitarily in between such hypersurfaces in the absence of measurements and by state vector reduction on them when a measurement occurs. The inner product is defined by integrals over fields on a spacelike hypersurface. Nevertheless, at the quantum realm, spacetime geometry is not fixed, but a dynamical variable fluctuating and without definite value. It is not possible to determine whether two given nearby points on a spacetime manifold are spacelike separated or not. Instead, the amplitudes for predictions are sums over different metrics on the manifold. Additionally, points separated by a spacelike intervals in one metric could be timelike separated in another metric, that contributes just as significantly to the sum. Moreover, quantum theory does not provide a natural time parameter and the quantum constraints of general relativity do not contain any time parameter. For this reason, standard quantum mechanics needs to be generalized to accommodate quantum spacetime.

On the other hand, the application of quantum mechanics to quantum cosmology also requires another kind of generalization of the standard formulation. Standard quantum mechanics predicts the outcome of measurements carried out on a system by another system outside it. However, in cosmology there is no outside. Therefore, quantum cosmology requires a quantum mechanics for closed systems, i.e., a generalization of the standard quantum theory.

All the attempts to implement the canonical quantization procedure to quantize systems in which time is not Newtonian do not provide a reasonable description of the corresponding quantum system. The quantization of general relativity has been an open problem for more than 70 years and the leading present approaches, string theory and loop quantum gravity, are far from providing an ultimate solution, although many technical problems have been attacked and partially solved in the past 20 years. Nevertheless, it seems that the main conceptual problems, especially the one related to time, are still not well understood. In our opinion, it is not possible to reconciliate and integrate into a common scheme the absolute and non–dynamical character of Newtonian time of canonical quantization with the relativistic and dynamical character of time in general relativity. What is needed is a radical change of perspective either in general relativity or in quantum mechanics. That is to say, we need either a theory of gravity with an non–dynamical Newtonian time or a quantum theory with a dynamical time in its construction. We believe that what requires radical changes is the canonical quantization procedure in such a way that the concept of time enters it in a more flexible manner. The issue of time remains as an open problem.

Acknowledgements

One of us (A.M.) thanks Claus Lämmerzahl and Bertfried Fauser for enlightening discussions in Bremen and Blaubeuren, respectively. This work was supported by CONACyT grants 48601–F, and 48404–F.

References

[1] G. Amelino–Camelia, C. Lämmerzahl, A. Macías, and H. Müller: *"The Search for Quantum Gravity Signals"*, in: *Gravitation and Cosmology*, eds. A. Macías, C. Lämmerzahl, and D. Núñez (American Institute of Physics, 2005) pp. 30–80.

[2] C. Kiefer, *Quantum Gravity* (Clarendon Press, Oxford, 2004).

[3] H.D. Zeh, *The physical basis of the direction of time* (Springer–Verlag, Berlin, 2001).

[4] J.J. Halliwell, *Derivation of the Wheeler–DeWitt equation from a path integral for minisuperspace models*, Phys. Rev. **D38** (1988), 2468.

[5] K. Kuchař, *"Time and interpretation of quantum gravity"*, in: *Proc. 4th Canadian Conf. on General Relativity and Relativistic Astrophysics*, eds. G. Kunstatter, D. Vincent, and J. Williams (World Scientific, Singapore, 1992).

[6] N. M. J. Woodhouse, *Geometric Quantization* (Clarendon Press, Oxford, 1980).

[7] C. J. Isham, *"Topological and global aspects of quantum theory"*, in: *Relativity, Groups and Topology II: Les Houches 1983*, eds. B. S. DeWitt and R. Stora (North–Holland, Amsterdam, 1984) pp. 1059.

[8] A. Ashtekar, *Lectures on Non–Perturbative Canonical Gravity* (World Scientific, Singapore, 1991).

[9] R. Wald, *Quantum Field Theory in Curved Spacetime and Black Hole Thermodynamics* (University of Chicago Press, Chicago, 1994).

[10] R. Arnowitt, S. Deser, and C.W. Misner, *"The dynamics of general relativity"*, in: *Gravitation: An introduction to current research*. ed. L. Witten (Wiley, New York, 1962).

[11] J. W. York, *Role of conformal three–geometry in the dynamics of gravitation*, Phys. Rev. Lett. **28** (1972) 1082.

[12] B.S. DeWitt, *"The space–time approach to quantum field theory"*, in: *Relativity, groups, and topology II* eds. B.S. DeWitt and R. Stora (North–Holland, Amsterdam, 1984) pp. 381.

[13] J. Butterfield and C. J. Isham, *Spacetime and philosophical challenge of quantum gravity*, arXiv: qr-qc/9903072, Ch. 4.5

[14] S. Carlip, *Quantum Gravity: a progress report*, Rep. Prog. Phys. **64** (2001) 885.

[15] C. Rovelli, *Quantum mechanics: A model*, Phys. Rev. **D42** (1990) 2638.

[16] C. Rovelli, *Time in quantum gravity: An hypotesis*, Phys. Rev. **D43** (1991) 442.

[17] C. Rovelli, *Quantum Gravity* (Cambridge University Press, Cambridge, UK, 2004).

[18] M. Bojowald, *Loop quantum gravity: II. Volume operators*, Class. Quantum Grav. **17** (2000) 1509.

[19] M. Bojowald, *Loop quantum gravity: II. Discrete time evolution*, Class. Quantum Grav. **18** (2001) 1071.

[20] A. Sen, *Time and Tachyon*, Int. J. Mod. Phys. **A18** (2003) 4869.

[21] B.S. DeWitt, *Quantum Theory of Gravity I. The canonical theory*, Phys. Rev. **160** (1967) 1113.

[22] C.W. Misner, *"Minisuperspace"*, in: *Magic without Magic: John Archibald Wheeler*, ed. J. R. Klauder (Freeman, San Francisco, 1972) pp. 441.

[23] J.A. Wheeler, *"Geometrodynamics and the issue of the final state"* in: *Relativity, Groups and Topology*, ed. C. DeWitt and B. DeWitt. (Gordon and Breach, New York, 1964), pp. 459 ff.

[24] J. B. Hartle and S.W. Hawking, *Wave function of the universe*, Phys. Rev. **D28** (1983) 2690.

[25] A. Macías, O. Obregón and M. P. Ryan, *Quantum Cosmology: The Supersymmetric Square Root*, Class. Quantum Grav. **4** (1987) 1477.

[26] P.D. D'Eath, D.I. Hughes, *Supersymmetric minisuperspace*, Phys. Lett. **B214** (1988) 498.

[27] P.D. D'Eath, D.I. Hughes, *Minisuperspace with local supersymmetry*, Nucl. Phys. **B378** (1992) 381.

[28] P. D. D'Eath, *Quantization of the supersymmetric Bianchi I model with a cosmological constant*, Phys. Lett. **B320** (1994) 12.

[29] A.D.Y. Cheng, P.D. D'Eath, and P.R.L.V. Moniz, *Quantization of the Bianchi IX model in supergravity with a cosmological constant*, Phys. Rev. **D49** (1994) 5246.

[30] A.D.Y. Cheng, P.D. D'Eath, and P.R.L.V. Moniz, *Quantization of a Friedmann–Robertson–Walker model in N=1 supergravity with gauged supermatter*, Class. Quantum Grav. **12** (1995) 1343.

[31] A. Macías, E.W. Mielke, and J. Socorro, *Supersymmetric quantum cosmology: the physical states*, Phys. Rev. **D57** (1998) 1027.

[32] A. Macías, E.W. Mielke, and J. Socorro, *Supersymmetric quantum cosmology for Bianchi class A models*, Int. J. Mod. Phys. **D7** (1998) 701.

[33] S.M. Carroll, D. Z. Freedman, M.E. Ortiz, and D. Page, *Physical states in canonically quantized supergravity*, Nucl. Phys. **B423** (1994) 661.

[34] K. V. Kuchař and M. P. Ryan, *Is minisuperspace quantization valid? Taub and Mixmaster*, Phys. Rev. **D40** (1989) 3982.

[35] M.P. Ryan, *Hamiltonian Cosmology* (Springer-Verlag, New York, 1972); M.P. Ryan and L.C. Shepley: *Homogeneous Relativistic Cosmologies* (Princeton University Press, New Jersey, 1975).

[36] A. Macías, H. Quevedo and A. Sánchez, *Gowdy T^3 Cosmological Models in $N = 1$ Supergravity*, arXiv: grqc/0505013.

[37] A. Ashtekar and M. Pierri, *Probing quantum gravity through exactly soluble midisuperspaces*, J. Math. Phys. **37** (1996) 6250.

[38] G. A. Mena–Marugan, *Canonical quantization of the Gowdy model*, Phys. Rev. **D56** (1997) 908.

[39] A. Corichi, J. Cortez, and H. Quevedo, *On unitary time evolution in Gowdy T^3 models*, Int. J. Mod. Phys. **D11** (2002) 1451.

[40] A. Einstein, *"Uber die spezielle und allgemeine Relativitätstheorie* (Vieweg, Braunschweig, 1920).

[41] D. Hilbert, *Grundlagen der Physik*, 2Mitt., Nachr.Eiss.Göttinger **53** (1917).

[42] A. Einstein, *Die Grundlagen der allgemeinen Relativitätstheorie*, Ann. Phys. (Leipzig), **49** (1916) 769.

[43] W.G. Unruh, *Unimodular Theory of Canonical Quantum Gravity*, Phys. Rev. **D40** (1989) 1048.

[44] R. Sorkin, *On the role of time in the sum over histories framework for gravity*, Int. J. Theor. Phys. **33** (1994) 523.

[45] R. Sorkin, *Forks on the road, on the way to quantum gravity*, Int. J. Theor. Phys. **36** (1997) 2759.

[46] M. Henneaux and C. Teitelboim, *The Cosmological Constant and General Covariance*, Phys. Lett. **B222** (1989) 195.

[47] W.G. Unruh and R.M. Wald, *Time and the interpretation of canonical quantum gravity*, Phys. Rev. **D40** (1989) 2598.

[48] K. Kuchař, *Does an Unspecified Cosmological Constant Solve the Problem of Time in Quantum Gravity?* Phys. Rev. **D43** (1991) 3332.

[49] D.N. Page, *Density matrix of the universe*, Phys. Rev. **D34** (1986) 2267.

[50] D.N. Page and W.K. Wootters, *Evolution without evolution: Dynamics decribed by stationary observables*, Phys Rev. **D27** (1983) 2885.

[51] J.B. Hartle, *General Relativity and Quantum Mechanics*, Int. J. Mod. Phys. **A16** (2001) 1.

Alfredo Macías
Departamento de Física,
Universidad Autonoma Metropolitana–Iztapalapa,
A.P. 55–534, México D.F. 09340, México
e-mail: amac@xanum.uam.mx

Hernando Quevedo
Instituto de Ciencias Nucleares,
Universidad Nacional Autónoma de México
A.P. 70-543, México D.F. 04510, México
e-mail: `quevedo@nuclecu.unam.mx`

Quantum Gravity
B. Fauser, J. Tolksdorf and E. Zeidler, Eds., 61–75
© 2006 Birkhäuser Verlag Basel/Switzerland

Differential Geometry in Non-Commutative Worlds

Louis H. Kauffman

Abstract. Aspects of gauge theory, Hamiltonian mechanics and quantum mechanics arise naturally in the mathematics of a non-commutative framework for calculus and differential geometry. A variant of calculus is built by defining derivations as commutators (or more generally as Lie brackets). We embed discrete calculus into this context and use this framework to discuss the pattern of Hamilton's equations, discrete measurement, the Schrodinger equation, dynamics and gauge theory, a generalization of the Feynman–Dyson derivation of electromagnetic theory, and differential geometry in a non-commutative context.

Mathematics Subject Classification (2000). Primary 39A12; Secondary 81T75, 46L87.

Keywords. Discrete differential geometry, Feynman–Dyson derivation of electromagnetism, noncommutative alegbra, Hamiltonian mechanics.

1. Introduction to non-commutative worlds

Aspects of gauge theory, Hamiltonian mechanics and quantum mechanics arise naturally in the mathematics of a non-commutative framework for calculus and differential geometry. This chapter consists in three sections including the introduction. The introduction sketches our general results in this domain. The introduction is based on the paper [15]. The second section discusses relationships with differential geometry. The third section discusses, in more depth, relationships with gauge theory and differential geometry.

Constructions are performed in a Lie algebra $\mathcal{A}$. One may take $\mathcal{A}$ to be a specific matrix Lie algebra, or abstract Lie algebra. If $\mathcal{A}$ is taken to be an abstract Lie algebra, then it is convenient to use the universal enveloping algebra so that the Lie product can be expressed as a commutator. In making general constructions of operators satisfying certain relations, it is understood that one can always begin with a free algebra and make a quotient algebra where the relations are satisfied.

On $\mathcal{A}$, a variant of calculus is built by defining derivations as commutators (or more generally as Lie products). For a fixed N in $\mathcal{A}$ one defines

$$\nabla_N : \mathcal{A} \longrightarrow \mathcal{A}$$

by the formula

$$\nabla_N F = [F, N] = FN - NF.$$

∇_N is a derivation satisfying the Leibniz rule.

$$\nabla_N(FG) = \nabla_N(F)G + F\nabla_N(G).$$

There are many motivations for replacing derivatives by commutators. If $f(x)$ denotes (say) a function of a real variable x, and $\tilde{f}(x) = f(x + h)$ for a fixed increment h, define the *discrete derivative* Df by the formula $Df = (\tilde{f} - f)/h$, and find that the Leibniz rule is not satisfied. One has the basic formula for the discrete derivative of a product:

$$D(fg) = D(f)g + \tilde{f}D(g).$$

Correct this deviation from the Leibniz rule by introducing a new non-commutative operator J with the property that

$$fJ = J\tilde{f}.$$

Define a new discrete derivative in an extended non-commutative algebra by the formula

$$\nabla(f) = JD(f).$$

It follows at once that

$$\nabla(fg) = JD(f)g + J\tilde{f}D(g) = JD(f)g + fJD(g) = \nabla(f)g + f\nabla(g).$$

Note that

$$\nabla(f) = (J\tilde{f} - Jf)/h = (fJ - Jf)/h = [f, J/h].$$

In the extended algebra, discrete derivatives are represented by commutators, and satisfy the Leibniz rule. One can regard discrete calculus as a subset of non-commutative calculus based on commutators.

In $\mathcal{A}$ there are as many derivations as there are elements of the algebra, and these derivations behave quite wildly with respect to one another. If one takes the concept of *curvature* as the non-commutation of derivations, then $\mathcal{A}$ is a highly curved world indeed. Within $\mathcal{A}$ one can build a tame world of derivations that mimics the behaviour of flat coordinates in Euclidean space. The description of the structure of $\mathcal{A}$ with respect to these flat coordinates contains many of the equations and patterns of mathematical physics.

The flat coordinates X_i satisfy the equations below with the P_j chosen to represent differentiation with respect to X_j:

$$[X_i, X_j] = 0,$$
$$[P_i, P_j] = 0,$$
$$[X_i, P_j] = \delta_{ij}.$$

Derivatives are represented by commutators.

$$\partial_i F = \partial F/\partial X_i = [F, P_i],$$

$$\hat{\partial}_i F = \partial F/\partial P_i = [X_i, F].$$

Temporal derivative is represented by commutation with a special (Hamiltonian) element H of the algebra:

$$dF/dt = [F, H].$$

(For quantum mechanics, take $i\hbar dA/dt = [A, H]$.) These non-commutative coordinates are the simplest flat set of coordinates for description of temporal phenomena in a non-commutative world. Note that *Hamilton's equations are a consequence of these definitions.* The very short proof of this fact is given below.

Hamilton's Equations.

$$dP_i/dt = [P_i, H] = -[H, P_i] = -\partial H/\partial X_i$$

$$dX_i/dt = [X_i, H] = \partial H/\partial P_i.$$

These are exactly Hamilton's equations of motion. The pattern of Hamilton's equations is built into the system.

Discrete Measurement. Consider a time series $\{X, X', X'', \cdots\}$ with commuting scalar values. Let

$$\dot{X} = \nabla X = JDX = J(X' - X)/\tau$$

where τ is an elementary time step (If X denotes a times series value at time t, then X' denotes the value of the series at time $t + \tau$.). The shift operator J is defined by the equation $XJ = JX'$ where this refers to any point in the time series so that $X^{(n)}J = JX^{(n+1)}$ for any non-negative integer n. Moving J across a variable from left to right, corresponds to one tick of the clock. This discrete, non-commutative time derivative satisfies the Leibniz rule.

This derivative ∇ also fits a significant pattern of discrete observation. Consider the act of observing X at a given time and the act of observing (or obtaining) DX at a given time. Since X and X' are ingredients in computing $(X' - X)/\tau$, the numerical value associated with DX, it is necessary to let the clock tick once, Thus, if one first observe X and then obtains DX, the result is different (for the X measurement) if one first obtains DX, and then observes X. In the second case, one finds the value X' instead of the value X, due to the tick of the clock.

1. Let $\dot{X}X$ denote the sequence: observe X, then obtain $\dot{X}$.
2. Let $X\dot{X}$ denote the sequence: obtain $\dot{X}$, then observe X.

The commutator $[X, \dot{X}]$ expresses the difference between these two orders of discrete measurement. In the simplest case, where the elements of the time series are commuting scalars, one has

$$[X, \dot{X}] = X\dot{X} - \dot{X}X = J(X' - X)^2/\tau.$$

Thus one can interpret the equation

$$[X, \dot{X}] = Jk$$

(k a constant scalar) as

$$(X' - X)^2/\tau = k.$$

This means that the process is a (possibly random) walk with spatial step

$$\Delta = \pm\sqrt{k\tau}$$

where k is a constant. In other words, one has the equation

$$k = \Delta^2/\tau.$$

This is the diffusion constant for a Brownian walk. Any walk with spatial step size Δ and time step τ will satisfy the commutator equation above exactly when the square of the spatial step divided by the time step remains constant. This shows that the diffusion constant of a Brownian process is a structural property of that process, independent of considerations of probability and continuum limits.

Heisenberg/Schrödinger Equation. Here is how the Heisenberg form of Schrödinger's equation fits in this context. Let $J = (1 + H\Delta t/i\hbar)$. Then $\nabla\psi = [\psi, J/\Delta t]$, and we calculate

$$\nabla\psi = \psi[(1 + H\Delta t/i\hbar)/\Delta t] - [(1 + H\Delta t/i\hbar)/\Delta t]\psi = [\psi, H]/i\hbar.$$

This is exactly the form of the Heisenberg equation.

Dynamics and Gauge Theory. One can take the general dynamical equation in the form

$$dX_i/dt = \mathcal{G}_i$$

where $\{\mathcal{G}_1, \cdots, \mathcal{G}_d\}$ is a collection of elements of $\mathcal{A}$. Write $\mathcal{G}_i$ relative to the flat coordinates via $\mathcal{G}_i = P_i - A_i$. This is a definition of A_i and $\partial F/\partial X_i = [F, P_i]$. The formalism of gauge theory appears naturally. In particular, if

$$\nabla_i(F) = [F, \mathcal{G}_i],$$

then one has the curvature

$$[\nabla_i, \nabla_j]F = [R_{ij}, F]$$

and

$$R_{ij} = \partial_i A_j - \partial_j A_i + [A_i, A_j].$$

This is the well-known formula for the curvature of a gauge connection. Aspects of geometry arise naturally in this context, including the Levi-Civita connection (which is seen as a consequence of the Jacobi identity in an appropriate non-commutative world).

With $\dot{X}_i = P_i - A_i$, the commutator $[X_i, \dot{X}_j]$ takes the form

$$[X_i, \dot{X}_j] = [X_i, P_j - A_j] = [X_i, P_j] - [X_i, A_j] = \delta_{ij} - [X_i, A_j] = g_{ij}.$$

Thus we see that the "gauge field" A_j provides the deviation from the Kronecker delta in this commutator. We have $[\dot{X}_i, \dot{X}_j] = R_{ij}$, so that these commutators represent the curvature.

One can consider the consequences of the commutator $[X_i, \dot{X}_j] = g_{ij}$, deriving that

$$\ddot{X}_r = G_r + F_{rs}\dot{X}^s + \Gamma_{rst}\dot{X}^s\dot{X}^t,$$

where G_r is the analogue of a scalar field, F_{rs} is the analogue of a gauge field and Γ_{rst} is the Levi-Civita connection associated with g_{ij}. This decompositon of the acceleration is uniquely determined by the given framework. We shall give this derivation in Section 3.

In regard to the commutator $[X_i, \dot{X}_j] = g_{ij}$, It is worth noting that this equation is a consequence of the right choice of Hamiltonian. By this I mean, that in a given non-commutative world we choose an H in the algebra to represent the total (or discrete) time derivative so that $\dot{F} = [F, H]$ for any F. Suppose we have elements g_{ij} such that

$$[g_{ij}, X_k] = 0$$

and

$$g_{ij} = g_{ji}.$$

We choose

$$H = (g_{ij}P_iP_j + P_iP_jg_{ij})/4.$$

This is the non-commutative analog of the classical $H = (1/2)g_{ij}P_iP_j$. In Section 2, we show that this choice of Hamiltonian implies that $[X_i, \dot{X}_j] = g_{ij}$.

Feynman–Dyson Derivation. One can use this context to revisit the Feynman–Dyson derivation of electromagnetism from commutator equations, showing that most of the derivation is independent of any choice of commutators, but highly dependent upon the choice of definitions of the derivatives involved. Without any assumptions about initial commutator equations, but taking the right (in some sense simplest) definitions of the derivatives one obtains a significant generalization of the result of Feynman–Dyson.

Electromagnetic Theorem. (See [15]) With the appropriate [see below] definitions of the operators, and taking

$$\nabla^2 = \partial_1^2 + \partial_2^2 + \partial_3^2, \; B = \dot{X} \times \dot{X} \text{ and } E = \partial_t\dot{X}, \text{ one has}$$

1. $\ddot{X} = E + \dot{X} \times B$
2. $\nabla \bullet B = 0$
3. $\partial_t B + \nabla \times E = B \times B$
4. $\partial_t E - \nabla \times B = (\partial_t^2 - \nabla^2)\dot{X}$

The key to the proof of this Theorem is the definition of the time derivative. This definition is as follows

$$\partial_t F = \dot{F} - \Sigma_i \dot{X}_i\partial_i(F) = \dot{F} - \Sigma_i\dot{X}_i[F, \dot{X}_i]$$

for all elements or vectors of elements F. The definition creates a distinction between space and time in the non-commutative world. In the non-commutative world, we are give the process derivative $\dot{F} = [F, H]$, conceived originally as a discrete difference ratio. The elements of the non-commutative world are subject to this temporal variation, but they are not functions of a "time variable" t. The concept of a time variable is a classical notion that we bring partially into the non-commutative context by defining the notion of a partial derivative $\partial_t F$.

A calculation reveals that

$$\ddot{X} = \partial_t \dot{X} + \dot{X} \times (\dot{X} \times \dot{X}).$$

This suggests taking $E = \partial_t \dot{X}$ as the electric field, and $B = \dot{X} \times \dot{X}$ as the magnetic field so that the Lorentz force law

$$\ddot{X} = E + \dot{X} \times B$$

is satisfied.

This result is applied to produce many discrete models of the Theorem. These models show that, just as the commutator $[X, \dot{X}] = Jk$ describes Brownian motion (constant step size processes) in one dimension, a generalization of electromagnetism describes the interaction of triples of time series in three dimensions.

Remark. While there is a large literature on non-commutative geometry, emanating from the idea of replacing a space by its ring of functions, work discussed herein is not written in that tradition. Non-commutative geometry does occur here, in the sense of geometry occuring in the context of non-commutative algebra. Derivations are represented by commutators. There are relationships between the present work and the traditional non-commutative geometry, but that is a subject for further exploration. In no way is this chapter intended to be an introduction to that subject. The present summary is based on [6, 7, 8, 9, 10, 11, 12, 13, 14, 15] and the references cited therein.

The following references in relation to non-commutative calculus are useful in comparing with the present approach [2, 3, 4, 17]. Much of the present work is the fruit of a long series of discussions with Pierre Noyes, influenced at critical points by Tom Etter and Keith Bowden. Paper [16] also works with minimal coupling for the Feynman–Dyson derivation. The first remark about the minimal coupling occurs in the original paper by Dyson [1], in the context of Poisson brackets. The paper [5] is worth reading as a companion to Dyson. It is the purpose of this summary to indicate how non-commutative calculus can be used in foundations.

2. Differential geometry and gauge theory in a non-commutative world

We take the dynamical law in the form

$$dX_i/dt = \dot{X}_i = P_i - A_i = \mathcal{G}_i.$$

This gives rise to new commutation relations

$$[X_i, \dot{X}_j] = [X_i, P_j] - [X_i, A_j] = \delta_{ij} - \partial A_j/\partial P_i = g_{ij}$$

where this equation defines g_{ij}, and

$$[\dot{X}_i, \dot{X}_j] = R_{ij} = \partial_i A_j - \partial_j A_i + [A_i, A_j].$$

We define the "covariant derivative"

$$\nabla_i F = [F, P_i - A_i] = \partial_i(F) - [F, A_i] = [F, \dot{X}_i],$$

while we can still write

$$\hat{\partial}_i F = [X_i, F].$$

It is natural to think that g_{ij} is analogous to a metric. This analogy is strongest if we *assume* that $[X_i, g_{jk}] = 0$. By assuming that the spatial coordinates commmute with the metric coefficients we have that

$$[\dot{X}_i, g_{jk}] + [X_i, \dot{g}_{jk}] = 0.$$

Hence

$$\nabla_i g_{jk} = \hat{\partial}_i \dot{g}_{jk}.$$

Here, we shall assume from now on that

$$[X_i, g_{jk}] = 0.$$

A stream of consequences then follows by differentiating both sides of the equation

$$g_{ij} = [X_i, \dot{X}_j].$$

We will detail these consequences in section 3. For now, we show how the form of the Levi-Civita connection appears naturally.

In the following we shall use D as an abbreviation for d/dt.

The Levi-Civita connection

$$\Gamma_{ijk} = (1/2)(\nabla_i g_{jk} + \nabla_j g_{ik} - \nabla_k g_{ij})$$

associated with the g_{ij} comes up almost at once from the differentiation process described above. To see how this happens, view the following calculation where

$$\hat{\partial}_i \hat{\partial}_j F = [X_i, [X_j, F]].$$

We apply the operator $\hat{\partial}_i \hat{\partial}_j$ to the second time derivative of X_k.

Lemma 2.1. *Let* $\Gamma_{ijk} = (1/2)(\nabla_i g_{jk} + \nabla_j g_{ik} - \nabla_k g_{ij})$. *Then*

$$\Gamma_{ijk} = (1/2)\hat{\partial}_i \hat{\partial}_j \ddot{X}_k.$$

Proof. Note that by the Leibniz rule

$$D([A, B]) = [\dot{A}, B] + [A, \dot{B}],$$

we have

$$\dot{g}_{jk} = [\dot{X}_j, \dot{X}_k] + [X_j, \ddot{X}_k].$$

Therefore

$$\hat{\partial}_i \hat{\partial}_j \ddot{X}_k = [X_i, [X_j, \ddot{X}_k]]$$
$$= [X_i, \dot{g}_{jk} - [\dot{X}_j, \dot{X}_k]]$$
$$= [X_i, \dot{g}_{jk}] - [X_i, [\dot{X}_j, \dot{X}_k]]$$
$$= [X_i, \dot{g}_{jk}] + [\dot{X}_k, [X_i, \dot{X}_j]] + [\dot{X}_j, [\dot{X}_k, X_i]]$$
$$= -[\dot{X}_i, g_{jk}] + [\dot{X}_k, [X_i, \dot{X}_j]] + [\dot{X}_j, [\dot{X}_k, X_i]]$$

$$= \nabla_i g_{jk} - \nabla_k g_{ij} + \nabla_j g_{ik}$$
$$= 2\Gamma_{kij}.$$

This completes the proof. $\square$

It is remarkable that the form of the Levi-Civita connection comes up directly from this non-commutative calculus without any apriori geometric interpretation.

The upshot of this derivation is that it confirms our interpretation of

$$g_{ij} = [X_i, \dot{X}_j] = [X_i, P_j] - [X_i, A_j] = \delta_{ij} - \partial A_j/\partial P_i$$

as an abstract form of metric (in the absence of any actual notion of distance in the non-commutative world). *This calls for a re-evaluation and reconstruction of differential geometry based on non-commutativity and the Jacobi identity.* This is differential geometry where the fundamental concept is no longer parallel translation, but rather a non-commutative version of a physical trajectory. This approach will be the subject of a separate paper.

At this stage we face the mystery of the appearance of the Levi-Civita connection. There is a way to see that the appearance of this connection is not an accident, but rather quite natural. We are thinking about the commutator $[X_i, \dot{X}_j] = g_{ij}$. It is worth noting that this equation is a consequence of the right choice of Hamiltonian. By this I mean, that in a given non-commutative world we choose an H in the algebra to represent the total (or discrete) time derivative so that $\dot{F} = [F, H]$ for any F. Suppose we have elements g_{ij} such that

$$[g_{ij}, X_k] = 0$$

and

$$g_{ij} = g_{ji}.$$

We choose

$$H = (g_{ij}P_iP_j + P_iP_jg_{ij})/4.$$

This is the non-commutative analog of the classical $H = (1/2)g_{ij}P_iP_j$. In the non-commutative case, there is no reason for the metric coefficients and the momenta P_i to commute since the metric coefficients are dependent on the positions X_j.

We now show that this choice of Hamiltonian implies that $[X_i, \dot{X}_j] = g_{ij}$. Once we see this consequence of the choice of the Hamiltonian, the appearance of the Levi-Civita connection is quite natural, since the classical case of a particle moving in generalized coordinates under Hamilton's equations implies geodesic motion in the Levi-Civita connection.

Lemma 2.2. *Let g_{ij} be given such that $[g_{ij}, X_k] = 0$ and $g_{ij} = g_{ji}$. Define*

$$H = (g_{ij}P_iP_j + P_iP_jg_{ij})/4$$

(where we sum over the repeated indices) and

$$\dot{F} = [F, H].$$

Then

$$[X_i, \dot{X}_j] = g_{ij}.$$

Proof. Consider

$$[X_k, g_{ij}P_iP_j] = g_{ij}[X_k, P_iP_j]$$
$$= g_{ij}([X_k, P_i]P_j + P_i[X_k, P_j])$$
$$= g_{ij}(\delta_{ki}P_j + P_i\delta_{kj}) = g_{kj}P_j + g_{ik}P_i$$
$$= 2g_{kj}P_j.$$

Then

$$[X_r, \dot{X}_k] = [X_r, [X_k, H]] = [X_r, [X_k, (g_{ij}P_iP_j + P_iP_jg_{ij})/4]]$$
$$= [X_r, [X_k, (g_{ij}P_iP_j)/4]] + [X_r, [X_k, (P_iP_jg_{ij})/4]]$$
$$= 2[X_r, 2g_{kj}P_j/4] = [X_r, g_{kj}P_j] = g_{kj}[X_r, P_j] = g_{kj}\delta_{rj}$$
$$= g_{kr} = g_{rk}.$$

This completes the proof. □

It is natural to extend the present analysis to a discussion of general relativity. A joint paper on general relativity from this non-commutative standpoint is in preparation (joint work with Tony Deakin and Clive Kilmister).

3. Consequences of the metric

In this section we shall follow the formalism of the metric commutator equation

$$[X_i, \dot{X}_j] = g_{ij}$$

very far in a semi-classical context. That is, we shall set up a non-commutative world, *and* we shall make assumptions about the non-commutativity that bring the operators into close analogy with variables in standard calculus. In particular we shall regard an element F of the Lie algebra to be a "function of the X_i" if F commutes with the X_i, and we shall assume that if F and G commute with the X_i, then F and G commute with each other. We call this the *principle of commutativity*. With these background assumptions, it is possible to get a very sharp result about the behaviour of the theory.

We assume that

$$[X_i, \dot{X}_j] = g_{ij}$$
$$[X_i, X_j] = 0$$
$$[X_i, g_{jk}] = 0$$
$$[g_{ij}, g_{kl}] = 0.$$

We assume that there exists a g^{ij} with

$$g^{ij}g_{jk} = \delta^i_k = g_{ij}g^{jk} = \delta^k_i.$$

We also assume that if

$$[A, X_i] = 0$$

and

$$[B, X_i] = 0$$

for all i, then
$$[A, B] = 0$$
for all expressions A and B in the algebra under consideration. To say that $[A, X_i] = 0$ is to say the analogue of the statement that A is a function only of the variables X_i and not a function of the $\dot{X}_j$. This is a stong assumption about the algebraic structure, and it will not be taken when we look at strictly discrete models. It is, however, exactly the assumption that brings the non-commutative algebra closest to the classical case of functions of positions and momenta.

The main result of this section will be a proof that
$$\ddot{X}_r = G_r + F_{rs}\dot{X}^s + \Gamma_{rst}\dot{X}^s\dot{X}^t,$$
and that this decompositon of the acceleration is uniquely determined by the given framework. Since
$$F^{rs} = [\dot{X}^r, \dot{X}^s] = g^{ri}g^{sj}F_{ij},$$
we can regard this result as a description of the motion of the non-commutative particle influenced by a scalar field G_r, a qauge field F^{rs}, and geodesic motion with respect to the Levi-Civita connection corresponding to g_{ij}. Let us begin.
Note that, as before, we have that $g_{ij} = g_{ji}$ by taking the time derivative of the equation $[X_i, X_j] = 0$.
Note also that the Einstein summation convention (summing over repeated indices) is in effect when we write equations, unless otherwise specified.

As before, we define
$$\partial_i F = [F, \dot{X}_i]$$
and
$$\hat{\partial}_i F = [X_i, F].$$
We also make the definitions
$$\dot{X}^i = g^{ij}\dot{X}_j$$
and
$$\partial^i F = [F, \dot{X}^i].$$
Note that we do not assume the existence of a variable X^j whose time derivative is $\dot{X}^j$. Note that we have
$$\dot{X}_k = g_{ki}\dot{X}^i.$$
Note that it follows at once that
$$\hat{\partial}_i \dot{g}_{jk} = \partial_i g_{jk}$$
by differentiating the equation $[X_i, g_{jk}] = 0$.

We assume the following postulate about the time derivative of an element F with $[X_i, F] = 0$ for all k :
$$\dot{F} = (\partial_i F)\dot{X}^i.$$
This is in accord with the concept that F is a function of the variables X_i. Note that in one interpretation of this formalism, one of the variables X_i could be itself a time variable. Here there is no restriction on the number of independent variables X_i.

We have the following Lemma.

Lemma 3.1.

1. $[X_i, \dot{X}^j] = \delta_i^j$.
2. $\partial_r(g^{ij})g_{jk} + g^{ij}\partial_r(g_{jk}) = 0$.
3. $[X_r, \partial_i g_{jk}] = 0$.

Proof.

$$[X_i, \dot{X}^j] = [X_i, g^{jk}\dot{X}_k] = [X_i, g^{jk}]\dot{X}_k + g^{jk}[X_i, \dot{X}_k]$$
$$= g^{jk}[X_i, \dot{X}_k] = g^{jk}g_{ik} = g^{jk}g_{ki} = \delta_i^j.$$

The second part of the proposition is an application of the Leibniz rule:

$$0 = \partial_r(\delta_k^i) = \partial_r(g^{ij}g_{jk}) = \partial_r(g^{ij})g_{jk} + g^{ij}\partial_r(g_{jk}).$$

Finally,

$$[X_r, \partial_i g_{jk}] = [X_r, [g_{jk}, \dot{X}_i]] = -[\dot{X}_i, [X_r, g_{jk}]] - [g_{jk}, [\dot{X}_i, X_r]]$$
$$= -[\dot{X}_i, 0] + [g_{jk}, g_{ir}] = 0 + 0 = 0.$$

This completes the proof of the Lemma. $\qquad\square$

It follows from this lemma that ∂^i can be regarded as $\partial/\partial X_i$.

We have seen that it is natural to consider the commutator of the velocities $R_{ij} = [\dot{X}_i, \dot{X}_j]$ as a field or curvature. For the present analysis, we would prefer the field to commute with all the variables X_k in order to identify it as a "function of the variables X_k". We shall find, by a computation, that R_{ij} does not so commute, but that a compensating factor arises naturally. The result is as follows.

Proposition 3.2. *Let* $F_{rs} = [\dot{X}_r, \dot{X}_s] + (\partial_r g_{ks} - \partial_s g_{kr})X^k$ *and* $F^{rs} = [\dot{X}^r, \dot{X}^s]$. *Then*

1. F_{rs} *and* F^{rs} *commute with the variables* X_k.
2. $F^{rs} = g^{ri}g^{sj}F_{ij}$.

Proof. We begin by computing the commutator of X_i and $R_{rs} = [\dot{X}_r, \dot{X}_s]$ by using the Jacobi identity.

$$[X_i, [\dot{X}_r, \dot{X}_s]] = -[\dot{X}_s, [X_i, \dot{X}_r]] - [\dot{X}_r, [\dot{X}_s, X_i]] = \partial_s g_{ir} - \partial_r g_{is}.$$

Note also that

$$[X_i, \partial_r g_{ks}] = [X_i, [g_{ks}, \dot{X}_r]] = -[\dot{X}_r, [X_i, g_{ks}]] - [g_{ks}, [\dot{X}_r, X_i]]$$
$$= -[\dot{X}_r, [X_i, g_{ks}]] + [g_{ks}, g_{ir}] = 0.$$

Hence

$$[X_i, (\partial_r g_{ks} - \partial_s g_{kr})X^k] = \partial_r g_{is} - \partial_s g_{ir}.$$

Therefore

$$[X_i, F_{rs}] = [X_i, [\dot{X}_r, \dot{X}_s] + (\partial_r g_{ks} - \partial_s g_{kr})X^k] = 0.$$

This, and an a similar computation that we leave for the reader, proves the first part of the proposition. We prove the second part by direct computation: Note the following identity:

$$[AB, CD] = [A, C]BD + A[B, C]D + C[A, D]B + CA[B, D].$$

Using this identity we find

$$[\dot{X}^r, \dot{X}^s] = [g^{ri}\dot{X}_i, g^{sj}\dot{X}_j]$$

$$= [g^{ri}, g^{sj}]\dot{X}_i\dot{X}_j + g^{ri}[\dot{X}_i, g^{sj}]\dot{X}_j + g^{sj}[g^{ri}, \dot{X}_j]\dot{X}_i + g^{sj}g^{ri}[\dot{X}_i, \dot{X}_j]$$

$$= -g^{ri}\partial_i(g^{sj})\dot{X}_j + g^{sj}\partial_j(g^{ri})\dot{X}_i + g^{sj}g^{ri}[\dot{X}_i, \dot{X}_j]$$

$$= -g^{ri}\partial_i(g^{sj})g_{jl}\dot{X}^l + g^{sj}\partial_j(g^{ri})g_{il}\dot{X}^l + g^{sj}g^{ri}[\dot{X}_i, \dot{X}_j]$$

$$= g^{ri}g^{sj}\partial_i(g_{jl})\dot{X}^l - g^{sj}g^{ri}\partial_j(g_{il})\dot{X}^l + g^{sj}g^{ri}[\dot{X}_i, \dot{X}_j]$$

$$= g^{ri}g^{sj}(\partial_i(g_{jl})\dot{X}^l - \partial_j(g_{il})\dot{X}^l + [\dot{X}_i, \dot{X}_j])$$

$$= g^{ri}g^{sj}F_{ij}.$$

This completes the proof of the proposition.   □

We now consider the full form of the acceleration terms $\ddot{X}_k$. We have already shown that

$$\hat{\partial}_i\hat{\partial}_j\ddot{X}_k = \partial_i g_{jk} + \partial_j g_{ik} - \partial_k g_{ij}.$$

Letting

$$\Gamma_{kij} = (1/2)(\partial_i g_{jk} + \partial_j g_{ik} - \partial_k g_{ij}),$$

we *define* G_r by the formula

$$\ddot{X}_r = G_r + F_{rs}\dot{X}^s + \Gamma_{rst}\dot{X}^s\dot{X}^t.$$

Proposition 3.3. *Let Γ_{rst} and G_r be defined as above. Then both Γ_{rst} and G_r commute with the variables X_i.*

Proof. Since we know that $[X_l, \partial_i g_{jk}] = 0$, it follows at once that $[X_l, \Gamma_{rst}] = 0$. It remains to examine the commutator $[X_l, G_r]$. We have

$$[X_l, G_r] = [X_l, \ddot{X}_r - F_{rs}\dot{X}^s - \Gamma_{rst}\dot{X}^s\dot{X}^t]$$

$$= [X_l, \ddot{X}_r] - [X_l, F_{rs}\dot{X}^s] - [X_l, \Gamma_{rst}\dot{X}^s\dot{X}^t]$$

$$= [X_l, \ddot{X}_r] - F_{rs}[X_l, \dot{X}^s] - \Gamma_{rst}[X_l, \dot{X}^s\dot{X}^t]$$

(since F_{rs} and Γ_{rst} commute with X_l). Note that

$$[X_l, \dot{X}^s] = \delta_l^s$$

and that

$$[X_l, \dot{X}^s\dot{X}^t] = [X_l, \dot{X}^s]\dot{X}^t + \dot{X}^s[X_l, \dot{X}^t]$$

$$= \delta_l^s\dot{X}^t + \dot{X}^s\delta_l^t.$$

Thus

$$[X_l, G_r] = [X_l, \ddot{X}_r] - F_{rs}\delta_l^s - \Gamma_{rst}(\delta_l^s\dot{X}^t + \dot{X}^s\delta_l^t)$$

$$= [X_l, \ddot{X}_r] - F_{rl} - \Gamma_{rlt}\dot{X}^t - \Gamma_{rsl}\dot{X}^s.$$

It is easy to see that $\Gamma_{rlt}\dot{X}^t = \Gamma_{rsl}\dot{X}^s$. Hence

$$[X_l, G_r] = [X_l, \ddot{X}_r] - F_{rl} - 2\Gamma_{rlt}\dot{X}^t.$$

On the other hand,

$$[X_l, \dot{X}_r] = g_{lr}.$$

Hence

$$[X_l, \ddot{X}_r] = \dot{g}_{lr} - [\dot{X}_l, \dot{X}_r].$$

Therefore

$$[X_l, G_r] = \dot{g}_{lr} - [\dot{X}_l, \dot{X}_r] - F_{rl} - 2\Gamma_{rlt}\dot{X}^t$$
$$= \dot{g}_{lr} - (\partial_r g_{kl} - \partial_l g_{kr})\dot{X}^k - 2\Gamma_{rlt}\dot{X}^t.$$

(since $F_{rl} = [\dot{X}_r, \dot{X}_l] + (\partial_r g_{kl} - \partial_l g_{kr})\dot{X}^k$) Hence

$$[X_l, G_r] = \dot{g}_{lr} - (\partial_r g_{tl} - \partial_l g_{tr})\dot{X}^t - (\partial_l g_{tr} + \partial_t g_{lr} - \partial_r g_{lt})\dot{X}^t$$
$$= \dot{g}_{lr} - (\partial_t g_{lr})\dot{X}^t = 0.$$

This completes the proof of the proposition. □

We now know that G_r, F_{rs} and Γ_{rst} commute with the variables X_k. As we now shall see, the formula

$$\ddot{X}_r = G_r + F_{rs}\dot{X}^s + \Gamma_{rst}\dot{X}^s\dot{X}^t$$

allows us to extract these functions from $\ddot{X}_r$ by differentiating with respect to the dual variables. We already know that

$$\hat{\partial}_i\hat{\partial}_j\ddot{X}_k = 2\Gamma_{kij},$$

and now note that

$$\hat{\partial}_i(\ddot{X}_r) = [X_i, \ddot{X}_r] = [X_i, G_r + F_{rs}\dot{X}^s + \Gamma_{rst}\dot{X}^s\dot{X}^t]$$
$$= F_{rs}[X_i, \dot{X}^s] + \Gamma_{rst}[X_i, \dot{X}^s\dot{X}^t]$$
$$= F_{ri} + 2\Gamma_{rit}\dot{X}^t.$$

We see now that the decomposition

$$\ddot{X}_r = G_r + F_{rs}\dot{X}^s + \Gamma_{rst}\dot{X}^s\dot{X}^t$$

of the acceleration is uniquely determined by these conditions. Since

$$F^{rs} = [\dot{X}^r, \dot{X}^s] = g^{ri}g^{sj}F_{ij},$$

we can regard this result as a description of the motion of the non-commutative particle influenced by a scalar field G_r, a qauge field F^{rs}, and geodesic motion with respect to the Levi-Civita connection corresponding to g_{ij}. The structural appearance of all of these physical aspects is a mathematical consequence of the choice of non-commutative framework.

Remark. It follows from the Jacobi identity that

$$F_{ij} = g_{ir}g_{js}F^{rs}$$

satisfies the equation

$$\partial_i F_{jk} + \partial_j F_{ki} + \partial_k F_{ij} = 0,$$

identifying F_{ij} as a non-commutative analog of a gauge field. G_i is a non-commutative analog of a scalar field. The derivation in this section generalizes the Feynman–Dyson derivation of non-commutative electromagnetism [1] where $g_{ij} = \delta_{ij}$. The results of this section sharpen considerably an approach of Tanimura [18]. In Tanimura's paper, normal ordering techniques are used to handle the algebra. In the derivation given above, we have used straight non-commutative algebra, just as in the original Feynman–Dyson derivation. It will be of interest to return to the normal ordering techniques in the light of our work here.

Remark. It is interesting to note that we can rewrite the equation

$$\ddot{X}_r = G_r + F_{rs}\dot{X}^s + \Gamma_{rst}\dot{X}^s\dot{X}^t$$

as

$$\ddot{X}_r = G_r + [\dot{X}_r, \dot{X}_s]\dot{X}^s + \Gamma_{srt}\dot{X}^s\dot{X}^t.$$

(Just substitute the expression for F_{rs} and recollect the terms.) The reader may enjoy trying her hand at other ways to reorganize this data. It is important to note that in the first form of the equation, the basic terms G_r, F_{rs} and Γ_{rst} commute with the coordinates X_k. It is this decomposition into parts that commute with the coordinates that guides the structure of this formula in the non-commutative context.

Acknowledgements

It gives the author great pleasure to acknowledge support from NSF Grant DMS-0245588 and to thank Pierre Noyes, Keith Bowden and Tony Deakin for continuing conversations related to the contents of this paper.

References

[1] Dyson, F. J. [1990], Feynman's proof of the Maxwell Equations, *Am. J. Phys.* 58 (3), March 1990, 209-211.

[2] Connes, Alain [1990], *Non-commutative Geometry* Academic Press.

[3] Dimakis, A. and Müller-Hoissen F. [1992], Quantum mechanics on a lattice and q-deformations, *Phys. Lett.* 295B, p.242.

[4] Forgy, Eric A. [2002] Differential geometry in computational electromagnetics, PhD Thesis, UIUC.

[5] Hughes, R. J. [1992], On Feynman's proof of the Maxwell Equations, *Am. J. Phys.* 60, (4), April 1992, 301-306.

[6] Kauffman, Louis H.[1991,1994], *Knots and Physics,* World Scientific Pub.

[7] Kauffman, Louis H. and Noyes, H. Pierre [1996], Discrete Physics and the Derivation of Electromagnetism from the formalism of Quantum Mechanics, *Proc. of the Royal Soc. Lond. A*, **452**, pp. 81-95.

[8] Kauffman, Louis H. and Noyes, H. Pierre [1996], Discrete Physics and the Dirac Equation, *Physics Letters A*, 218 ,pp. 139-146.

[9] Kauffman, Louis H. and Noyes, H.Pierre (In preparation)

[10] Kauffman, Louis H.[1996], Quantum electrodynamic birdtracks, *Twistor Newsletter Number 41*

[11] Kauffman, Louis H. [1998], Noncommutativity and discrete physics, *Physica D* 120 (1998), 125-138.

[12] Kauffman, Louis H. [1998], Space and time in discrete physics, *Intl. J. Gen. Syst.* Vol. 27, Nos. 1-3, 241-273.

[13] Kauffman, Louis H. [1999], A non-commutative approach to discrete physics, in *Aspects II - Proceedings of ANPA 20*, 215-238.

[14] Kauffman, Louis H. [2003], Non-commutative calculus and discrete physics, in *Boundaries- Scientific Aspects of ANPA 24*, 73-128.

[15] Kauffman, Louis H. [2004], Non-commutative worlds, *New Journal of Physics 6*, 2-46.

[16] Montesinos, M. and Perez-Lorenzana, A., [1999], Minimal coupling and Feynman's proof, arXiv:quant-phy/9810088 v2 17 Sep 1999.

[17] Müller-Hoissen, Folkert [1998], Introduction to non-commutative geometry of commutative algebras and applications in physics, in *Proceedings of the 2nd Mexican School on Gravitation and Mathematical Physics*, Kostanz (1998) ¡http://kaluza.physik.uni-konstanz.de/2MS/mh/mh.html¿.

[18] Tanimura, Shogo [1992], Relativistic generalization and extension to the non-Abelian gauge theory of Feynman's proof of the Maxwell equations, *Annals of Physics, vol. 220*, pp. 229-247.

Louis H. Kauffman
Department of Mathematics, Statistics and Computer Science
University of Illinois at Chicago
851 South Morgan Street
Chicago, IL, 60607-7045
U.S.A.
e-mail: kauffman@uic.edu

Quantum Gravity
B. Fauser, J. Tolksdorf and E. Zeidler, Eds., 77–100
© 2006 Birkhäuser Verlag Basel/Switzerland

Algebraic Approach to Quantum Gravity III: Non-Commmutative Riemannian Geometry

S. Majid

Abstract. This is a self-contained introduction to quantum Riemannian geometry based on quantum groups as frame groups, and its proposed role in quantum gravity. Much of the article is about the generalisation of classical Riemannian geometry that arises naturally as the classical limit; a theory with nonsymmetric metric and a skew version of metric compatibilty. Meanwhile, in quantum gravity a key ingredient of our approach is the proposal that the differential structure of spacetime is something that itself must be summed over or 'quantised' as a physical degree of freedom. We illustrate such a scheme for quantum gravity on small finite sets.

Mathematics Subject Classification (2000). Primary 83C65; Secondary 58B32; 20C05; 58B20; 83C27.

Keywords. Poisson geometry, generalised Riemannian geometry, quantum groups, noncommutative geometry, quantum gravity.

1. Introduction

Why is quantum gravity so hard? Surely it is because of its nonrenormalisable nature leading to UV divergences that cannot be tamed. However, UV divergences in quantum field theory arise from the assumption that the classical configurations being summed over are defined on a continuum. This is an assumption that is not based on observation but on mathematical constructs that were invented in conjunction with the classical geometry visible in the 19th century. *A priori* it would therefore be more reasonable to expect classical continuum geometry to play a role only in a macroscopic limit and not as a fundamental ingredient. While this problem might not be too bad for matter fields in a fixed background, the logical nonsensicality of putting the notion of a classical manifold that is supposed to emerge from quatum gravity as the starting point for quantum gravity inside the

---

The work was completed on leave at the Mathematical Institute, Oxford, UK.

functional path integral, is more severe and this is perhaps what makes gravity special. The idea of course is to have the quantum theory centred on classical solutions but also to take into account nearby classical configurations with some weight. However, taking that as the actual definition is wishful and rather putting 'the cart before the horse'. Note that string theory also assumes a continuum for the strings to move in, so does not address the fundamental problem either.

If the problems of quantum gravity are indeed such an artifact of a false continuum assumption then what can we do about dropping? Noncommutative geometry (NCG) is a more general, algebraic, framework for geometry that includes the classical continuum case but goes much beyond it. In the last 20 years it has developed into a systematic computable framework and that is capable of making quantum gravity predictions as well as being a possible basis for a complete theory. The 'coordinate algebra' here can be noncommutative as it is on a quantum phase space and hence conjecturally on quantum spacetime as well, at least as an effective theory, or it can be finite-dimensional or infinite-dimensional but reflecting a discrete or poor topology of the underlying space.

Most well-known among frameworks for NCG are Connes' 'spectral triples' [Co]. However, the axioms for these are too closely modelled on the classical case or objects very close to classical ones (such as the noncommutative torus or Heisenberg algebra *aka* θ-spacetimes)[1] and do not hold for many other examples of noncommutative geometries such as coming from genuine quantum groups. In this article we cover our alternative 'quantum groups approach' [BM1, M6, M5, M7, M11] where quantum groups, as the analogues of Lie groups, play a fundamental role not only as key examples but as gauge groups in quantum principal bundles. Therefore part of this article is a systematic account of quantum bundles.

Note also that whereas two decades ago most physicists knew only two or three noncommutative algebras (the Heisenberg algebra, symmetry algebras like the angular momentum algebra, and the algebra of all matrices (or all operators on a Hilbert space)), by now it is accepted that noncommutative 'algebras' *per se* have as rich a structure as that of Riemannian manifolds and indeed rich geometric content when geometry is expressed algebraically. To reach this point of acceptance of noncommutative algebras as having their own structure and geometry is perhaps in the long term the most important legacy of quantum groups. These *are* algebras and force one to take algebras seriously, while at the same time they are analogues of Lie groups with an analogous geometry. Beyond them are category-theory based 'functorial' constructions also coming in part out of quantum groups and their use in constructing 3-manifold invariants but not limited to that. Such methods eventually could be expected to tie up with other approaches to quantum gravity such as spin networks and causal sets that also give up the continuum (but for these there is not yet a fully developed alternative geometry limiting to our familiar continuum one). We will touch upon some aspects of such approaches.

---

[1] Annotation by the Eds: Compare the chapter by Grosse and Wulkenhar in this volume.

This article is complementary to [MII] where we cover the use of NCG in weak-gravity effective theories that might come out of quantum gravity in the form of noncommutative flat spacetime. This covers explicit models of 'deformed special relativity' and the (several) issues regarding making physical predictions that might be actually tested (such as variable speed of light) and for which there is a large literature on the 'quantum gravity phenomenology' side. If there is one important general lesson from these models for the theoretical side, it is that in highly noncommutative cases there is generally an *anomaly* whereby the classical differential calculus cannot be quantised covariantly and forces extra dimensions either in the spacetime or in the Poincaré group to neutralise the anomaly.

By contrast, the present article is concerned with a general formalism for NCG at the same level as general relativity but with possibly noncommutative coordinate algebras. We will include at least one example with nontrivial cotangent bundle, from [M11]. If one is of the view that gravity does not need to be quantised as long as it is suitably extended to include quantized matter, then this may be as far as one needs. Alternatively if one assumes that quantum gravity does need a sum over all geometries, NCG allows the geometries inside the summation to be already more general which is likely needed for self-consistency and finiteness as explained above. Also in this case, having a better algebraic control of the geometry we can and will do such things as sum over all (noncommutative) differential structures. The article begins with a reprise of these. At the semiclassical level the classical data for a quantum calculus in the symplectic case is a certain type of symplectic connection and we are therefore saying that this is a new field in physics that in conventional terms we should 'integrate over'. There is also planned a first article in our series of three, which will deal with the philosophical basis of the approach as introduced in [M1].

2. Reprise of quantum differential calculus

The theory we provide throws out conventional topology and analysis, as founded too closely in classical mechanics. Instead, we demand only a unital algebra A over a field k. The latter, with care, could be a commutative ring (for example if one were to work over $\mathbb{Z}$ one could safely say that this assumption had been relegated to something unavoidable). For a conventional picture it should be $\mathbb{C}$.

A differential structure on A means (Ω^1, d) where

1. Ω^1 is an $A - A$-bimodule (so one can associatively multiply 1-forms by functions from the left and the right).
2. $\mathrm{d} : A \to \Omega^1$ obeys the Leibniz rule $\mathrm{d}(fg) = (\mathrm{d}f)g + f\mathrm{d}g$ for all $f, g \in A$.
3. $\Omega^1 = \mathrm{span}_k\{f\mathrm{d}g\}$
4. (Optional connectedness condition) $\ker \mathrm{d} = k.1$

It is important that this is just about the absolute minimum that one could require in an associative context, but we shall see that it is adequate for Riemannian geometry.

In general a given algebra can have zillions of differential structures, just as can a topological space in classical geometry; we have to focus on those with some symmetry. To describe symmetry in noncommutative geometry the most reasonable notion is that of a quantum group H. Much has been written on such objects and we refer to [Ma]. In brief, the minimal notion of a 'group multiplication' is, in terms of the coordinate algebra, an algebra homomorphism $\Delta : H \to H \otimes H$. We will use the shorthand $\Delta f = f_{(1)} \otimes f_{(2)}$. Associativity of group multiplication corresponds to 'coassociativity' of this 'coproduct' in the sense $(\Delta \otimes \mathrm{id})\Delta = (\mathrm{id} \otimes \Delta)\Delta$. The group identity corresponds to a map $\epsilon : H \to k$ (it evaluates the function at the group identity in the classical case) characterised by $(\mathrm{id} \otimes \epsilon)\Delta = (\epsilon \otimes \mathrm{id})\Delta = \mathrm{id}$. Finally, the group inverse appears as an 'antipode' map $S : H \to H$ characterised by $\cdot(\mathrm{id} \otimes S)\Delta = \cdot(S \otimes \mathrm{id})\Delta = 1\epsilon$. Most relevant for us at the moment, the action of a group on a vector space V appears at the level of H as a 'coaction' $\Delta_R : V \to V \otimes H$ (here a right coaction) characterised by

$$(\Delta_R \otimes \mathrm{id})\Delta_R = (\mathrm{id} \otimes \Delta)\Delta_R, \quad (\mathrm{id} \otimes \epsilon)\Delta_R = \mathrm{id}.$$

There is a similar notion for a left coaction. The nicest case for a quantum differential calculus on an algebra is the *bicovariant* case when there are both left and right coactions of a Hopf algebra H on A extending to Ω^1 according to

$$\Delta_R(f\,dg) = f_{(1)}dg_{(1)} \otimes f_{(2)}g_{(2)}, \quad \Delta_L(f\,dg) = f_{(1)}g_{(1)} \otimes f_{(2)}dg_{(2)}$$

where we use $\Delta_R f = f_{(1)} \otimes f_{(2)}$, $\Delta_L f = f_{(1)} \otimes f_{(2)}$ also for the given coaction on A in the two cases.

When a quantum group acts on an algebra we require the coaction to be an algebra homomorphism. In the case of a differential calculus, we require the functions and 1-forms to generate an entire exterior algebra (Ω, d) with a wedge product, d of degree 1 and $d^2 = 0$, and in the covariant case we require the exterior algebra to be covariant. A lot is known about the construction and classification (by certain ideals) of covariant differential structures on many algebras, but relatively little is known about their extension to an entire exterior algebra; there is a universal extension (basically let the 1-forms generate it and impose the stated requirements) which is always covariant, but in practice it gives too large an answer for the geometry and cohomology to be realistic; one has to quotient it further and there will be many ways to do this (so the higher geometry can involve more data).

Example [Wo]: The quantum group $\mathbb{C}_q[SU_2]$ has a matrix $t^1{}_1 = a, t^1{}_2 = b$ etc. of four generators with q-commutation relations, a unitary $*$-algebra structure and a q-determinant relation $ad - q^{-1}bc = 1$. This is a quantum group with $\Delta t^i{}_j = t^i{}_k \otimes t^k{}_j$. We take $\Delta_L = \Delta$ (left translation on the quantum group). There is a left-covariant calculus

$$\Omega^1 = \mathbb{C}_q[SU_2].\{e_0, e_\pm\}, \quad e_\pm f = q^{\deg(f)} f e_\pm, \quad e_0 f = q^{2\deg(f)} f e_0$$

where $\deg(f)$ the number of a, c minus the number of b, d in a monomial f. The $e_\pm, e_0$ will later be a dreibein on the quantum group. At the moment they are a basis of left-invariant 1-forms. Right multiplication on 1-forms is given via the commutation relations shown. The exterior derivative is

$$\mathrm{d}a = ae_0 + qbe_+, \quad \mathrm{d}b = ae_- - q^{-2}be_0, \quad \mathrm{d}c = ce_0 + qde_+, \quad \mathrm{d}d = ce_- - q^{-2}de_0.$$

The natural extension to an entire exterior algebra is

$$\mathrm{d}e_0 = q^3 e_+ \wedge e_-, \quad \mathrm{d}e_\pm = \mp q^{\pm 2}[2]_{q^{-2}} e_\pm \wedge e_0, \quad (e_\pm)^2 = (e_0)^2 = 0$$

$$q^2 e_+ \wedge e_- + e_- \wedge e_+ = 0, \quad e_0 \wedge e_\pm + q^{\pm 4} e_\pm \wedge e_0 = 0$$

where $[n]_q \equiv (1 - q^n)/(1 - q)$ denotes a q-integer. This means that there are the same dimensions as classically, including a unique top form $e_- \wedge e_+ \wedge e_0$.

Warning: in most interesting noncommutative examples the requirement of full covariance is so restrictive that a differential structure does not exist with the classical dimensions. For example, for the above quantum group, the smallest bicovariant calculus is 4-dimensional with generators similar to $e_\pm, e_0$ but some different relations, and a new non-classical generator Θ (see below).

2.1. Symplectic connections: a new field in physics

If an algebra for spacetime is noncomutative due to quantum gravity effects, we can suppose it has the form

$$f \bullet g - g \bullet f = \lambda\{f, g\} + O(\lambda^2)$$

with respect to some deformation parameter λ, where we write $\bullet$ to stress the noncommutative product. On the right is some Poisson bracket on the manifold M which is supposed to be found in the classical limit. In this sense a Poisson bracket will have to arise in the semiclassical limit of quantum gravity. However, Darboux' theorem says that all nondegenerate (symplectic) Poisson brackets are equivalent so we do not tend to give too much thought to this; we take it in a canonical form. If one does the same thing for a non(super)commutative exterior algebra Ω it is obvious that one has a graded (super)Poisson bracket. This has been observed by many authors, most recently in the form of a flat connection of some form [Haw, BeM2].

This is not much use, however, since as we have mentioned there is often no calculus of the correct dimensions for Ω to be a flat deformation. If one wants to stay in a deformation setting one must therefore, and we shall, temporarily, relax the full axioms of a differential structure. Specifically, we shall suppose them only at the lowest order in λ [BeM2]. Now define

$$f \bullet \tau - \tau \bullet f = \lambda \hat{\nabla}_f \tau + O(\lambda^2)$$

for all functions f and 1-forms τ. Morally speaking, one can think of $\hat{\nabla}_f$ as a covariant derivative $\nabla_{\hat{f}}$ where $\hat{f} = \{f, \ \}$ is the Hamiltonian vector field generated by a function f, and this will be true in the symplectic case. We therefore call it in general a 'preconnection'. It further obeys the 'Poisson-compatibility' condition

$$\hat{\nabla}_f \mathrm{d}g - \hat{\nabla}_g \mathrm{d}f = \mathrm{d}\{f, g\}$$

and its curvature (defined in the obvious way) appears in

$$[f, [g, dh]_\bullet]_\bullet + \text{cyclic} = \lambda^2 R_{\hat{\nabla}}(f, g)dh + O(\lambda^3).$$

If the differential structure was strictly associative, this would imply that the jacobiator on the left would vanish and we see that this would correspond to the preconnection being flat. In the symplectic case the assumption $[f, \omega] = O(\lambda^2)$ (where ω is the symplectic 2-form) is equivalent to ∇ being a torsion free symplectic connection [BeM2]. So this is the new classical data that should somehow emerge from quantum gravity in the semiclassical limit. If it emerges with zero curvature it might be expected to be the leading part of an associative noncommutative differential calculus, but if it emerges with nonzero curvature, which is surely the more likely generic case, there will be no noncommutative associative differential structure of classical dimensions. This is exactly what we typically find in NCG.

Without knowing the full theory of quantum gravity, we can also classify covariant poisson-compatible preconnections in classical geometry, which classifies possibilities in the quantum gravity theory. For example, the theory in [BeM2, BeM3] finds an essentially unique such object for all simple Lie group manifolds G for the type of Poisson-bracket that appears from quantum groups $\mathbb{C}_q[G]$ and it has curvature given by the Cartan 3-form (in the case of SU_n, $n > 2$ one has a 1-parameter family but also with curvature). Hence if these quantum groups arise from quantum gravity we know the data $\hat{\nabla}$. In general, however, it is a new field in physics that has to come out of the quantum gravity theory along with the metric and other fields.

2.2. Differential anomalies and the orgin of time

When $\hat{\nabla}$ does not have zero curvature we say that there is a 'differential anomaly': even the minimal axioms of a differential structure do not survive on quantisation. We have said that it can typically be neutralised by increasing the dimension of the cotangent bundle beyond the classical one. But then we have cotangent directions not visible classically at all; these are purely 'quantum' directions even more remarkable than the unobservable compact ones of string theory.

This is tied up with another natural feature of NCG; in many models there is à 1-form Θ such that

$$[\Theta, f] = \lambda df \tag{2.1}$$

where we assume for the sake of discussion a noncommutativity parameter λ (it also applies in finite non-deformation cases). This equation has no classical meaning, as both sides are zero when $\lambda \to 0$; it says that the geometry for $\lambda \neq 0$ is sufficiently noncommutative to be 'inner' in this sense. Now, if Θ is part of a basis of 1-forms along with (say) dx_i where x_i are some suitable coordinate functions in the noncommutative algebra, we can write df in that basis. The coefficients in that basis are 'quantum partial derivative' operators ∂^i, ∂^0 on A defined by

$$df \equiv (\partial^i f)dx_i + (\partial^0 f)\Theta. \tag{2.2}$$

The quantum derivatives ∂^i would have their classical limits if x_i become usual classical coordinates, but ∂^0/λ will typically limit to some other differential operator which we call the *induced Hamiltonian*.

Example [M12]: The simplest example is $U(su_2)$ regarded as a quantisation of su^*. The smallest connected covariant calculus here is 4D with an extra direction Θ as well as the usual $\mathrm{d}x_i$. The associated ∂^0 operator is

$$\partial^0 = \frac{\imath\mu}{2\lambda^2}\left(\sqrt{1 + \lambda^2\partial_i\partial^i} - 1\right)$$

where $\mu = 1/m$ is a free (but nonzero) parameter inserted (along with $-\imath$) into the normalisation of Θ. If the effect is a quantum gravity one we would expect $\mu \sim \lambda$ as in (2.1) but in principle the noncommutativity scales of the algebra A and its calculus are independent. More details of this 'flat space' example are in [M12, MII]. If we go a little further and explicitly adjoin time by extending our algebra by a central element t with $\Theta = \mathrm{d}t$, then we obtain a natural description of Schroedinger's equation for a particle of mass m on the noncommutative spacetime. The induced calculus on t in this model is actually a finite-difference one.

We call this general mechanism *spontaneous time generation* [M12] arising intrinsically from the noncommutive differential geometry of the spatial algebra. In effect, any sufficiently noncommutative differential algebra 'evolves itself' and this is the reason in our view for equations of motion in classical physics in the first place. Note that this is unrelated to 'time' defined as the modular automorphism group of a von-Neuman algebra with respect to a suitable state [CR].

Example: For a 'curved space' example, we consider the bicovariant calculus on $\mathbb{C}_q[SU_2]$ mentioned in the warning above. The phenomoneon in this case is just the same (it is in fact a 'cogravity' phenomenon dual to gravity and independent of it). Thus, the 4D calculus [Wo], which is the smallest bicovariant one here, has basis $\{e_a, e_b, e_c, e_d\}$ and relations:

$$e_a\begin{pmatrix} a & b \\ c & d \end{pmatrix} = \begin{pmatrix} qa & q^{-1}b \\ qc & q^{-1}d \end{pmatrix} e_a$$

$$[e_b, \begin{pmatrix} a & b \\ c & d \end{pmatrix}] = q\lambda\begin{pmatrix} 0 & a \\ 0 & c \end{pmatrix} e_d, \quad [e_c, \begin{pmatrix} a & b \\ c & d \end{pmatrix}] = q\lambda\begin{pmatrix} b & 0 \\ d & 0 \end{pmatrix} e_a$$

$$[e_d, \begin{pmatrix} a \\ c \end{pmatrix}]_{q^{-1}} = \lambda\begin{pmatrix} b \\ d \end{pmatrix} e_b, \quad [e_d, \begin{pmatrix} b \\ d \end{pmatrix}]_q = \lambda\begin{pmatrix} a \\ c \end{pmatrix} e_c + q\lambda^2\begin{pmatrix} b \\ d \end{pmatrix} e_a,$$

where $[x, y]_q \equiv xy - qyx$ and $\lambda = 1 - q^{-2}$. The exterior differential has the inner form of a graded anticommutator $\mathrm{d} = \lambda^{-1}[\Theta, \}$ where $\Theta = e_a + e_d$. By iterating the above relations one may compute [GM1, M9]:

$$\begin{aligned} \mathrm{d}(c^k b^n d^m) = {}& \lambda^{-1}(q^{m+n-k} - 1)\,c^k b^n d^m\,e_d + q^{n-k+1}[k]_{q^2}\,c^{k-1}b^n\,d^{m+1}\,e_b \\ &+ q^{-k-n}\,([m+n]_{q^2}\,c^{k+1}b^n d^{m-1} + q[n]_{q^2}\,c^k b^{n-1}d^{m-1})e_c \\ &+ \lambda\,q^{-k-m-n+2}([k+1]_{q^2}\,[m+n]_{q^2}\,c^k b^n d^m + q[n]_{q^2}\,[k]_{q^2}\,c^{k-1}b^{n-1}d^m)e_a \\ &+ \lambda^{-1}(q^{-m-n+k} - 1)\,c^k b^n d^m\,e_a \end{aligned}$$

using our previous notations. This is in a coordinate 'patch' where d is invertible so that $a = (1 + q^{-1}bc)d^{-1}$; there are similar formulae in the other patch where a is inverted. Now to extract the 'geometry' of this calculus let us change to a basis $\{e_b, e_c, e_z, \Theta\}$ where $e_z = q^{-2}e_a - e_d$. Then the first three become in the classical limit the usual left-invariant 1-forms on SU_2. Writing

$$\mathrm{d}f \equiv (\partial^b f)e_b + (\partial^c f)e_c + (\partial^z f)e_z + (\partial^0 f)\Theta$$

we compute from the above on $f = c^k b^n d^m$ that

$$\frac{(2)_q}{q^2\lambda}\partial^0 f = q^{-k}(k)_q(n)_q c^{k-1}b^{n-1}d^m + \left(\frac{k+n+m}{2}\right)_q \left(\frac{k+n+m}{2}+1\right)_q f \equiv \Delta_q f$$

where $(n)_q \equiv (q^n - q^{-n})/(q - q^{-1})$ is the 'symmetrized q-integer' (so the first term can be written as $q^{-k}\partial^c_q\partial^b_q f$ in terms of symmetrized q-derivatives that bring down q-integers on monomials, see [Ma] for notations). The right hand side here is exactly a q-deformation of the classical Laplace Beltrami operator Δ on SU_2. To see this, let us note that this is given by the action of the Casimir $x_+ x_- + \frac{h^2}{4} - \frac{h}{2}$ in terms of the usual Lie algebra generators of su_2 where $[x_+, x_-] = h$. To compute the action of the vector fields for these Lie algebra generators in the coordinate patch above we let $\partial^b, \partial^c, \partial^d$ denote partials keeping the other two generators constant but regarding $a = (1 + bc)d^{-1}$. Then

$$\partial^b = \frac{\partial}{\partial b} + d^{-1}c\frac{\partial}{\partial a}, \quad \partial^c = \frac{\partial}{\partial c} + d^{-1}b\frac{\partial}{\partial a}, \quad \partial^d = \frac{\partial}{\partial d} - d^{-1}a\frac{\partial}{\partial a}$$

if one on the right regards the a, b, c, d as independent for the partial derivations. Left-invariant vector fields are usually given in the latter redundant form as $\tilde{x} = t^i{}_j x^j{}_k \frac{\partial}{\partial t^i{}_k}$ for x in the representation associated to the matrix of coordinates. Converting such formulae over for our coordinate system, we find

$$\tilde{h} = c\partial^c - b\partial^b - d\partial^d, \quad \tilde{x}_+ = a\partial^b + c\partial^d, \quad \tilde{x}_- = d\partial^c$$

$$\Rightarrow \quad \Delta = \partial^b\partial^c + \frac{1}{4}(c\partial^c + b\partial^b + d\partial^d)^2 + \frac{1}{2}(c\partial^c + b\partial^b + d\partial^d)$$

which is indeed what we have above when $q \to 1$. The other coordinate patch is similar. It is worth noting that the 4D (braided) Lie algebra (see below) of q-vector fields defined by this calculus [M4] is irreducible for generic $q \neq 1$ so one sees also from this point of view that one cannot avoid an 'extra dimension'.

Also, by the same steps as in [M12] we can manifest this extra dimension by adjoining a new central variable t with $\Theta = \mathrm{d}t$. By applying d to $[f, t] = 0$ we deduce that

$$[e_b, t] = \lambda e_b, \quad [e_c, t] = \lambda e_c, \quad [e_z, t] = \lambda e_z, \quad [\mathrm{d}t, t] = \lambda \mathrm{d}t$$

where the last implies again that the induced calculus on the t variable is a finite difference one: $\mathrm{d}g(t) = \frac{g(t+\lambda)-g(t)}{\lambda}\mathrm{d}t \equiv (\partial_\lambda g)\mathrm{d}t$, which is the unique form of noncommutative covariant calculus on the algebra of functions $\mathbb{C}[t]$ with its additive Hopf algebra structure. We also deduce that $(\mathrm{d}f)g(t) = g(t + \lambda)\mathrm{d}f$ for any $f = f(a,b,c,d)$ and hence compute ∂^i, ∂^0 (where $i = b,c,z$) in the extended calculus from

$$\mathrm{d}(fg) \equiv \partial^i(fg)e_i + \partial^0(fg)\Theta = (\mathrm{d}f)g + f\mathrm{d}g = \partial^i(f)e_ig(t) + (\partial^0 f)\Theta g(t) + f\partial_\lambda g\Theta.$$

In these terms Schroedinger's equation or the heat equation (depending on normalisation) appears in a natural geometrical form on the spacetime algebra fields $\psi(a,b,c,d,t)$, regarded as $\psi(t) \in \mathbb{C}_q[SU_2]$, as the condition that $\mathrm{d}\psi$ is purely spatial, i.e. $\partial^0\psi = 0$. Explicitly, this means

$$\frac{\psi(t) - \psi(t-\lambda)}{\lambda} + \frac{q^2\lambda}{(2)_q}\Delta_q(\psi(t)) = 0.$$

Finally, one may change the normalisation of Θ above to introduce the associated mass parameter and also $\imath$ according to the reality structures in the model.

For completeness, the rest of the bicovariant exterior algebra here is with e_b, e_c behaving like usual forms or Grassmann variables and

$$e_z \wedge e_c + q^2 e_c \wedge e_z = 0, \quad e_b \wedge e_z + q^2 e_z \wedge e_b = 0, \quad e_z \wedge e_z = (1 - q^{-4})e_c \wedge e_b.$$

$$\mathrm{d}\Theta = 0, \quad \mathrm{d}e_c = q^2 e_c \wedge e_z, \quad \mathrm{d}e_b = q^2 e_z \wedge e_b, \quad \mathrm{d}e_z = (q^{-2} + 1)e_b \wedge e_c.$$

3. Classical weak Riemannian geometry

In this section we will set aside NCG and prepare Riemannian geometry for quantisation by recasting it in a suitable and slightly weaker form. This theory is due to the author in the first half of [M6]. The first thing to note is that in classical Riemannian geometry one defines covariant derivatives ∇_X for the action of vector fields X. Similarly the Riemann curvature $R(X,Y)$ as an operator on vector fields and the torsion $T(X,Y)$. But in NCG we work more naturally with 1-forms which are dual to vector fields. Thus we instead think of a covariant derivative in the same spirit as for a left coaction Δ_L, namely

$$\nabla : \Omega^1 \to \Omega^1 \bar{\otimes} \Omega^1$$

where $\bar{\otimes}$ means tensor product over the algebra of (say smooth) functions on the classical manifold M and Ω^1 means $\Omega^1(M)$. The left hand output of ∇ is 'waiting' for a vector field X to be evaluated against it, which would give ∇_X as usual. Similarly the standard formulae convert over to [M6]

$$R_\nabla = (\mathrm{id} \wedge \nabla - \mathrm{d} \otimes \mathrm{id})\nabla : \Omega^1 \to \Omega^2 \bar{\otimes} \Omega^1, \quad T_\nabla = \nabla \wedge -\mathrm{d} : \Omega^1 \to \Omega^2 \quad (3.1)$$

where $\wedge$ means to apply the product $\Omega^1 \bar{\otimes} \Omega^1 \to \Omega^2$ in the indicated place. Thus curvature is a 2-form valued operator on 1-forms and torsion is a measure of the failure of the projection of ∇ to coincide with the exterior derivative. In our view the structure of classical Riemannian geometry is much more cleanly expressed in these terms and in a coordinate free manner.

3.1. Cotorsion and weak metric compatibility

So far, these remarks apply for any covariant derivative. In Riemannian geometry we also need a metric. In noncommutative geometry it is not reasonable to assume a naively symmetric metric (for example in our q-deformed example it will be q-symmetric as dictated by the stringencies of $\mathbb{C}_q[S0_3]$-invariance). Non-symmetric metrics are also suggested in other contexts such as for T-duality in string theory or Poisson-Lie T-duality [BeM1]. The idea is that semiclassical corrections to classical Riemannian geometry may entail antisymmetric terms given, for example, by the Poisson bivector that also has to come out of quantum gravity (see above) and one has to work in this weaker setting to study certain semiclassical phenomena. Therefore our first bit of generalisation is to define a metric as simply a bundle isomorphism $TM \cong T^*M$ or in our sectional terms $\Omega^{-1} \cong \Omega^1$ (the former denotes the space of vector fields) as a module over the algebra of functions. In plain English it means a nondegenerate tensorial bilinear map $g(X, Y)$ on vector fields, or a nondegenerate element $g \in \Omega^1 \bar\otimes \Omega^1$. The metric is symmetric if $\wedge(g) = 0$ and we are free to impose symmetry in this form in NCG if we want.

Next, we define an adjoint connection ∇^* such that

$$X(g(Y, Z)) = g(\nabla^*_X Y, Z) + g(Y, \nabla_X Z)$$

or in dual form

$$(\nabla^* \bar\otimes \mathrm{id} + \mathrm{id} \bar\otimes \nabla)g = 0 \quad \Leftrightarrow \quad ((\nabla^* - \nabla) \bar\otimes \mathrm{id})g + \nabla g = 0$$

where in the 2nd term of the first expression, the left-hand output of ∇ is understood to be positioned to the far left. Note that ∇g is defined similarly by extending ∇ as a derivation but keeping its left-hand output to the far left (where it could be safely evaluated against a vector field X).

Proposition 3.1. ∇^* *is a connection and*

$$(R_{\nabla^*} \bar\otimes \mathrm{id})g + (\mathrm{id} \bar\otimes R_\nabla)g = 0, \quad ((T_{\nabla^*} - T_\nabla) \bar\otimes \mathrm{id})g = (\nabla \wedge \bar\otimes \mathrm{id} - \mathrm{id} \wedge \nabla)g$$

where the left hand (2-form valued) output of R_∇ is understood to the far left.

The direct proof is tedious but elementary and will therefore be omitted. One also has to check that the constructions are indeed well defined over $\bar\otimes$ which is not completely obvious. (The result is implicitly proven a different way using frame bundles, see below, in [M6].) The quantity T_{∇^*} is called the *cotorsion* of ∇ and we see that a *torsion free and cotorsion free* connection ∇ is the same thing as torsion free and

$$(\nabla \wedge \bar\otimes \mathrm{id} - \mathrm{id} \wedge \nabla)g = 0. \tag{3.2}$$

We call such a connection a 'weak Levi-Civita' connection (it is no longer unique). Note that (3.2) can also be written as $(\wedge \bar\otimes \mathrm{id})\nabla g = 0$ or in components, $\nabla_\mu g_{\nu\rho} - \nabla_\nu g_{\mu\rho} = 0$ and we call this 'weak metric compatibility'. Why do we need this weakening? Quite simply in most NCG examples that have been computed the weaker version of the Levi-Civita condition actually determines ∇ uniquely within some reasonable context, and the ∇ that arise this way simply do not obey the full $\nabla g = 0$. So Riemannian geometry in its usual form does not generalise to most key examples but this weaker version does and indeed suffices. We will give some examples later. As is not untypical in NCG (as we saw for the 1-form Θ), the classical theory is degenerate and in that limit (3.2) does not suffice, and this in our view is why we have grown artificially used to the stronger form.

3.2. Framings and coframings

Next, it is quite well-known that the notion of a vector bundle may be expressed in algebraic terms simply as the sections $\mathcal{E}$ being a finitely-generated projective module (the Serre-Swann theorem). This is the line taken for example in [Co]. However, it is important in Riemannian geometry that the vector bundles that arise are not simply vector bundles but are associated to a principal frame bundle. This ensures that the spinor bundle, the cotangent bundle etc. all have compatible structures induced from a single connection on the frame bundle. It is also important in physics to have the 'moving frame' picture for calculations. Our first problem in noncommutative geometry is, if we replace GL_n or O_n by a quantum group, what flavour of quantum group should we use? There are in fact many different types of Hopf algebras that deform even GL_n whereas we should like to have a general and not specific theory. The answer is to generalise the classical notion of frame bundle to allow for a general choice of frame group.

Thus, let $G \subset X \to M$ be a principal G-bundle over M (where G acts on X from the right, say) and V a representation of G. There is an associated bundle $E = P \times_G V \to M$ and its sections may be identified as the space $\mathcal{E} = C_G(X, V)$ of V-valued G-equivariant functions on X. The following lemma is known to experts:

Lemma 3.2. *Bundle maps $E \to T^*M$ are in 1-1 correspondence with $C(M)$-module maps $C_G(X, V) \to \Omega^1$ and these in turn are in correspondence with $\theta \in \Omega^1_{\text{tensorial}}(X, V^*)$, i.e. with G-equivariant horizontal 1-forms on X.*

A proof is in [M6] in some other notations. One way is easy: given θ we multiply it by any V-valued function on X and evaluate the V^* against the V to obtain a 1-form on X which actually is the pull-back of a 1-form on M due to the equivariance and horizontality assumptions. Using this lemma we define:

1. A framing of a manifold M is (G, X, V, θ) such that $E \cong T^*M$ or equivalently $\mathcal{E} \cong \Omega^1$ by the above maps.
2. A framed weak Riemannian manifold structure on M means framed as above and *also* framed by (G, X, V^*, θ^*) (we call this the coframing).

Here the framing implies that the dual bundle is $E^* \cong TM$. Given this, a coframing (which is similarly equivalent to $E^* \cong T^*M$) is equivalent to $E^* \cong E$ or $TM \cong T^*M$ given the framing. In other words, it is equivalent to a metric g in the generalised sense above. The latter is given explicitly by

$$g = \langle \theta^*, \theta \rangle \in \Omega^1 \bar{\otimes} \Omega^1 \tag{3.3}$$

where the angular brackets denote evaluation of V^*, V and where the result is the pull back of forms on M due to equivariance and horizontality of θ, θ^*.

Next, any connection ω on the principal bundle X induces a covariant derivative $D : \mathcal{E} \to \Omega^1 \bar{\otimes} \mathcal{E}$ on any associated bundle with sections $\mathcal{E}$. Given the framing isomorphism this becomes $\nabla : \Omega^1 \to \Omega^1 \bar{\otimes} \Omega^1$. One also finds that $D \wedge \theta$ corresponds to the torsion T_∇ and that $F(\omega) = d\omega + \omega \wedge \omega$ acting on the sections corresponds to R_∇. As shown in [M6], none of this actually needs the bundle X to be the usual frame bundle, we have instead 'abstracted' the necessary properties in the notion of a framing. Similarly, we can regard θ^* in the role of framing and have an induced ∇^* with torsion T_{∇^*} corresponding to $D \wedge \theta^*$ and an induced curvature R_{∇^*} which is merely adjoint to that for ∇. One may easily verify that ∇^* is adjoint to ∇ in the sense of Proposition 3.1 with respect to g defined by (3.3).

In the case of a parallelisable manifold (or a local coordinate patch, for example) one does not need to work globally 'upstairs' in terms of θ, θ^*. In this case θ is equivalent to a basis of 1-forms $\{e_a\}$ such that $\Omega^1 = C(M).\{e_a\}$ (a basis over $C(M)$) that transform among themselves under the action of G, in other words a G-covariant n-vielbein. The first part means that the sections of the cotangent bundle are a free module rather than the general case of a projective one. Similarly a coframing is the choice of another collection $\Omega^1 = \{e^a\}.C(M)$ which we call an 'n-covielbein' and which transforms in the dual representation. The corresponding metric (given a framing) is $g = e_a \bar{\otimes} e^a$ or in indices,

$$g_{\mu\nu} = e_{a\mu} e^a_\nu.$$

Our decision not to assume that the metric is symmetric is reflected in the fact that the vielbein and covielbein are treated independently. We can of course write $e^a = \eta^{ab} e_b$ and call η^{ab} the frame metric but it need not be of a fixed or symmetric form (it can vary over the manifold), but at each point should be G-invariant.

Finally, in order to do the minimum of gravitational physics we need to be able to define the Ricci tensor and scalar. To do this we need to explicate a 'lifting map'

$$\imath : \Omega^2 \to \Omega^1 \bar{\otimes} \Omega^1$$

that splits the surjection $\wedge$ going the other way (so that $i \circ \wedge$ is a projection operator on $\Omega^1 \bar{\otimes} \Omega^1$). In classical differential geometry this is a trivial map since 2-forms are already defined by antisymmetric tensors so that $i(\alpha^{\mu\nu} dx_\mu \wedge dx_\nu) = \alpha^{\mu\nu} dx_\mu \bar{\otimes} x_\nu$ does the job. Evaluating i against a vector field X, it is equivalent to the interior product of vector fields against 2-forms, but as before we prefer to take the 'coaction' point of view. In NCG we will need to specify this data in the course of constructing the 2-forms rather than to take it for granted. Then $(i \bar{\otimes} \mathrm{id}) R_\nabla : \Omega^1 \to \Omega^1 \bar{\otimes} \Omega^1 \bar{\otimes} \Omega^1$ and we can now take the trace (at least in the local or parallelizable case, but it seems to work in practice globally) by feeding the middle Ω^1 (say) back into the input. This defines

$$\mathrm{Ricci} = \mathrm{Tr}(i \bar{\otimes} \mathrm{id}) R_\nabla = i(F_j)^{ab} e_b \bar{\otimes} f^j \triangleright e_a \in \Omega^1 \bar{\otimes} \Omega^1$$

from the general and from the framed points of view. Here $i(F)$ is expressed in the vielbein basis and $\triangleright$ denotes the action of the Lie algebra $\mathfrak{g}$ of G with basis $\{f^i\}$ on $V = \mathrm{span}_k \{e_a\}$. Notice that Ricci does not depend exactly on the metric or covielbein when set up in this way, rather just on the framing which is half way to a metric. Also note that in these conventions the Ricci tensor is minus the usual one. We then define the Ricci scalar by applying g^{-1} as a map and taking the trace $R = \mathrm{Tr}g^{-1}\mathrm{Ricci}$ again at least in the parallelizable case. It is fair to say that a completely abstract and more conceptual picture of the Ricci curvature and scalar is missing both in classical geometry in this setup and in the NCG case, but the above does give more or less reasonable results in examples including nonparallelizable ones.

The same remarks apply to the Dirac operator where there is a practical definition as follows: we require some other representation W of G such that the associated bundle with sections $\mathcal{S}$ can serve as the spinor bundle. What is required for this is a bundle map $\Gamma : \Omega^1 \to \mathrm{End}(\mathcal{S})$ which at least in the parallelizable case is given by a G-equivariant map $\gamma : V \to \mathrm{End}(W)$. The latter is our notion of 'generalised Clifford structure'. The Dirac operator is defined by $\slashed{\nabla} = \circ(\Gamma\bar{\otimes}\mathrm{id})D$ where $D : \mathcal{S} \to \Omega^1\bar{\otimes}\mathcal{S}$ is the covariant derivative from the connection on X and $\circ$ is the application of $\mathrm{End}(\mathcal{S})$ on $\mathcal{S}$. This constructive approach gives reasonable answers in NCG examples including non-parallelizable ones but a conceptual picture and axiomatization based on it are missing at the time of writing (it does not usually obey Connes' axioms for a spectral triple, for example, but may obey some generalised version of them).

Note also that the above reformulation and weakening of classical Riemannian geometry is completely symmetric between 1-forms and vector fields or vielbeins and covielbeins up until the Ricci curvature. In other words the basic geometrical set-up is self-dual and should therefore be better adapted to microlocal analysis or local Fourier transform ideas. This is part of our 'vision' of Einstein's equation as a self-duality equation, see Section 6, when both sides are understood properly (in NCG for example). At a more practical level it may be interesting to take these weaker axioms seriously and ask if there are interesting weak classical solutions, such as new kinds of weak black holes etc. At the moment most attention has been given to quantising the geometry which means by intention landing back on a conventional configuration in the classical limit $\lambda \to 0$, rather than this question for classical weak Riemannian geometry.

4. Quantum bundles and Riemannian structures

The above reformulation and weakenig of classical Riemannian geometry is now NCG-ready. All notions in the previous section make sense over any algebra A with differential structure. Moreover, the extra rigidity provided by a high degree of noncommutativity of the geometry tends to compensate for the weakening and yield canonical answers, for example for Riemannian structures on quantum groups and homogeneous spaces.

First, a quantum bundle over an algebra A means [BM1, BM2]:

1. H a Hopf algebra coacting $\Delta_R : P \to P \otimes H$ covariantly on an algebra P with $A = P^H \equiv \{p \in P \mid \Delta_R p = p \otimes 1\} \subseteq P$ as the fixed subalgebra.
2. Compatible differential structures where $\Omega^1(H) = H.\Lambda_H^1$ is bicovariant, $\Omega^1(P)$ is H-covariant and $\Omega^1(A) = A(dA)A \subseteq \Omega^1(P)$.
3. $0 \to P\Omega^1(A)P \to \Omega^1(P) \to P \otimes \Lambda_H^1 \to 0$ is exact.

Here any quantum group left (or bi)-covariant calculus is a free module of the form shown with basic left-invariant 1-forms spanning a vector space Λ_H^1. This can be explicitly constructed as a quotient of $\ker \epsilon \subset H$ by some Ad-invariant right ideal $\mathcal{I}_H$ and should be thought of as (and typically is) the dual of some finite-dimensional (braided) Lie algebra underlying H [GM2]. The compatibility of the calculi includes the stated requirement that the calculus on P restricted to A gives the desired calculus on A and a further condition on the ideal $\mathcal{I}_H$ under Δ_R reflecting 'smoothness' of this coaction. The third item is key and says that the kernel of the 'left-invariant vector fields map' $\Omega^1(P) \to P \otimes \Lambda_H^1$ is precisely the space of horizontal forms $P\Omega^1(A)P$, where the former is an infinitesimal version of the coaction Δ_R and determined by that. If one evaluates Λ_H^1 against an element x of the (braided) Lie algebra of H, one has a map $\Omega^1(P) \to P$ which is the noncommutative vector field generated by the action of x. This notion is now about 14 years old and quite well explored by many authors. A special case is to take so-called universal calculi on A, H, P (so for example the ideal $\mathcal{I}_H = 0$) and in this case a quantum bundle is equivalent to an algebraic notion of 'Hopf-Galois extension'. This is in some sense the 'topological' level of the theory maximally far from classical differential geometry.

Next, if V is a right H-comodule we have $\mathcal{E} = (P \otimes V)^H$ the associated bundle in NCG given by the fixed subspace under the tensor product coaction (it can also be set up as a cotensor product). We work directly with the sections as there is not necessarily any underlying classical total space. Then a framed manifold structure on A means (H, P, V, θ) where we have the above and an equivariant map

$$\theta : V \to P\Omega^1(A)$$

(the right hand side here is the space of 'left-horizontal' forms) such that the induced map $\cdot(\mathrm{id} \otimes \theta) : \mathcal{E} \to \Omega^1(A)$ is an isomorphism . Being a map from V is the same as having values in V^* as we had before (at least in the finite-dimensional case of main interest).

Finally, a coframing means similarly (H, P, V^*, θ^*) where $\theta^* : V^* \to \Omega^1(A)P$. A framed and coframed algebra is the same as a framed algebra with metric g given as in (3.3). A (strong) connection ω on H, which we call 'spin connection' with respect to the framing induces a covariant derivative D on associated bundles and we define ∇ in the same way as before. In the framed and coframed case we also have ∇^* and we say that the connection is 'weakly metric compatible' if the torsion and cotorsion vanish. The latter is equivalent as before to metric-compatibilty in a skew form (3.2). The torsion and curvature are given by the same formulae as before in (3.1) and correspond to a certain $\bar{D} \wedge \theta$ and $F(\omega)$ respectively but now on the quantum bundle (this requires certain flatness conditions on Ω^2 but these hold in practice). Similarly for the Dirac operator in terms of equivariant $\gamma : V \to \text{End}(W)$ at least in the parallelizable case or in terms of Γ in the general case.

Note that in the parallelizable case ω 'upstairs' is determined by $\alpha : \Lambda^1_H \to \Omega^1(A)$ which one should think of as a (braided) Lie algebra-valued connection $\alpha = \alpha_i f^i$ with $\{f^i\}$ a basis of $\Lambda^{1*}_H \subset H^*$. The torsion and cotorsion equations then become

$$\mathrm{d}e_a + \sum \alpha_i \wedge f^i \triangleright e_a = 0, \quad \mathrm{d}e_a + S^{-1}f^i \triangleright e_a \wedge \alpha_i = 0$$

while the curvature 'downstairs' is

$$F(\alpha) = \mathrm{d}\alpha_i + c_i^{jk}\alpha_j \wedge \alpha_k$$

where $f^j f^k = f^i c_i^{jk}$ expresses the product of H^* (or the coproduct of H).

All of this would be pie in the sky if not for the fact that over the years a lot has been computed and while there are still some rough edges to the abstract theory, it does apply to a wide range of examples. Still more examples are covered by a generalisation to coalgebras in place of Hopf algebras [BM3].

Example [M6]: Just to set the scene with something familiar, let G be a compact Lie group with Lie algebra $\mathfrak{g}$. We use G to frame itself, with $X = G \times G$ trivial. This choice of frame group restricts what kind of ∇ can be induced but is adequate for many purposes. We take $\alpha = \frac{1}{2}e$ where $e \in \Omega^1(G, \mathfrak{g})$ is the Maurer-Cartan form obeying $\mathrm{d}e + e \wedge e = 0$. We let $V = \mathfrak{g}^*$ with the coadjoint action and basis $\{f_a\}$ say, and we take vielbein $e_a = f_a(e)$ given by the components of the Maurer-Cartan form. In short, e defines the framing and $e/2$ defines the spin connection. The torsion vanishes since $De = \mathrm{d}e + [\alpha, e] = \mathrm{d}e + \frac{1}{2}[e, e] = \mathrm{d}e + e \wedge e = 0$ but its curvature $F(\alpha) = -\frac{1}{4}e \wedge e$ does not. If we take the coveilbein to be related to the vielbein by the Killing form (so the local metric η is the Killing form) then the cotorsion automatically also vanishes. The corresponding ∇ is in fact the usual Levi-Civita connection for the Killing metric (i.e. $\nabla g = 0$ as it happens) but we have constructed it in a novel way.

Example [M9]: The above example works just as well for quantum groups. For $A = \mathbb{C}_q[SU_2]$ with its 4D bicovariant calculus as studied in Section 2, we take $H = A$ and $P = A \otimes A$. There is a Maurer-Cartan form e and a (braided)-Killing form on the 4D cotangent space which is invariant and used to define the coframing as well. Remarkably, this metric is a q-deformation of the Minkowksi metric when all the reality constraints are taken into account, i.e. the extra direction Θ in Λ^1_H enters with a negative signature in the limit. We then solve for a spin connection α (it is no longer just $e/2$ but is uniquely determined) and find the resulting torsion free cotorsion free ∇. One finds $\nabla g = O(\lambda)$ (where $q = \exp(\lambda/2)$) but not zero. Finally, the exterior algebra is given by braided-skew symmetrization and hence, as classically, there is a canonical lifting i given from this braiding. The resulting Ricci tensor is

$$\text{Ricci} = -\frac{2q^2}{[4]_{q^2}}\left(g + \frac{q^4}{1+q^2}\Theta \bar{\otimes} \Theta\right).$$

In other words, apart from the mysterious non-classical extra direction Θ, the noncommutative space is 'Einstein'.

Example [M7]: The nice thing is that algebras of functions on finite groups G are also perfectly good Hopf algebras. Even though the algebra $A = \mathbb{C}(G)$ is commutative, its differential calculus is necessarily noncommutative (there is no non-zero classical differential structure on a finite set). The bicovariant calculi have invariant forms e_a labelled by Ad-stable subsets (e.g. conjugacy classes) in G not containing the group identity element 1. For S_3 there is a natural calculus $\Omega^1 = \mathbb{C}(S_3)\{e_u, e_v, e_w\}$ labeled by the 2-cycles $u = (12), v = (23), w = (13)$. There is an element $\Theta = \sum_a e_a$ which makes the calculus inner as we said was typical of a strictly noncommutative geometry. The (braided) Killing form comes out as the standard $g = \sum_a e_a \otimes e_a$ and again one can solve for torsion free cotorsion free α. Under a regularity condition this is unique and given by

$$\nabla e_u = -e_u \bar{\otimes} e_u - e_v \bar{\otimes} e_w - e_w \bar{\otimes} e_v + \frac{1}{3}\Theta \bar{\otimes} \Theta$$

One can verify that $\nabla g \neq 0$ but that (3.2) does hold as it must by construction. Again there is a nontrivial braiding Ψ that defines the exterior algebra and hence a canonical lifting map i. The Ricci curvature is then

$$\mathrm{Ricci} = \frac{2}{3}(-g + \Theta \bar{\otimes} \Theta).$$

In other words, apart form this mysterious extra dimension we again have an Einstein space with (what would be positive in usual conventions) constant curvature. Indeed, the canonical NCG of S_3 is more like quantum S^3 (the quantum group SU_2). If one does the same thing for the alternating group A_4 one has again a unique spin connection and this time Ricci = 0.

Example [M11]: To convince the reader we have to give a nontrivial non-parallel-izable example. The one that has been fully worked out is $\mathbb{C}_q[S^2]$ the standard quantum sphere. We define on $P = \mathbb{C}_q[SU_2]$ the coaction $\Delta_R f = f \otimes t^{\deg(f)}$ of the Hopf algebra $H = \mathbb{C}[t, t^{-1}]$ with coproduct $\Delta t = t \otimes t$. The latter is the coordinate algebra of S^1 and we are giving the q-analogue of $S^2 = SU_2/S^1$. Then $A = \mathbb{C}_q[S^2] \equiv \mathbb{C}_q[SU_2]_0$, the degree zero subspace. It inherits an algebra structure generated by $b_+ = cd$, $b_- = ab$, $b_0 = bc$ and differential calculus from that of $\mathbb{C}_q[SU_2]$, where we use the 3D calculus given in Section 2. Finally, we need to take a calculus $\Omega^1(H) = \mathbb{C}[t, t^{-1}]dt$ with relations $dt.t = q^2 t.dt$ and $dt^n = [n]_{q^2}t^{n-1}$, which is the unique form of noncommutive differential structure on a classical circle for some parameter (here q^2). It is known that all this gives a quantum bundle, essentially it was found in [BM1] in some form along with a canonical connection α, the q-monopole.

Next, a main theorem of [M11] is that any quantum homogeneous bundle built from quantum groups $\pi : P \to H$ like this makes the base algebra A a framed quantum manifold. Here

$$V = \frac{P^+ \cap A}{\mathcal{I}_P \cap A}, \quad \Theta(v) = S\tilde{v}_{(1)}d\tilde{v}_{(2)}, \quad \Delta_R(v) = \tilde{v}_{(2)} \otimes S\pi(\tilde{v}_{(1)})$$

where $\mathcal{I}_P$ is the ideal that describes the left-covariant $\Omega^1(P)$ and $P^+ = \ker \epsilon \subseteq P$. Here $\tilde{v}$ denotes any representative of $v \in V$ in the larger space. Applying this theorem in the above example gives

$$V = \langle b_\pm \rangle / \langle b_\pm^2, b_0 \rangle = \mathbb{C} \oplus \mathbb{C}, \quad \Delta_R b_\pm = b_\pm \otimes t^{\pm 2}$$

which means that

$$\Omega^1(\mathbb{C}_q[S^2]) = \mathcal{E}_{-2} \oplus \mathcal{E}_2$$

as the direct sum of associated quantum bundles of monopole charges ± 2. We identify them with the holomorphic and antiholomorphic differentials $\Omega^{1,0}$, $\Omega^{0,1}$ respectively in a double complex. This defines a q-deformed complex structure on the q-sphere, with several applications. Meanwhile, the invariant metric is

$$g = q^2 db_- \bar{\otimes} db_+ + db_+ \bar{\otimes} db_- - (2)_q db_0 \bar{\otimes} db_0$$

where $(2)_q = q + q^{-1}$, and the q-monopole connection induces a ∇ which is torsion free and cotorsion free. It is given by

$$\nabla db_\pm = (2)_q b_\pm g, \quad \nabla db_0 = (1 + (2)_q) b_0 g.$$

Its curvature is a multiple of $\text{Vol} \otimes \text{id}$ on each of the two parts, where $\text{Vol} = e_+ \wedge e_-$ is the top form on the q-sphere. The Ricci curvature for the simplest choice of lifting map is

$$\text{Ricci} = \frac{q^{-1}(1 + q^4)}{2} g + (2)_q \frac{(1 - q^4)}{2} i(\text{Vol}).$$

We see that Ricci acquires what in classical geometry would be an antisymmetric part as a result of the deformation. There is also a spinor bundle $\mathcal{S} = \mathcal{E}_{-1} \oplus \mathcal{E}_1$ and a canonical Clifford map with resulting gravitational Dirac operator $\slashed{\nabla}$.

Since our machinery applies also to finite groups it would be nice to similarly have a finite quantum homogeneous space analogous to the above example, using finite groups and calculi on them. At the time of writing we lack a non-parallelizable such example. Finally, working with finite geometries obviously improves the divergences coming from the continuum, but so does (for example) q-deformation. For example, the natural categorically defined dimension of the left-regular representation

$$\dim_q(\mathbb{C}_q[SU_2]) = \frac{\sum_{n \in \mathbb{Z}} q^{-\frac{n^2}{2}}}{1 - q^{-2}}$$

is finite for $q < 1$. Similar formulae exist for all $\mathbb{C}_q[G]$ and have links to number theory and quantum mechanics in bounded domains [MS].

5. Quantum gravity on finite sets

So far we have shown that the general theory is quite able to construct canonical metrics and connections in a range of examples not possible in ordinary differential geometry. However, for quantum gravity we need to work with the entire moduli of geometries. This has been investigated in detail for small finite sets in [MR2] and we discuss the results now.

Here $A = \mathbb{C}(M)$ is the algebra of functions on a finite set M. A (necessarily noncommutative) differential structure is given by a subset $E \subset M \times M$ which we indicate as 'arrows' $x \to y$ for $x, y \in M$. Here $x \to x$ are not allowed. To keep things simple let us suppose that the calculus is bidirectional so if $x \to y$ then $y \to x$. Even such unoriented graphs have not been classified, it is a wild problem. To simplify it let us stick to the parallelizable case where we assume that the graph is n-regular for a fixed n, i.e. for each x the cardinality of the set of arrows out of x is fixed. The simplest class of vielbeins in this context boils down to a combinatorial part and a continuous part at each point. The former is the choice at each x of a numbering $1, \cdots, n$ of the arrows out of x, i.e., $x \to s_x(a)$ for some bijection s_x at each x. The continuous degree of freedom is a non-vanishing normalisation which we denote $\lambda_a(x)$. The differential structure and vielbein are of the form

$$\Omega^1(\mathbb{C}(M)) = \mathbb{C}(M).\{e_a \ : \ a = 1, \cdots n\}, \quad e_a = \sum_{x \in M} \lambda_a(x)\delta_x \mathrm{d}\delta_{s_a(x)}$$

in terms of the differentials of Kronecker delta-functions. The relations of the calculus are

$$e_a f = f(s.(a))e_a, \quad \mathrm{d}f = (\partial^a f)e_a, \quad (\partial^a f)(x) = \frac{f(s_x(a)) - f(x)}{\lambda_a(x)}$$

where we see that the partial derivatives are finite differences. These calculi are all inner with $\Theta = \sum_a \lambda_a^{-1} e_a$. The finite group example in the previous section was of the above form with $s_x(a) = xa$ right translation in the group and $\lambda_a \equiv 1$.

Again for simplicity we take the Euclidean frame metric so that the vielbein and covielbein are the same and $g = \sum_a e_a \bar{\otimes} e_a$. We also have to fix a frame group G, its calculus and the action of its (braided) Lie algebra on the vielbein, all of which will generally be dictated by the geometrical picture we want to allow. If $n = 2$ and we have in mind a curve then a natural choice for G is $\mathbb{Z}_2$ flipping the e_1, e_2 considered at angle π (so the quantum cotangent bundle has an extra dimension). If for $n = 2$ we have in mind a surface then a natural choice might be the group $\mathbb{Z}_4$ acting by $\pi/2$ frame rotations (if we consider e_1, e_2 at right angles), and so forth. We also have at some point to specify the $\wedge$ product for the higher exterior algebra through the choice of a G-invariant projector (this also gives the lifting i). In the simplest cases it could just be the usual antisymmetrization. In this way we fix the 'type' of data for the context of the model.

To do quantum gravity with fixed cotangent dimension n we then have to sum over all $m = |M|$ the number of points in the NCG, all n-regular graphs, all

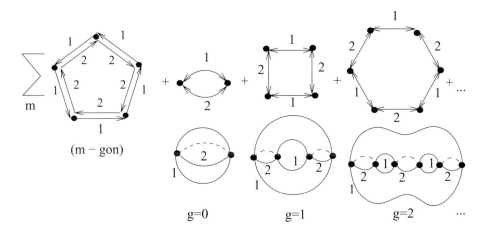

FIGURE 1. Combinatorial part of sum for finite $n = 2$ quantum gravity includes $2(g + 1)$-gons as models of genus g surfaces

numberings s_x on them, and integrate over all the functions λ_a and all compatible connections α_i for the given framing and wedge product (probably we should sum over choices for the latter as well). If we take a (perhaps naive) path integral approach we should do all this with some action such as given by R, the Ricci scalar computed in the NCG after all the choices stated. So, a partition function

$$\mathcal{Z} = \sum_m \sum_{n-\text{graphs}} \sum_s \int \prod_a \mathcal{D}\lambda_a \int \prod_i \mathcal{D}\alpha_i \; e^{-\beta \sum_{x \in M} R(x)}.$$

Note that the choice s is a *colouring* of the oriented graph into loops labelled from $\{1, \cdots, n\}$. This is because one can follow each number label from vertex to vertex; since the graph is finite it has to come back to itself at some point and every arrow must be coloured (in each direction). In addition to this the gravitational modes are allowed to set the physical length of each arrow which is the continuous part of the metric or vielbein data. For example, if we fix $n = 2$ the graphs are polygons and the colourings are either (i) colours 1,2 running oppositely around the polygon (ii) when m is even; colours 1,2 alternating and with the two directions of each edge having the same colour. This gives the combinatorial part of the sum of configurations for $n = 2$ quantum gravity as shown schematically in Figure 1. The type (i) cases are the groups $\mathbb{Z}_m$ with their natural 2D calculus, which we view as approximations of a circle S^1. The type (ii) cases we propose correspond to surfaces of genus $(m-2)/2$ where each $\leftrightarrow$ approximates a circle, so we have m geodesic circles intersecting at right angles (since the labels alternate and describe orthogonal zweibeins), as indicated in Figure 1. (Different weights λ_a will stretch these pictures about.) Thus the square of type (ii) is the natural geometry for $\mathbb{Z}_2 \times \mathbb{Z}_2$ as a model of a torus, while the square of type (i) is the natural geometry for $\mathbb{Z}_4$ as a model of S^1. We have a similar choice for $n = 3$ and higher; we can think of a trivalent graph as a triangulation of a surface (note the extra dimension, linked to the existence of Θ) or as the geometry of a 3-manifold. For example a cube with one kind of colouring can be a model of a sphere, or, with a different colouring, $\mathbb{Z}_2^3$ as a model of a 3-torus.

Then for each combinatorial configuraton we have to integrate over the continuous degrees of freedom, preferably with a source term in the action as we also want expectation values of operators, not just the partition function. Such an approach has been carried out in the $U(1)$ not gravitational case in [M8]; the integrals over α, while still divergent, are now ordinary not usual functional integrals and the theory is renormalisable. The gravitational case has not been carried so far but the moduli of α and its curvature have been investigated in the simplest $n = 2$ cases [MR2]. For example, consider the square with type (ii) colouring. We take zweibeins e_1, e_2 anticommuting as usual and $G = \mathbb{Z}_4$ with its 2D calculus generating $\pm\pi/2$ rotations in the group. The actions are therefore

$$f^1 \triangleright e_1 = e_2 - e_1, \quad f^1 \triangleright e_2 = -e_1 - e_2, \quad f^2 \triangleright e_1 = -e_2 - e_1, \quad f^2 \triangleright e_2 = e_1 - e_2$$

(the Lie algebra elements are of the form $g - 1$ in the group algebra, where g is a group element). To focus on the connections, let us fix $\lambda_a = 1$, which is of course not the full story. Then $\mathrm{d}e_a = 0$ and the torsion-free equation becomes

$$\alpha_1 \wedge (e_2 - e_1) = \alpha_2 \wedge (e_2 + e_1), \quad \alpha_1 \wedge (e_1 + e_2) = \alpha_2 \wedge (e_1 - e_2).$$

Let us also assume unitarity constraints in the form $e_a^* = e_a$ and $\alpha_1^* = \alpha_2$ ('antihermitian') which makes the cotorsion-free equation automatic as the conjugate of the torsion-free one. This is solved with α_i in basis e_1, e_2 given by functions a, b subject to $\bar{a} = R_2 a = -R_1 a$, $\bar{b} = R_1 b = -R_2 b$, where R_a are translations in each $\mathbb{Z}_2$ of $\mathbb{Z}_2 \times \mathbb{Z}_2$. The resulting covariant derivative and Ricci scalar are

$$\nabla e_1 = (ae_1 + be_2)\bar{\otimes}e_1 + (be_1 - ae_2)\bar{\otimes}e_2, \quad \nabla e_2 = (-be_1 + ae_2)\bar{\otimes}e_1 + (ae_1 + be_2)\bar{\otimes}e_2$$

$$R = -2(|a|^2 + |b|^2), \quad \sum_x R(x) = -8(|A|^2 + |B|^2),$$

where A, B are the values of a, b are a basepoint (the values at the other 3 points being determined). The full picture when we include varying λ_a has additional terms involving their derivatives and a regularity condition cross-coupling the a, b systems (otherwise there is a kind of factorisation), see [MR2] for details but with some different notations. Note that the model in the type (ii) case is very different from the simplicial interpretation in the type (i) models, and indeed we don't appear to have a Gauss-Bonnet theorem in the naive sense of $\sum_x R(x) = 0$ for a torus. Also note that the torus case is too 'commutative' for the weak Riemannian geometry to determine ∇ from the metric and we see the larger moduli of connections. If we impose full $\nabla g = 0$ then we do get a, b determined from derivatives of λ_a but this significantly constrains the allowed λ_a and is not very natural. Our naive choice of Ω^2 also constrains the λ_a, so the above is perhaps not the last word on this model.

Looking at the diagrams in Figure 1 we see a similarity with matrix models and their sum over genus. Also note that the sum over graphs is not unlike Feynman rules for a ϕ^n interacting scaler theory in flat space. There vertices are the interactions, while graph arrows are propagators (doubled up in the type (i) colouring cf. in matrix models). As with Feynman rules, we follow the momentum (=colouring in our case) around in loops and sum over all values. The weights are different and given in our case by NCG from the continuous degrees of freedom. In general terms, however, we see that finite set quantum gravity at fixed cotangent dimension n has some kind of duality with flat space interacting ϕ^n scaler theory of some (strange) kind. There are likewise similarities with group field theory coming out of spin foams when expressed in a suitable way.[2]

We also have obvious links with the causal poset approach to spacetime [RS]. The difference is that NCG is a general framework that embraces both classical and these new approaches to geometry in a uniform manner, which addresses the main problem in say the poset as well as spin foam approaches. The main problem is to see how ordinary spacetime and geometry will emerge from such models. In NCG this is a matter of the algebra and the calculus. As $m \to \infty$ but with the calculus in a controlled limit, the finite differences will become usual differentials and, modulo some issues of extra dimensions, we will see the classical geometry emerge within the uniform NCG framework. The specific comparison with causal sets is nevertheless interesting and suggests that a causal structure is 'half' of a bidirectional differential structure in which if $x \to y$ then the other way is not allowed. Given a poset we can double the arrows so that they are both ways, and proceed as above. Or indeed NCG also works with unidirectional calculi so we can work just on the poset. One can also have a mixture of arrows, for example to model a black hole. In NCG one seeks a Hodge $*$-operator on the algebra of differential forms (it is known for all the examples above [MR1, GM1, M11]) after which one can do Maxwell and matter field theory as well as the gravitational case discussed above.

6. Outlook: Monoidal functors

We have offered NCG as a more general framework definitely useful as an effective theory and conjecturally better as a foundation for geometry to avoid divergences. But is it the ulimate theory? In my view there is a deeper philosophical picture for quantum gravity that requires us to address the nature of physical reality itself, with NCG as a key stage of development. This was expounded many years ago in [M1] so I shall be brief. Basically, the claim is that what is real are some measured outcomes $f(x)$ but whether x is real and being measured by f or f is real and being measured by x (what I have called observable-state duality) is like a gauge choice and not absolute as long as the choice is made coherently. This sets up a duality between different bits of physics, basically elevating Born reciprocity

---

[2]Annotation by the Eds.: See the chapter by Qriti on this subject in thsi volume.

to a deeper conceptual level that applies not only to position and momentum but to geometry and quantum theory. In this setting Einstein's equations $G = 8\pi T$ should be a self-duality equation identifying a quantum part of the theory with a classical geometry part, when both sides have been expressed in the same language, such NCG. Quantum group toy models [M3] demonstrate the idea in terms of Hopf algebra duality and T-duality [BeM1], where indeed a non-symmetric metric is inverted in the dual model (micro-macro duality). Quantum group Fourier transform implements the Hopf algebra duality as explored for quantum spacetime models in our previous work, see [MII] for a review.

But what about in the other direction, beyond Hopf algebras and NCG? In [M2] we showed that the next self-dual type of object beyond Hopf algebras and admitting such duality was: a pair of monoidal categories $F : \mathcal{C} \to \mathcal{V}$ with a functor between then. To such a triple we constructed a dual $F^\circ : \mathcal{C}^\circ \to \mathcal{V}$ of the same type and a map $\mathcal{C} \subset \mathcal{C}^{\circ\circ}$. Therefore the next more complex theory to fulfill our self-duality requirement of observer-observed symmetry after toy quantum group models should be a self-dual such object in some sense. There is a bit more going on here as unless $\mathcal{V}$ is trivial (such as vector spaces) the duality operation extends the category (for example it takes a quantum group to its quantum double) so something should be projected out before we can speak about self-dual objects. That being said, it is interesting that 15 years later the notion of QFT on curved spaces has been nicely set-up by Fredenhagen and coworkers precisely as a monoidal functor[3]

$$F : \{\text{Globally hyperbolicmanifolds}\} \to C^* - \text{Algebras}.$$

This is a noncommutative version of the functor $C(\ )$ that assigns to a manifold its commutative algebra of functions, but is only half the story and not self-dual. However, deformations of F could lead to a self-dual functor in the same way as one has genuine self-dual quantum groups. These would be systems with both quantum theory and gravity along the lines of our experience with quantum group toy models. Moreover, the NCG constructions such as quantum bundles etc. extend to monoidal categories [M5], so this line can be explored.

References

[BeM1] E.J. Beggs and S. Majid, Poisson-Lie T-Duality for Quasitriangular Lie Bialgebras. *Commun. Math. Phys.* 220:455-488, 2001.

[BeM2] E.J. Beggs and S. Majid. Semiclassical Differential Structures, to appear *Pac. J. Math.*, 2006.

[BeM3] E.J. Beggs and S. Majid. Quantization by Cochain Twists and Nonassociative Differentials. Preprint Math.QA/0506450.

[BM1]   T. Brzeziński and S. Majid. Quantum Group Gauge Theory on Quantum Spaces. *Commun. Math. Phys.*, 157:591–638, 1993. Erratum 167:235, 1995.:

---

[3]Annotation by the Eds.: See the chapter by Brunetti and Fredenhagen in this volume dwelling on this point.

[BM2]   T. Brzeziński and S. Majid. Quantum Differentials and the q-Monopole Revisited. *Acta Appl. Math.*, 54:185–232, 1998.

[BM3]   T. Brzezinski and S. Majid. Quantum Geometry of Algebra Factorisations and Coalgebra Bundles. *Commun. Math. Phys* 213 (2000) 491-521.

[Co]   A. Connes. *Noncommutative Geometry.* Academic Press, 1994.

[CR]   A Connes and C. Rovelli. Von Neumann Algebra Automorphisms and Time-Thermodynamics Relation in Generally Covariant Quantum Theories. *Class. Quantum Grav.* 11: 2899-2917, 1994.

[GM1]   X. Gomez and S.Majid. Noncommutative Cohomology and Electromagnetism on Cq[SL2] at Roots of Unity. *Lett. Math. Phys.* 60:221-237, 2002.

[GM2]   X. Gomez and S.Majid. Braided Lie Algebras and Bicovariant Differential Calculi over Coquasitriangular Hopf Algebras. *J. Algebra* 261: 334-388, 2003.

[HM]   P. Hajac and S. Majid. Projective Module Description of the q-Monopole. *Commun. Math. Phys.*, 206:246–464, 1999.

[Haw]   E. Hawkins. Noncommutative Rigidity. *Commun. Math. Phys.* 246:211-235, 2004.

[NML]   F. Ngakeu, S. Majid and D. Lambert. Noncommutative Riemannian Geometry of the Alternating Group A_4. *J. Geom. Phys.* 42: 259-282, 2002.

[MII]   S. Majid. Algebraic Approach to Quantum Gravity II: Noncommutative Spacetime. To appear in D. Oriti, ed., Cambridge Univ. Press; hep-th/0604130.

[Ma]   S. Majid. *Foundations of Quantum Group Theory.* Cambridge Univeristy Press, 1995.

[M1]   S. Majid. The Principle of Representation-theoretic Self-duality. *Phys. Essays.* 4:395-405, 1991.

[M2]   S. Majid. Representations, Duals and Quantum Doubles of Monoidal Categories. *Suppl. Rend. Circ. Mat. Palermo., Series II* 26, 197-206, 1991.

[M3]   S. Majid. Hopf Algebras for Physics at the Planck Scale. *J. Classical and Quantum Gravity,* 5:1587–1606, 1988.

[M4]   S. Majid. Quantum and Braided Lie-algebras. *J. Geom. Phys.* 13:307-356, 1994.

[M5]   S. Majid. Diagrammatics of Braided Group Gauge Theory. *J. Knot Th. Ramif.* 8:731-771, 1999.

[M6]   S. Majid. Quantum and Braided Group Riemannian Geometry. *J. Geom. Phys.,* 30:113–146, 1999.

[M7]   S. Majid. Riemannian Geometry of Quantum Groups and Finite Groups with Nonuniversal Differentials. *Commun. Math. Phys.* 225:131–170, 2002.

[M8]   S. Majid. Noncommutative Physics on Lie Algebras, $\mathbb{Z}_2^n$ Lattices and Clifford Algebras. In *Clifford Algebras: Application to Mathematics, Physics, and Engineering*, ed. R. Ablamowicz, pp. 491-518. Birkhauser, 2003.

[M9]   S. Majid. Ricci Tensor and Dirac Operator on $\mathbb{C}_q[SL_2]$ at Roots of Unity. *Lett. Math. Phys.*, 63:39-54, 2003.

[M10]   S. Majid. Classification of Differentials on Quantum Doubles and Finite Noncommutative Geometry, *Lect. Notes Pure Appl. Maths* 239:167-188, 2004. Marcel Dekker.

[M11]   S. Majid. Noncommutative Riemannian and Spin Geometry of the Standard q-Sphere. *Commun. Math. Phys.* 256:255-285, 2005.

[M12]   S. Majid. Noncommutative Model with Spontaneous Time Generation and Planckian Bound. *J. Math. Phys.* 46:103520, 18pp, 2005.

[MR1]   S. Majid and E. Raineri. Electromagnetism and Gauge Theory on the Permutation Group S_3. *J. Geom. Phys.* 44:129-155, 2002.

[MR2]   S. Majid and E. Raineri. Moduli of Quantum Riemannian Geometries on ¡= 4 Points. *J. Math. Phys.* 45: 4596-4627, 2004.

[MS]   S. Majid and Ya. S. Soibelman. Rank of Quantized Universal Enveloping Algebras and Modular Functions. *Comm. Math. Phys.* 137: 249-262, 1991.

[RS]   D. Rideout and R. Sorkin. A Classical Sequential Growth Dynamics for Causal Sets. *Phys. Rev. D* 61:024002, 2000.

[Wo]   S.L. Woronowicz. Differential Calculus on Compact Matrix Pseudogroups (Quantum Groups). *Commun. Math. Phys.*, 122:125–170, 1989.

S. Majid
School of Mathematical Sciences
Queen Mary, University of London
327 Mile End Rd, London E1 4NS, UK
& Perimeter Institute for Theoretical Physics
31 Caroline St N., Waterloo, ON N2L 2Y5, Canada
e-mail: `s.majid@qmul.ac.uk`

Quantum Gravity
B. Fauser, J. Tolksdorf and E. Zeidler, Eds., 101–126

Quantum Gravity as a Quantum Field Theory of Simplicial Geometry

Daniele Oriti

Abstract. This is an introduction to the group field theory approach to quantum gravity, with emphasis on motivations and basic formalism, more than on recent results; we elaborate on the various ingredients, both conceptual and formal, of the approach, giving some examples, and we discuss some perspectives of future developments.

Mathematics Subject Classification (2000). Primary 81R99; Secondary 81S99.

Keywords. Group field theory, spin networks, spin foams, canonical quantum gravity, third quantization, simplicial geometries.

1. Introduction: Ingredients and motivations for the group field theory approach

Our aim in this paper is to give an introduction to the group field theory (GFT) approach to non-perturbative quantum gravity. We want especially to emphasize the motivations for this type of approach, the ideas involved in its construction, and the links with other approaches to quantum gravity, more than reviewing the results that have been obtained up to now in this area. For other introductory papers on group field theory, see [1], but especially [2], and for a review of the state of the art see [3]. No need to say, the perspective on the group field theory approach we provide is a *personal* one and we do not pretend it to be shared or fully agreed upon by other researchers in the field, although of course we hope this is the case. First of all what do we mean by 'quantizing gravity' in the GFT approach? What kind of theory are we after? The GFT approach seeks to construct a theory of quantum gravity that is non-perturbative and background independent. By this we mean that we seek to describe at the quantum level *all* the degrees of freedom of the gravitational field and thus obtain a quantum description of the full spacetime geometry; in other words no perturbative expansion around any given gravitational background metric is involved in the definition of the theory, so on the one hand states and observables of the theory will not carry any dependence on such background structure, on the other hand the theory will not include only the gravitational configurations that are obtainable perturbatively starting from a given geometry. Also, let us add a (maybe not necessary) note: we are not after unification of fundamental forces; it cannot be excluded that a group field theory formulation of quantum gravity would be best phrased in terms of unified structures, be it the group manifold used or the field, but it is not a necessary condition of the formalism nor among the initial aims of the approach. So what are group field theories? In a word: group field theories are particular field theories on group manifolds that (aim to) provide a background independent third quantized formalism for simplicial gravity in any dimension and signature, in which both geometry and topology are thus dynamical, and described in purely algebraic and combinatorial terms. The Feynman diagrams of such theories have the interpretation of simplicial spacetimes and the theory provides quantum amplitudes for them, in turn interpreted as discrete, algebraic realisation of a path integral description of gravity. Let us now motivate further the various ingredients entering the formalism (for a similar but a bit more extensive discussion, see [20]), and at the same time discuss briefly other related approaches to quantum gravity in which the same ingredients are implemented.

1.1. Why path integrals? The continuum sum-over-histories approach

Why to use a description of quantum gravity on a given manifold in terms of path integrals, or sum-over-histories? The main reason is its generality: the path integral formulation of quantum mechanics, let alone quantum gravity, is more general than the canonical one in terms of states and Hamiltonians, and both problems of interpretation and of recovering of classicality (via decoherence) benefit from such a generalisation [4]. Coming to quantum gravity in particular, the main advantages follow from its greater generality: one does not need a canonical formulation or a definition of the space of states of the theory to work with a gravity path integral, the boundary data one fixes in writing it down do not necessarily correspond to canonical states nor have to be of spacelike nature (one is free to consider timelike boundaries), nor the topology of the manifold is fixed to be of direct product type with a *space* manifold times a time direction (no global hyperbolicity is required). On top of this, one can maintain manifest diffeomorphism invariance, i.e. general covariance, and does not need any $(n-1)+1$ splitting, nor the associated enlargement of spacetime diffeomorhism symmetry to the symmetry group of the canonical theory [5]. Finally, the most powerful non-perturbative techniques of quantum field theory are based on path integrals and one can hope for an application of some of them to gravity. So how would a path integral for continuum gravity look like? Consider a compact four manifold (spacetime) with trivial topology $\mathcal{M}$ and all the possible geometries (spacetime metrics up to diffeomorphisms) that are compatible with it. The partition function of the theory would then be defined [4] by an integral over all possible 4-geometries with a diffeomorphism invariant measure and weighted by a quantum amplitude given by the exponential of ('i' times) the action of the classical theory one wants to quantize, General Relativity. For computing transition amplitudes for given boundary configurations of the field, one would instead consider a manifold $\mathcal{M}$ again, of trivial topology, with two disjoint boundary components S and S' and given boundary data, i.e. 3-geometries, on them: $h(S')$ and $h'(S')$, and define the transition amplitude by:

$$Z_{QG}\left(h(S), h'(S')\right) = \int_{g(\mathcal{M}|h(S),h'(S'))} \mathcal{D}g \, e^{i\, S_{GR}(g,\mathcal{M})} \tag{1}$$

i.e. by summing over all 4-geometries inducing the given 3-geometries on the boundary, with the amplitude possibly modified by boundary terms if needed. The expression above is purely formal: first of all we lack a rigorous definition of a suitable measure in the space of 4-geometries, second the expectation is that the oscillatory nature of the integrand will make the integral badly divergent. To ameliorate the situation somehow, a 'Wick rotated' form of the above expression was advocated with the definition of a "Euclidean quantum gravity" where the sum would be only over Riemannian metrics with a minus sign in front of the action in the definition of the integral [6]. This however was not enough to make rigorous sense of the theory and most of the related results were obtained in semiclassical approximations [6]. Also, the physical interpretation of the above quantities presents several challenges, given that the formalism seems to be bound to a cosmological setting, where our usual interpretations of quantum mechanics are not applicable. We do not discuss this here, but it is worth keeping this issue in mind, given that a good point about group field theory is that it seems to provide a rigorous version of the above formulas (and much more than that) which is also *local* in a sense to be clarified below.

1.2. Why topology change? Continuum 3rd quantization of gravity

In spite of the difficulties in making sense of a path integral quantization of gravity on a fixed spacetime, one can think of doing even more and treat not only geometry but also topology as a dynamical variable in the theory. One would therefore try to implement a sort of "sum over topologies" alongside a sum over geometries, thus extending this last sum to run over *all* possible spacetime geometries and not only those that can live on a given topology. Again therefore the main aim in doing this is to gain in generality: there is no reason to *assume* that the spacetime topology is fixed to be trivial, so it is good not to assume it. Of course this has consequences on the type of geometries one can consider, in the Lorentzian case, given that a nontrivial spacetime topology implies spatial topology change [7] and this in turn forces the metric to allow either for closed timelike loops or for isolated degeneracies (i.e. the geometry may be degenerate, have zero volume element, at isolated points). While in a first order or tetrad formulation of gravity one can thus avoid the first possibility by allowing for the second, in the second order metric formulation one is bound to include metrics with causality violations. This argument was made stronger by Horowitz [8] to the point of concluding that *if* degenerate metrics are included in the (quantum) theory, then topology change is not only possible but *unavoidable* and non-trivial topologies therefore *must* be included in the quantum theory. However, apart from greater generality, there are various results that hint to the *need* for topology change in quantum gravity. Work on topological geons [9], topological configurations with particle-like properties, suggest that spatial topology change (the equivalent of pair creation for geons) is needed in order for them to satisfy a generalisation of the spin-statistics theorem. Work in string theory [10] indicates that different spacetime topology can be equivalent with respect to stringy probes. Wormholes, i.e. spatial topology changing spacetime configurations, have been advocated as a possible mechanism that turn off the cosmological constant decreasing its value toward zero [11], and the possibility has been raised that *all* constants of nature can be seen as vacuum *parameters*, thus in principle can be computed, in a theory in which topology is allowed to fluctuate [12]. This last idea, together with the analogy with string perturbation theory and the aim to solve some problems of the canonical formulation of quantum gravity, prompted the proposal of a "third quantization" formalism for quantum gravity [21, 22]. The idea is to define a (scalar) field in superspace $\mathcal{H}$ for a given choice of basic spatial manifold topology, i.e. in the space of all possible 3-geometries (3-metrics $^3h_{ij}$ up to diffeos) on, say, the 3-sphere, essentially turning the wave function of the canonical theory into an operator: $\phi(^3h)$, whose dynamics is defined by an action of the type:

$$S(\phi) = \int_{\mathcal{H}} \mathcal{D}^3h \, \phi(^3h) \Delta \phi(^3h) + \lambda \int_{\mathcal{H}} \mathcal{D}^3h \, \mathcal{V}\left(\phi(^3h)\right) \tag{2}$$

with Δ being the Wheeler-DeWitt operator of canonical gravity here defining the kinetic term (free propagation) of the theory, while $\mathcal{V}(\phi)$ is a generic, e.g. cubic, and generically non-local (in superspace) interaction term for the field, governing the topology changing processes. Notice that because of the choice of basic spatial topology needed to define the 3rd quantized field, the topology changing processes described here are those turning X copies of the 3-sphere into Y copies of the same. The quantum theory is defined by the partition function $Z = \int \mathcal{D}\phi e^{-S(\phi)}$, that produces the sum over histories outlined above, including a sum over topologies with definite weights, as a dynamical process, in its perturbative expansion in Feynman graphs:

The quantum gravity path integral for each topology will represent the Feynman amplitude for each 'graph', with the one for trivial topology representing a sort of one particle propagator, thus a Green function for the Wheeler-DeWitt equation. Some more features of this (very) formal setting are worth mentioning: 1) the full classical equations of motions for the scalar field will be a non-linear extension of the Wheeler-DeWitt equation of canonical gravity, due to the interaction term in the action, i.e. of the inclusion of topology change; 2) the perturbative 3rd quantized vacuum of the theory will be the "no spacetime" state, and not any state with a semiclassical geometric interpretation in terms of a smooth geometry, say a Minkowski state. We will see shortly how these ideas are implemented in the group field theory approach.

1.3. Why going discrete? Matrix models and simplicial quantum gravity

However good the idea of a path integral for gravity and its extension to a third quantized formalism may be, there has been no definite success in the attempt to realise them rigorously, nor in developing the formalism to the point of being able to do calculations and then obtaining solid predictions from the theory. A commonly held opinion is that the main reason for the difficulties encountered is the use of a *continuum* for describing spacetime, both at the topological and at the geometrical level. One can indeed advocate the use of *discrete structures* as a way to regularize and make computable the above expressions, to provide a more rigorous definition of the theory, with the continuum expressions and results emerging only in a continuum *limit* of the corresponding discrete quantities. This was in fact among the motivations for discrete approaches to quantum gravity as matrix models, or dynamical triangulations or quantum Regge calculus. At the same time, various arguments can be and have been put forward for the point of view that discrete structures instead provide a more *fundamental* description of spacetime. These arguments come from various quarters. On the one hand there is the possibility, suggested by various approaches to quantum gravity such as string theory or loop quantum gravity, that in a more complete description of space and time there should be a fundamental length scale that sets a least bound for measurable distances and thus makes the notion of a continuum loose its physical meaning, at least as a fundamental entity. Also, one can argue on both philosophical and mathematical grounds [13] that the very notion of "point" can correspond at most to an idealization of the nature of spacetime due to its lack of truly operational meaning, i.e. due to the impossibility of determining with absolute precision the location in space and time of any event (which, by the way, is implemented mathematically very precisely in non-commutative models of quantum gravity, see the contribution by Majid in this volume). Spacetime points are indeed to be replaced, from this point of view, by small but finite regions corresponding to our finite abilities in localising events, and a more fundamental (even if maybe not ultimate [13]) model of spacetime should take these local regions as basic building blocks. Also, the results of black hole thermodynamics seem to suggest that there should be a discrete number of fundamental spacetime degrees of freedom associated to any region of spacetime, the apparent continuum being the result of the microscopic (Planckian) nature of them. This means that the continuum description of spacetime will replace a more fundamental discrete one as an *approximation* only, as the result of a *coarse graining* procedure. In other words, a finitary topological space [14] would constitute a better model of spacetime than a smooth manifold. All these arguments against the continuum and in favor of a finitary substitute of it can be naturally seen as arguments in favor of a simplicial description of spacetime, with the simplices playing indeed the role of a finitary substitute of the concept of a point or fundamental event, or of a minimal spacetime region approximating it. Simplicial approaches to quantum gravity are matrix models, dynamical triangulations and quantum Regge calculus. The last one [15] is the straightforward translation of the path integral idea in a simplicial context. One starts from the definition of a discrete version of the Einstein-Hilbert action for General Relativity on a simplicial complex Δ, given by the Regge action S_R in which the basic geometric variables are the lengths of the edges of Δ, and then defines the quantum theory usually via Euclidean path integral methods, i.e. by:

$$Z(\Delta) = \int \mathcal{D}l \, e^{-S_R(l)}. \tag{3}$$

The main issue is the definition of the integration measure for the edge lengths, since it has to satisfy the discrete analogue of the diffeomorphism invariance of the continuum theory (the most used choices are the ldl and the dl/l measures) and then the proof that the theory admits a good continuum limit in which continuum general relativity is recovered, indeed the task that has proven to be the most difficult. Matrix models [23] can instead be seen as a surprisingly powerful implementation of the third quantization idea in a simplicial context, but in an admittedly simplified framework: 2d Rieammian quantum gravity. Indeed group field theories are a generalisation of matrix models to higher dimension and to Lorentzian signature. Consider the action

$$S(M) = \frac{1}{2}trM^2 - \frac{\lambda}{3!\sqrt{N}}trM^3 \tag{4}$$

for an $N \times N$ hermitian matrix M_{ij}, and the associated partition function $Z = \int dM e^{-S(M)}$. This in turn is expanded in perturbative expansion in Feynman diagrams; propagators and vertices of the theory can be expressed diagrammatically in Figure 1, and the corresponding Feynman diagrams, obtained as usual by gluing vertices with propagators, are given by *fat graphs* of **all** topologies. Moreover, propagators can be understood as topologically dual to edges and vertices to triangles, see Figure 2, of a 2-dimensional simplicial complex that is dual to the whole fat graph in which they are combined; this means that one can define a model for quantum gravity in 2d, via the perturbative expansion for the matrix model above, as sum over **all 2d triangulations** T of **all topologies**. Indeed the amplitude of each Feynman diagram for the above theory is related to the Regge action for classical simplicial gravity in 2 dimensions for fixed edge lengths equal to N and positive cosmological constant. More specifically, the partition function is:

$$Z = \int dM e^{-S(M)} = \sum_T \frac{1}{sym(T)} \lambda^{n_2(T)} N^{\chi(T)} \tag{5}$$

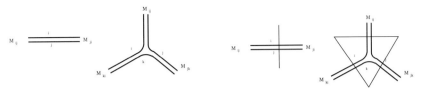

FIGURE 1. Propaga-
tor and vertex

FIGURE 2. Dual picture

where $sym(T)$ is the order of symmetries of the triangulation T, n_2 is the number
of triangles in it, and χ is the Euler characteristic of the same triangulation. Many
results have been obtained over the years for this class of models, for which we
refer to the literature [23]. Closely related to matrix models is the dynamical
triangulations approach [24], that extends the idea and results of defining a path
integral for gravity as a sum over equilateral triangulations of a given topology to
higher dimensions, weighted by the (exponential of the) Regge action for gravity:

$$Z(G, \lambda, a) = \sum_T \frac{1}{sym(T)} e^{iS_R(T,G,\Lambda,a)} \tag{6}$$

where G is the gravitational constant and Λ is a cosmological constant. In the
Lorentzian case one also distinguishes between spacelike edges (length square a^2)
and timelike ones (length square $-a^2$), and imposes some additional restrictions
on the topology considered and on the way the triangulations are constructed
via the gluing of d-simplices. In particular, one may then look for a continuum
limit of the theory, corresponding to the limit $a \to 0$ accompanied by a suitable
renormalisation of the constants of the theory Λ and G, and check whether in
this limit the structures expected from a continuum quantum gravity theory are
indeed recovered, i.e. the presence of a smooth phase with the correct macroscopic
dimensionality of spacetime. And indeed, the exciting recent results obtained in
this approach seem to indicate that, in the Lorentzian context and for trivial
topology, a smooth phase with the correct dimensionality is obtained even in 4
dimensions, which makes the confidence in the correctness of the strategy adopted
to define the theory grow stronger.

1.4. Why groups and representations? Loop quantum gravity and spin foams

We will see many of the previous ideas at work in the group field theory context.
There, however, a crucial role is played by the Lorentz group and its representations, as it is in terms of them that geometry is described. Another way to see
group field theories in fact is as a re-phrasing (in addition to a generalisation)
of the matrix model and simplicial quantum gravity formalism in an algebraic
language. Why would one want to do this? One reason is the physical meaning
and central role that the Lorentz group plays in gravity and in our description of
spacetime; another is that by doing this, one can bring in close contact with the
others yet another approach to quantum gravity: loop quantum gravity, through

spin foam models. But let us discuss one thing at the time. The Lorentz group enters immediately into play and immediately in a crucial role as soon as one passes from a description of gravity in terms of a metric field to a first order description in terms of tetrads and connections. Gravity becomes not too dissimilar from a gauge theory, and as such its basic observables (intended as correlations of partial observables [16]) are given by parallel transports of the connection itself along closed paths, i.e. holonomies, contracted in such a way as to be gauge invariant. Indeed these have a clear operational meaning [16]. The connection field is a $so(3,1)$ valued 1-form (in 4d) and therefore its parallel transports define elements of the Lorentz group, so that the above observables (in turn determining the data necessary to specify the states of a canonical formulation of a theory based on this variables) are basically given by collections of group elements associated to possible paths in spacetime organized in the form of networks. They are *classical spin networks*. In a simplicial spacetime, the valence of these networks will be constrained but they will remain the basic observables of the theory. A straightforward quantization of them would be obtained by the choice of a representation of the Lorentz group for each of the links of the network to which group elements are associated. Indeed, the resulting quantum structures are *spin networks*, graphs labeled by representations of the Lorentz group associated to their links, of the type characterizing states and observables of loop quantum gravity [16], the canonical quantization of gravity based on a connection formulation. A covariant path integral quantization of a theory based on spin networks will have as histories (playing the role of a 4-dimensional spacetime geometries) a higher-dimensional analogue of them: a spin foam [17, 18, 20], i.e. a 2-complex (collection of faces bounded by links joining at vertices) with representations of the Lorentz group attached to its faces, in such a way that any slice or any boundary of it, corresponding to a spatial hypersurface, will be indeed given by a spin network. Spin foam models [17, 18, 20] are intended to give a path integral quantization of gravity based on these purely algebraic and combinatorial structures.

In most of the current models the combinatorial structure of the spin foam is restricted to be topologically dual to a simplicial complex of appropriate dimension, so that to each spin foam 2-complex it corresponds a simplicial spacetime, with the representations attached to the 2-complex providing geometric information to the simplicial complex; in fact they are interpreted as volumes of the (n-2)-simplices topologically dual to the faces of the 2-complex. The models are then defined by an assignment of a quantum probability amplitude (here factorised in terms of face, edge, and vertex contributions)to each spin foam σ summed over, depending on the representations ρ labeling it, also being summed over, i.e. by the transition amplitudes for given boundary spin networks Ψ, Ψ' (which may include the empty spin network as well):

$$Z = \sum_{\sigma|\Psi,\Psi'} w(\sigma) \sum_{\{\rho\}} \prod_f A_f(\rho_f) \prod_e A_e(\rho_{f|e}) \prod_v A_v(\rho_{f|v});$$

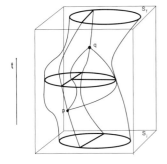

FIGURE 3. A spin foam

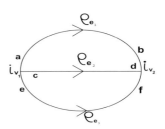

FIGURE 4.
A spin network

one can either restrict the sum over spin foams to those corresponding to a given fixed topology or try to implement a sum over topologies as well; the crucial point is in any case to come up with a well-motivated choice of quantum amplitudes, either coming from some sort of discretization of a classical action for gravity or from some other route. Whatever the starting point, one would then have an implementation of a sum-over-histories for gravity in a combinatorial-algebraic context, and the key issue would then be to prove that one can both analyse fully the quantum domain, including the coupling of matter fields, and at the same recover classical and semi-classical results in some appropriate limit. A multitude of results have been already obtained in the spin foam approach, for which we refer to [17, 18, 20]. We will see shortly that this version of the path integral idea is the one coming out naturally from group field theories.

2. Group field theory: What is it? The basic GFT formalism

Group field theories, as anticipated, are a new realization of the third quantization idea that we have outlined above, in a simplicial setting, and in which the geometry of spacetime as well as superspace itself are described in an algebraic language. As such, they bring together most of the ingredients entering the other approaches we have briefly discussed, thus providing hopefully a general encompassing framework for developing them, as we will try to clarify in the following. We describe the basic framework of group field theories and the rationale for its construction first, and then we will give an explicit (and classic) example of it so to clarify the details of the general picture.

2.1. A discrete superspace

The first ingredient in the construction of a third quantization theory of gravity in n dimensions is a definition of superspace, i.e. the space of (n-1)-geometries. In a simplicial setting, spacetime is discretized to a simplicial complex and thus it is built out of fundamental blocks represented by n-simplices; in the same way, an (n-1)-space, i.e. an hypersurface (not necessarily spacelike) embedded in it, is obtained gluing together along shared (n-2)-simplices a number of (n-1)-simplices in such a way as to reproduce through their mutual relations the topology of the hypersurface. In other words, a (n-1)-space is given by a (n-1)-dimensional triangulation and its geometry is given not by a metric field (thanks to which one can compute volumes, areas, lengths and so on) but by the geometric data assigned to the various elements of the complex: volumes, areas, lengths etc. There is some freedom in the choice of variables to use as basic ones for describing geometry and from which to compute the various geometric quantities. In Regge calculus, as we have seen, the basic variables are chosen to be the edge lengths of the complex; in group field theories [2, 25], as currently formulated, the starting assumption is that one can use as basic variables the volumes of (n-2)-simplices (edge lengths in 3d, areas of triangles in 4d, etc). The consequences and possible problems following from this assumption have not been fully investigated yet. These (n-2)-volumes are determined by unitary irreducible representations ρ of the Lorentz group, one for each $(n-2)$ face of the simplicial complex. Equivalently, one can take as basic variables appropriate Lorentz group elements g corresponding to the parallel transports of a Lorentz connection along dual paths (paths along the cell complex dual to the triangulation), one for each (n-2)-face of the complex. The equivalence between these two sets of variables is given by harmonic analysis on the group, i.e. by a Fourier-type relation between the representations ρ and the group elements g, so that they are interpreted as conjugate variables, as momenta and position of a particle in quantum mechanics [25]. Therefore, if we are given a collection of (n-1)-simplices together with their geometry in terms of associated representations ρ or group elements g, we have the full set of data we need to characterize our superspace. Now one more assumption enters the group field theory approach: that one can exploit the discreteness of this superspace in one additional way, i.e. by adopting a **local** point of view and considering as the fundamental superspace a single (n-1)-simplex; this means that one considers each (n-1)-simplex as a "one-particle state", and the whole (n-1)-d space as a "multiparticle" state, but with the peculiarity that these many "particles" (many (n-1)-simplices) can be glued together to form a collective extended structure, i.e. the whole of space. The truly fundamental superspace structure will then be given by a single (n-1)-simplex geometry, characterized by n Lorentz group elements or n representations of the Lorentz group, all the rest being reconstructed from it, either by composition of the fundamental superspace building blocks (extended space configurations) or by interactions of them as a dynamical process (spacetime configurations), as we will see. In the generalised group field theory formalism of [26], one uses an extended or parametrised formalism in which additional variables characterize the geometry of the fundamental (n-1)-simplices, so that the details of the geometric description are different, but the overall picture is similar, in particular the local nature of the description of superspace is preserved.

2.2. The field and its symmetries

Accordingly to the above description of superspace, the fundamental field of GFTs, as in the continuum a scalar field living on it, corresponds to the 2nd quantization of an (n-1)-simplex. The 1st quantization of a 3-simplex in 4d was studied in detail in [27] in terms of the algebraic set of variables motivated above, and the idea is that the field of the GFT is obtained promoting to an operator the wave function arising from the 1st quantization of the fundamental superspace building block. We consider then a complex scalar field over the tensor product of n copies of the Lorentz group in n dimensions and either Riemannian or Lorentzian signature,

$$\phi(g_1, g_2, ..., g_n) : G^{\otimes n} \to \mathbb{C}.$$

The order of the arguments in the field, each labeling one of its n boundary faces ((n-2)-simplices), corresponds to a choice of orientation for the geometric (n-1)-simplex it represents; therefore it is natural to impose the field to be invariant under even permutations of its arguments (that do not change the orientation) and to turn into its own complex conjugate under odd permutations. This ensures [28] that the Feynman graphs of the resulting field theory are given by orientable 2-complexes, while the use of a real field, with invariance under any permutation of its arguments, has as a result Feynman graphs including non-orientable 2-complexes as well. If the field has to correspond to an (n-1)-simplex, with its n arguments corresponding to an (n-2)-simplex each, one extra condition is necessary: a global gauge invariance condition under Lorentz transformations [27]. We thus require the field to be invariant under the global action of the Lorentz group, i.e. under the simultaneous shift of each of its n arguments by an element of the Lorentz group, and we impose this invariance through a projector operator: $P_g\phi(g_1; g_2; ...; g_n) = \int_G dg\, \phi(g_1g; g_2g; ...; g_ng)$[1]. Geometrically, this imposes that the n (n-2)-simplices on the boundary of the (n-1)-simplex indeed close to form it [27]; algebraically, this causes the field to be expanded in modes into a linear combination of Lorentz group invariant tensors (intertwiners). The mode expansion of the field takes in fact the form:

$$\phi^\alpha(g_i, s_i) = \sum_{J_i, \Lambda, k_i} \phi_{k_i}^{J_i \Lambda} \prod_i D_{k_i l_i}^{J_i}(g_i) C_{l_1..l_4}^{J_1..J_4 \Lambda},$$

with the J's being the representations of the Lorentz group, the k's vector indices in the representation spaces, and the C's are intertwiners labeled by an extra representation index Λ. In the generalised formalism of [26], the Lorentz group is extended to $(G \times \mathbb{R})^n$ with consequent extension of the gauge invariance one imposes and modification of the mode expansion. Note also that the timelike or spacelike nature of the (n-2)-simplices corresponding to the arguments of the field

---

[1]The Lorentzian case, with the use of the non-compact Lorentz group as symmetry group, will clearly involve, in the defintion of the symmetries of the field as well as in the definition of the action and of the Feynman amplitudes, integrals over a non-compact domain; this produces trivial divergences in the resulting expressions and care has to be taken in making them well-defined. However, this can be done quite easily in most cases with appropriate gauge fixing. We do not discuss issues of convergence here in order to simplify the presentation.

depends on the group elements or equivalently to the representations associated to them, and nothing in the formalism prevents us to consider timelike (n-1)-simplices thus a superspace given by a timelike (n-1)-geometry.

2.3. The space of states or a third quantized simplicial space

The space of states resulting from this algebraic third quantization is to have a structure of a Fock space, with N-particle states created out of a Fock vacuum, corresponding as in the continuum to the "no-spacetime" state, the absolute vacuum, not possessing any spacetime structure at all. Each field being an invariant tensor under the Lorentz group (in momentum space), labeled by n representations of the Lorentz group, it can be described by a n-valent spin network vertex with n links incident to it labeled by the representations. One would like to distinguish a 'creation' and an 'annihilation' part in the mode expansion of the field, as $\phi_{k_i}^{J_i \Lambda} = \varphi_{k_i}^{J_i \Lambda} + \left(\varphi_{k_i}^{J_i \Lambda} \right)^{\dagger}$, and then one would write something like: $\varphi_{k_i}^{J_i \Lambda} \mid 0 \rangle$ for a one particle state, $\varphi_{k_i}^{J_i \Lambda} \varphi_{\tilde{k}_i}^{\tilde{J}_i \tilde{\Lambda}} \mid 0 \rangle$ for a *disjoint* 2-particle state (two disjoint (n-1)-simplices), or $\varphi_{k_1 k_2 ... k_n}^{J_1 J_2 ... J_n \Lambda} \varphi_{\tilde{k}_1 k_2 ... \tilde{k}_n}^{\tilde{J}_1 J_2 ... \tilde{J}_n \tilde{\Lambda}} \mid 0 \rangle$ for a *composite* 2-particle state, made out of two (n-1)-simplices glued along one of their boundary (n-2)-simplices (the one labeled by J_2), and so on. Clearly the composite states will have the structure of a spin network of the Lorentz group. This way one would have a Fock space structure for a third quantized simplicial space of the same type as that of usual field theories, albeit with the additional possibility of creating or destroying at once composite structures made with more than one fundamental 'quanta' of space. At present this has been only formally realised [29] and a more complete and rigorous description of such a third quantized simplicial space is needed.

2.4. Quantum histories or a third quantized simplicial spacetime

In agreement with the above picture of (possibly composite) quanta of a simplicial space being created or annihilated, group field theories describe the evolution of these states in perturbation theory as a scattering process in which an initial quantum state (that can be either a collection of disjoint (n-1)-simplices, or spin network vertices, or a composite structure formed by the contraction of several such vertices, i.e. an extended (n-1)-dimensional triangulation) is transformed into another one through a process involving the creation or annihilation of a number of quanta. Being these quanta (n-1)-simplices, their interaction and evolution is described in terms of n-simplices, as fundamental interaction processes, in which D (n-1)-simplices are turned into $n + 1 - D$ ones (in each n-simplex there are n+1 (n-1)-simplices). Each of these fundamental interaction processes corresponds to a possible n-dimensional Pachner move, a sequence of which is known to allow the transformation of any given (n-1)-dimensional triangulation into any other. A generic scattering process involves however an arbitrary number of these fundamental interactions, with given boundary data, and each of these represents a possible quantum history of simplicial geometry, so our theory will appropriately sum over all these histories with certain amplitudes. The states being collections of suitably contracted spin network vertices, thus spin networks themselves labeled with representations of the Lorentz group (or equivalently by Lorentz group elements), dual to triangulations of a (n-1)-dimensional space, their evolution history will be given by 2-complexes labeled again by representations of the Lorentz group, dual to n-dimensional simplicial complexes. Spacetime is thus purely virtual in this context, as in the continuum third quantized formalism and as it should be in a sum over histories formulation of quantum gravity, here realised as a sum over labeled simplicial complexes or equivalently their dual labeled complexes, i.e. spin foams. We see immediately that we have here a formalism with the ingredients of the other discrete and algebraic approaches to quantum gravity we have outlined above.

2.5. The third quantized simplicial gravity action

The action of group field theories [2, 1, 17, 18, 26] is defined so to implement the above ideas, and it is given by:

$$
S_n(\phi, \lambda) = \frac{1}{2} \prod_{i=1,..,n} \int dg_i d\tilde{g}_i \phi(g_i) \mathcal{K}(g_i \tilde{g}_i^{-1}) \phi(\tilde{g}_i)
$$
$$
+ \frac{\lambda}{n+1} \prod_{i \neq j=1}^{n+1} \int dg_{ij} \phi(g_{1j})...\phi(g_{n+1j}) \mathcal{V}(g_{ij} g_{ji}^{-1}),
$$

where of course the exact choice of the kinetic and interaction operators is what defines the model. We see that indeed the interaction term in the action has the symmetries and the combinatorial structure of a n-simplex made out of $n+1$ (n-1)-simplices glued pairwise along common (n-2)-simplices, represented by their arguments, while the kinetic term represent the gluing of two n-simplices along a common (n-1)-simplex, i.e. the free propagation of the (n-1)-simplex between two interactions. λ is a coupling constant governing the strength of the interactions, and the kinetic and vertex operators satisfy the invariance property $\mathcal{K}(g_i \tilde{g}_i^{-1}) = \mathcal{K}(gg_i \tilde{g}_i^{-1} g')$ and $\mathcal{V}(g_{ij} g_{ji}^{-1}) = \mathcal{V}(g_i g_{ij} g_{ji}^{-1} g_j^{-1})$ as a consequence of the gauge invariance of the field itself. A complete analysis of the symmetries of the various group field theory actions has not been carried out yet, and in 3d for example it is known that there exist symmetries of the Feynman amplitudes (i.e. of the histories) of the theory that are not yet identified at the level of the GFT action. In the generalised models [26], the structure of the action is exactly the same, with the group extended to $G \times \mathbb{R}$. The simplest choice of action is given by $\mathcal{K} = \int dg \prod_{i=1}^{n} \delta(g_i g \tilde{g}_i^{-1})$ and $\mathcal{V} = \prod_{i=1}^{n+1} \int dg_i \prod_{i<j} \delta(g_{ij} g_i g_j^{-1} g_{ji}^{-1})$, that corresponds to a GFT formulation of topological BF theories in n dimensions, that gives gravity in 1st order formalism in 3d, as we will see shortly, while in dimension n=2 gives a sum over matrix models of increasing matrix dimension if one choses $SU(2)$ as group manifold [2]. Less trivial actions can be constructed [26], while a simple modification of the BF action gives much studied models of 4d quantum gravity [17, 18]. Unfortunately, we do not understand much at present of the classical theory described by these

actions, and paradoxically we understand better the (perturbative) quantum theory, thanks to the work done in the context of spin foam models [17, 18], and we turn now to this.

2.6. The partition function and its perturbative expansion

The partition function of the theory is then given by an integral over the field of the exponential of (minus) the GFT action. Our current understanding of the non-perturbative properties of the partition function, i.e. of the quantum theory, is quite poor, even if some work is currently in progress on instantonic calculations in GFTs [30]. More is known abour perturbative dynamics in terms of Feynman graphs, thanks to work on spin foam models. The perturbation expansion of the partition function is as usual given by the Schwinger-Dyson expansion in Feynman graphs:

$$ Z = \int \mathcal{D}\phi\, e^{-S[\phi]} = \sum_{\Gamma} \frac{\lambda^N}{sym[\Gamma]}\, Z(\Gamma), $$

where N is the number of interaction vertices in the Feynman graph Γ, $sym[\Gamma]$ is a symmetry factor for the graph and $Z(\Gamma)$ the corresponding Feynman amplitude. The Feynman amplitudes can be constructed easily after identification of the propagator, given by the inverse of the kinetic term in the action, and the vertex amplitude of the theory; each edge of the Feynman graph is made of n strands running parallel to each other, one for each argument of the field, and each is then re-routed at the interaction vertex, with the combinatorial structure of an n-simplex. Diagrammatically: Each strand in an edge of the Feynman graph goes

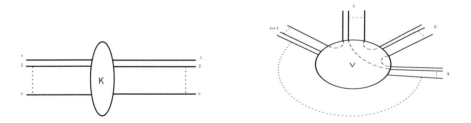

through several vertices and then comes back where it started, for closed Feynman graphs, and therefore identifies a 2-cell. The collection of 2-cells (faces), edges and vertices of the Feynman graph then characterizes a 2-complex that, because of the chosen combinatorics for the arguments of the field in the action, is dual to a n-dimensional simplicial complex. Each strand carries a field variable, i.e. a group element in configuration space or a representation label in momentum space. Therefore in momentum space each Feynman graph is given by a spin foam, and each Feynman amplitude (a complex function of the representation labels) of the GFT by a spin foam model. Indeed, one can show that the inverse is also true: any local spin foam model can be obtained from a GFT perturbative expansion [25, 2]. The sum over Feynman graphs for the partition function gives then a sum over spin foams (histories of the spin networks on the boundary in any scattering process considered), and equivalently a sum over triangulations, augmented by a sum over algebraic data (group elements or representations) with a geometric interpretation. This is true of course also for the generalised GFT models of [26]. This perturbative expansion of the partition function also allows (in principle) the explicit evaluation of expectation values of GFT observables; these are given [2] by gauge invariant combinations of the field operators that can be constructed using spin networks. In particular, the transition amplitude (probability amplitude for a certain scattering process) between certain boundary data represented by two spin networks can be expressed as the expectation value of field operators contracted as to reflect the combinatorics of the two spin networks [2].

2.7. GFT definition of the canonical inner product

Even though, as mentioned in the beginning, this is not meant to be a review of the results obtained so far in the GFT approach, for which we refer to [3], we would like to mention just one important recent result, because it shows clearly how useful this new formalism can be in addressing long-standing open problems of quantum gravity research. Also, it proves that the overall picture of GFTs that we have outlined above (and will summarize in the next subsection) is consistent and fertile. In ordinary QFT, the *classical* equations of motion for the (free) 2nd quantized field are also the *quantum* equation of motion for the 1st quantized relativistic particle wave function; in other words, the full dynamical content of the 1st quantized (non-interacting) theory is contained in the classical level of the 2nd quantized theory. If we buy the picture of GFTs as representing a 3rd quantization formalism of (simplicial) gravity, we expect a similar situation in which the classical GFT level encodes all the dynamical information about canonical quantum gravity (with fixed topology). We will mention in the final part of this paper some caveats to this perspective and some open issues. Given the limits in our understanding of the classical structure of GFT and of the non-perturbative level of the theory, at present we may hope to see these ideas realised explicitly only in the perturbative expansion of the partition function. Indeed, the hopes are fullfilled. It has been shown [2] that the restriction of the sum over Feynman graphs outlined above to *tree level*, thus neglecting all quantum correction and encoding the classical information only, gives a definition of the '2-point function', for fixed boundary spin networks, that is positive semidefinite and finite, including a sum over all triangulations and thus fully triangulation independent; the triangulations involved are of fixed trivial topology, as appropriate for dealing with canonical quantum gravity, and the sum over geometric data (representations) is well-defined. This prompts the interpretation of the resulting quantity as a well-posed and computable definition of the canonical inner product between quantum gravity states, represented by spin networks, the object that in canonical loop quantum gravity encodes the whole dynamical content of the theory (the action of the Hamiltonian constraint operator on spin network states). Even if more work is certainly needed to build up on this result, we see that the use of GFT techniques and ideas has an immediate and important usefulness even from the point of view of canonical quantum gravity: the GFT definition of the physical inner product for canonical spin network states provides finally a *solution* to the long standing issues of canonical quantum gravity on the definition of the Hamiltonian constraint operator, its action on kinematical states, the definition of the physical inner product trough some kind of projection operator formalism, the computation of physical observables, etc. Now it is time to put this concrete proposal to test. Finally, let us stress that this results also highlights the richness of the GFT formalism, and suggests that we have barely started to scratch the surface, with much more lying underneath. If the classical level of GFTs encodes already the full content of canonical quantum gravity, it is clear that a complete analysis of the quantum level of the GFTs will lead us even much further, in understanding a quantum spacetime, than we have hoped to do by studying canonical quantum gravity. This includes the physics of topology change, of course, and the use of quantum gravity sum over histories for other puroses than defining the canonical inner product, but probably much more than that.

2.8. Summary: GFT as a general framework for quantum gravity

Let us briefly summarise the nature of GFTs, before giving an explicit example of it, so to clarify the details of the formalism. We have a field theory over a group manifold that makes no reference to a physical spacetime (except implicitly in the combinatorial structure that one chooses for the GFT action), in which the field (thus the fundamental "particle" it describes) has a geometric interpretation of a quantized (n-1)-simplex. The states of the theory (given in momentum space by spin networks) are correspondingly interpreted as triangulations of (n-1)-dimensional (pseudo)manifolds [28]. The theory can be dealt with perturbatively through an expansion in Feynman graphs that describe the possible interactions of the field quanta, that can be created and annihilated as well as change intrinsic geometry (configuration variables and associated momenta), as a scattering process. Geometrically each possible interaction process for given boundary states is a possible simplicial spacetime with assigned geometry, and it is given by a spin foam, to which the theory assigns a precise quantum amplitude. The simplicial spacetimes summed over have arbitrary topology, as are constructed by all possible gluings of fundamental interaction vertices each corresponding to an n-simplex. GFTs are therefore a *third quantized formulation of simplicial geometry*. Interestingly, group field theories also have all the ingredients that enter other approaches to quantum gravity: boundary states given by spin networks, as in loop quantum gravity, a simplicial description of spacetime and a sum over geometric data, as in Quantum Regge calculus, a sum over triangulations dual to 2-complexes, as in dynamical triangulations, a sum over topologies like in matrix models, of which GFTs are indeed higher-dimensional analogues, as we said, an ordering of fundamental events (vertices of Feynman diagrams), given by the orientation of the 2-complex, which has similarities to that defining causal sets [19], and quantum amplitudes for histories that are given by spin foam models. Therefore one can envisage GFTs as a *general framework for non-perturbative quantum gravity*, that encompasses most of the current approaches. This is at present only a vague and quite optimistic point of view, not yet established nor strongly supported by rigorous results, but we feel that it can be a fruitful point of view both for the development of GFTs themselves and of these other approaches as well, by offering a new perspective on them and possibly new techniques that can be used to address the various open issues they still face.

3. An example: 3d Riemannian quantum gravity

For simplicity we consider explicitly in more detail only the 3d Riemannian quantum gravity case, whose group field theory formulation was first given by Boulatov [32]. Consider the real field: $\phi(g_1, g_2, g_3) : (SU(2))^3 \to \mathbb{R}$, with the symmetry: $\phi(g_1 g, g_2 g, g_3 g) = \phi(g_1, g_2, g_3)$, imposed through the projector: $\phi(g_1, g_2, g_3) = P_g \phi(g_1, g_2, g_3) = \int dg \, \phi(g_1 g, g_2 g, g_3 g)$ and the symmetry: $\phi(g_1, g_2, g_3) = \phi(g_{\pi(1)}, g_{\pi(2)}, g_{\pi(3)})$, with π an arbitrary permutation of its arguments, that one can realise through an explicit sum over permutations: $\phi(g_1, g_2, g_3) = \sum_\pi \phi(g_{\pi(1)}, g_{\pi(2)}, g_{\pi(3)})$. In this specific case, the interpretation is that of a 2nd quantized triangle with its 3 edges corresponding to the 3 arguments of the field; the irreps of $SU(2)$ labeling these edges in the mode expansion of the field have the interpretations of edge lengths. The classical theory is defined by the action:

$$
\begin{aligned}
S[\phi] \; = \; & \frac{1}{2} \int dg_1..dg_3 [P_g \phi(g_1, g_2, g_3)]^2 \; + \\
& + \; \frac{\lambda}{4} \int dg_1..dg_6 [P_{h_1} \phi(g_1, g_2, g_3)][P_{h_2} \phi(g_3, g_5, g_4)] \\
& \qquad [P_{h_3} \phi(g_4, g_2, g_6)][P_{h_4} \phi(g_6, g_5, g_1)].
\end{aligned}
$$

As we have discussed in the general case, we see that the structure of the action is chosen so to reflect the combinatorics of a 3d triangulation, with four triangles (fields) glued along their edges (arguments of the field) pairwise, to form a tetrahedron (vertex term) and two tetrahedra being glued along their common triangles (kinetic term $\to$ propagator). The quantum theory is given by the partition function, in turn again defined in terms of perturbative expansion in Feynman graphs:

$$
Z = \int d\phi \, e^{-S[\phi]} = \sum_\Gamma \frac{\lambda^N}{sym[\Gamma]} Z(\Gamma).
$$

Therefore, in order to construct explicitly the quantum amplitudes of the theory for each of its Feynman graphs, we need to identify their building blocks, i.e. propagator and vertex amplitude. These are to be read out from the action:

$$
S[\phi] = \frac{1}{2} \int dg_i d\tilde{g}_j \, \phi(g_i) \mathcal{K}(g_i, \tilde{g}_j) \phi(\tilde{g}_j) + \frac{\lambda}{4} \int dg_{ij} \mathcal{V}(g_{ij}) \phi(g_{1j}) \phi(g_{2j}) \phi(g_{3j}) \phi(g_{4j})
$$

For the propagator, starting from the kinetic term $P_g\phi(g_1, g_2, g_3)P_{\bar{g}}\phi(g_1, g_2, g_3)$, and considering the permutation symmetry, one gets immediately:

$$P = \mathcal{K}^{-1} = \mathcal{K} = \sum_{\pi} \int dg d\bar{g}\ \delta(g_1 g \bar{g}^{-1}\tilde{g}_{\pi(1)}^{-1})\delta(g_2 g \bar{g}^{-1}\tilde{g}_{\pi(2)}^{-1})\delta(g_3 g \bar{g}^{-1}\tilde{g}_{\pi(3)}^{-1}),$$

while the vertex is given by:

$$\mathcal{V} = \int dh_i\ \delta(g_1 h_1 h_3^{-1}\tilde{g}_1^{-1})\delta(g_2 h_1 h_4^{-1}\tilde{g}_2^{-1})\delta(g_3 h_1 h_2^{-1}\tilde{g}_3^{-1})$$
$$\delta(g_4 h_2 h_4^{-1}\tilde{g}_4^{-1})\delta(g_5 h_2 h_3^{-1}\tilde{g}_5^{-1})\delta(g_6 h_3 h_4^{-1}\tilde{g}_6^{-1})$$

We see that, as for BF theories in any dimensions, the vertex and propagator are given simply by products of delta functions over the group, represented simply by lines in the diagrams below, with boxes representing the integration over the group (following from the requirement of gauge invariance). The Feynman graphs

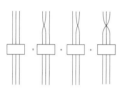

propagator

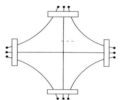

interaction vertex

are obtained as usual by gluing vertices with propagators. Let us see how these look like. As we explained for the general case, each line in a propagator goes through several vertices and for closed graphs it comes back to the original point, thus identifying a 2-cell, and these 2-cells, together with the set of lines in each propagator, and the set of vertics of the graph, identify a 2-complex. Each of these 2-complexes is dual to a 3d triangulation, with each vertex corresponding to a tetrahedron, each link to a triangle and each 2-cell to an edge of the triangulation.

tetrahedron

tetra + dual

dual

The sum over Feynman graphs is thus equivalent to a **sum over 3d triangulations of any topology**, as anticipated in the general case. Let us now identify the quantum amplitudes for these Feynman graphs. These are obtained the usual way using the above propagators and vertex amplitudes. In configuration space, where the variables being integrated over are group elements, the amplitude for each 2-complex is:

$$Z(\Gamma) = \left(\prod_{e \in \Gamma} \int dg_e \right) \prod_f \delta(\prod_{e \in \partial f} g_e)$$

which has the form of a lattice gauge theory partition function with simple delta function weights for each plaquette (face of the 2-complex) and one connection variable for each edge; the delta functions constraint the the curvature on any face to be zero, as we expect from 3d quantum gravity [31]. To have the corresponding expression in momentum space, one expands the field in modes $\phi(g_1, g_2, g_3) = \sum_{j_1, j_2, j_3} \phi^{j_1 j_2 j_3}_{m_1 n_1 m_2 n_2 m_3 n_3} D^{j_1}_{m_1 n_1}(g_1) D^{j_2}_{m_2 n_2}(g_2) D^{j_3}_{m_3 n_3}(g_3)$, where the j's are irreps of the group $SU(2)$ (the Lorentz group, local gauge group of gravity, for 3d and Riemannian signature) obtaining, for the propagator, vertex and amplitude:

$$\mathcal{P} = \delta_{j_1 \tilde{j}_1} \delta_{m_1 \tilde{m}_1} \delta_{j_2 \tilde{j}_2} \delta_{m_2 \tilde{m}_2} \delta_{j_3 \tilde{j}_3} \delta_{m_3 \tilde{m}_3}$$

$$\mathcal{V} = \delta_{j_1 \tilde{j}_1} \delta_{m_1 \tilde{m}_1} \delta_{j_2 \tilde{j}_2} \delta_{m_2 \tilde{m}_2} \delta_{j_3 \tilde{j}_3} \delta_{m_3 \tilde{m}_3}$$

$$\delta_{j_4 \tilde{j}_4} \delta_{m_4 \tilde{m}_4} \delta_{j_5 \tilde{j}_5} \delta_{m_5 \tilde{m}_5} \delta_{j_6 \tilde{j}_6} \delta_{m_6 \tilde{m}_6} \left\{ \begin{matrix} j_1 & j_2 & j_3 \\ j_4 & j_5 & j_6 \end{matrix} \right\}$$

$$Z(\Gamma) = \left(\prod_f \sum_{j_f} \right) \prod_f \Delta_{j_f} \prod_v \left\{ \begin{matrix} j_1 & j_2 & j_3 \\ j_4 & j_5 & j_6 \end{matrix} \right\}$$

where Δ_j is the dimension of the representation j and for each vertex of the 2-complex we have a so-called $6j - symbol$, i.e. a scalar function of the 6 representations meeting at that vertex. The amplitude for each 2-complex is given then by a spin foam model, the Ponzano-Regge model for 3d gravity without cosmological constant, about which a lot more is known [31]. This amplitude, after gauge fixing, gives a well-defined topological invariant of 3-manifolds, as one expects from 3d quantum gravity, and as such it is invariant under choice of triangulation. This means that it evaluates to the same number for any triangulation (2-complex) for given topology, so that the only dynamical degrees of freedom in the theory are indeed the topological. The full theory is then defined as we said by the sum over all Feynman graphs weighted by the above amplitudes:

$$Z = \sum_\Gamma \frac{\lambda^N}{sym[\Gamma]} \left(\prod_f \sum_{j_f} \right) \prod_f \Delta_{j_f} \prod_v \left\{ \begin{matrix} j_1 & j_2 & j_3 \\ j_4 & j_5 & j_6 \end{matrix} \right\}_v .$$

This gives a rigorous (after gauge-fixing of the symmetries of the theory [33] the above quantity is well-defined, even in the Lorentzian case, in spite of the non-compactness of the Lorentz group used and of the infinite dimensionality of the irreps used) and un-ambiguous (in the sense that every single element in the above formula has a known closed expression and therefore it can be computed exactly at least in principle) realisation, in purely algebraic and combinatorial terms, of the sum over both geometries and topologies, i.e. of the third quantization idea, in the 3-dimensional case. Issues about interpretation and about the convergence of the sum over complexes (but see [39]) of course remain, but we see a definite progress at least for what concerns the definition of the amplitude and measure for given 2-complex (i.e. given spacetime) with respect to the continuum path integral.

Strikingly, group field theory models for quantum gravity in 4 spacetime dimensions, that seem to have many of the right properties we seek, can be obtained by a very simple modification of the 3-dimensional model [17, 18]. Motivated by the classical formulation of gravity as a constrained BF theory [17, 18], one first generalises the above field to a 4-valent one with arguments living in the 4-dimensional Lorentz group $SO(3,1)$, modifying the combinatorial structure of the above action to mimic the combinatorics of a 4-simplex in the interaction term, with 5 tetrahedra (fields) glued along triangles, and then simply imposes a restriction on the arguments of the field to live in the homogeneous space $SO(3,1)/SO(3) \simeq H^3$, i.e. on the upper sheet of the timelike hyperboloid in Minkowski space. The resulting Feynman graphs expansion produces quantum amplitudes for them given by the Barrett-Crane model [34], that has been recently the focus of much work for which we refer to the literature [17, 18].

4. Assorted questions for the present, but especially for the future

We have outlined the general formalism of group field theories and the picture of a third quantized simplicial spacetime they suggest. We have organised this outline and presented these models from the perspective that sees GFTs as a candidate to and a proposal for a fundamental formulation of a theory of quantum gravity, as opposed to just a tool to be used to produce for example spin foam models, that is the way they have been used up to now. We have tried to highlight the features of the formalism and of the resulting picture that we find most appealing and fascinating. However, it is probably transparent that there is and there should be much more in the theory than what we have described. As a matter of fact, there is certainly in the GFTs much more than it is presently known, which is basically only their action, their perturbative expansion, and a few properties of their Feynman amplitudes, i.e. spin foam models. To tell the whole truth, even many of the details of the general picture we have presented are only tentative and there remains lots to be understood about them. Therefore we want to conclude by posing a (limited) set of assorted questions regarding GFTs and that future work should answer, if GFTs are to be taken seriously as candidates for a fundamental formulation of third quantized simplicial gravity, and thus of non-perturbative quantum gravity.

- **What about the classical theory?** The description of the classical simplicial superspace that we have briefly described is to be investigated further as the validity of the variables that GFTs use to describe classical geometry is not solidly established yet. Also, even given this for granted, not much is known about the classical theory behind the GFT action. Some work is in progress [30], but it is fair to say that we do not have a good understanding of the physical meaning of the classical equations of motion of the theory nor we know enough solutions of them. As for the physical menaing, we stress again that the *classical* GFT dynamics should already encode the canonical *quantum gravity* dynamics in full, as it happens in ordinary quantum field theories (where the classical field equations are the quantum Schroedinger equation for the one-particle theory). Also, in analogy to the continuum third quantization setting, one would expect them to describe a modification of the Wheeler-DeWitt equation, in a simplicial setting, due to the presence of topology change; however, unlike the continuum case it is not easy, and maybe not even possible, to distinguish (the simplicial analogue of) a Wheeler-DeWitt operator and a topology changing term as distinct contributions to the equation, due to the local nature of the description of superspace that we have in GFTs. As for obtaining solutions of the classical equations of motion, the non-local nature of the equations of motion (coming from that of the action) makes solving them highly non-trivial even in lower dimensions. The form of the action in the generalised GFT formalism [26] may probably facilitate this task somehow, due to a greater similarity with usual scalar quantum field theories. Similarly for the Hamiltonian analysis of the theory, not yet performed and that is quite non-trivial in the usual formalism, but may be easier to carry out in a covariant fashion thanks to the proper time variables in the generalised formalism. In this regard, the fact extablished in [26] that usual GFTs are the Static Ultra-local (SUL), and un-oriented, limit of well-defined generalised GFTs may help in that one may first tackle with standard methods the Hamiltonian analysis of such generalised models and then study the SUL limit of the corresponding Hamiltonian structures.

- **Fock space and +ve/-ve energy states?** The picture of creation and annihilation of quanta of simplicial geometry, the construction of quantum states of the field theory, and the Fock structure of the corresponding space of states built out of the 3rd quantized vacuum may be appealing and geometrically fascinating, but it is at present mostly based on intuition. A rigorous construction needs to be carried out, starting from a proper hamiltonian analysis, and a convincing identification of creation and annihilation operators from the mode expansion of the field. Having at hand the hamiltonian form and the harmonic decomposition of the field itself in modes (resulting from the harmonic analysis on group manifolds), one should investigate the quantum gravity analogue of the notion of positive and negative energy (particle/anti-particle) states of QFT. This has probably to do with the opposite orientations one can assign to the fundamental building blocks of our quantum simplicial spacetime, so maybe it is again best investigated in the generalised GFT formalism [26], in which the orientation data play a crucial role.

- **What is** λ**?** In [2], one consistent and convincing interpretation of the GFT coupling constant has been given: it was shown that a suitable power of the coupling constant can be interpreted as the parameter governing the sum over topologies in the perturbative expansion in Feynman graphs. This result was based on the analysis of the Schwinger-Dyson equations of the n-dimensional GFT. More work is now needed to expand on these result and elucidate all its implications. Another interpretation of the GFT coupling constant, although not as well extablished is suggested by the work [28]. If one considers only one fixed representation of the Lorentz group, thus reducing GFTs to simple tensor extensions of matrix models, then the coupling constant enters in the resulting amplitudes as the exponential of a cosmological constant, and the perturbative expansion is interpreted as an expansion in the "size" of spacetime [28]; this is beautiful, but one must check whether a similar picture applies for the full GFT with no restriction on the representations. Assuming this can be extablished, more work is then needed to prove the compatibility of the two interpretations. This last point has important consequences, in that it may lead to a simplicial and exact realisation of the old idea, put forward in the context of continuum formulations of 3rd quantised gravity [11], of a connection between cosmological constant and wormholes, i.e. topology change, in quantum gravity.

- **Where are the diffeos? Actually, where are all the continuum symmetries?** Another crucial point that needs to be addressed is a deeper understanding of GFT symmetries; in fact, already in the 3-dimensional case it is known [31] that the amplitudes generated by the GFT possess symmetries that are not immediately identified as symmetries of the GFT action. The most important symmetry of a gravity theory is diffeomorphism symmetry. One could argue that GFTs are manifestly diffeomorphism invariant in the sense that there is no structure in the theory corresponding to a continuum spacetime, but one should identify a discrete symmetry corresponding to diffeos in the appropriate continuum approximation of the theory. Since any continuum metric theory that is invariant under diffeomorphisms reproduces Einstein gravity (plus higher derivative terms), the identification of diffeomorphism symmetry in GFTs is crucial, it that it would support the idea that they possess a continuum approximation given by General Relativity. It was argued in [2] that diffeomorphisms are the origin of loop divergences in spin foam models, that in turn are just Feynman amplitudes for the GFT. This needs to be investigated. Another possibility to be investigated is that diffeomophisms originate from a non-trivial renormalisation group acting on the parameter space of GFTs.

- **A GFT 2-point functions zoo? where is causality?** In ordinary quantum field theory, one can define different types of N-point functions or transition amplitudes, with different uses and meanings; is this the case also for GFTs, and more generally for Quantum Gravity? if so, what is their respective use and interpretation? The difference between various N-point functions in QFT is in their different causal properties, so this question is related to the more general issue of causality in GFTs. Where is causality? How to implement causality restrictions? Recent work [26] seems to suggest that these issues can be dealt with satisfactorily, but much more work is certainly needed.

- **What is the exact relation with the canonical theory?** A canonical theory based on Lorentz spin networks, adapted to a simplicial spacetime, and thus a kind of covariant discretization of loop quantum gravity [16], has to be given by a subsector of the GFT formalism. The reduction to this subsector as well as the properties of the resulting theory are not yet fully understood. A canonical theory of quantum gravity needs a spacetime topology of product type, i.e. $\Sigma \times \mathbb{R}$, so that the only non-trivial topology can be in the topology of space Σ. and a positive definite inner product between quantum states. A precise and well-posed definition of such an inner product and a way to reduce to trivial topology in the perturbative expansion of the GFT was proposed in [2], as we discussed: it is given by the perturbative evaluation of the expectation value of appropriate spin network observables in a *tree level* truncation, that indeed generates only 2-complexes with trivial topology. The consequences of this proposal, that would amount to a complete definition of the canonical theory corresponding to the GFTs, need to be investigated in detail. It may be possible for example to extract a definition of an Hamiltonian constraint operator from the so-defined inner product, to study its properties, and to compare it with those proposed in the loop quantum gravity approach. Even when this has been done, it would remain to investigate the relation between covariant spin network structures based on the full Lorentz group, and the loop quantum gravity ones based on $SU(2)$, and to check how much of the many mathematical results obtained on the kinematical Hilbert space of $SU(2)$ spin networks can be reproduced for the covariant ones.

- **How to include matter?** The inclusion and the correct description of matter fields at the group field theory level is of course of crucial importance. Work on this has started only recently [35, 36, 37] for the 3-dimensional case, with very interesting results. The idea pursued there was that one could perform a 3rd quantization of gravity and a 2nd quantization of matter fields in one stroke, thus writing down a coupled GFT action for both gravity and matter fields that would produce, in perturbative expansion, a sum over simplicial complexes with dynamical geometry (quantum gravity histories) together with Feynman graphs for the matter fields living on the simplicial complexes (histories for the matter fields). The whole description of the coupled system would thus be purely algebraic and combinatorial. Indeed, this can be realised consistently for any type of matter field [37]. However, in 3d life is made easier by the topological nature of gravity and by the fact that one can describe matter as a topological defect. The difficult task that lies ahead is to extend these results to 4 dimensions. In this much more difficult context, some work is currently in progress regarding the coupling of gauge fields to quantum gravity at the GFT level [38].

- **Does the GFT perturbation theory make sense?** Even if the only thing we know about GFTs is basically their perturbative expansion in Feynman graphs, strangely enough we do not know for sure if this perturbative expansion makes sense. Most likely the perturbative series is not convergent, but this is not too bad, as it is what happens in ordinary field theories. One would expect (better, hope) it to be an asymptotic series to a non-perturbatively defined function, but this has been realised up to now only in a specific model in 3d [39], and more work is needed for what concerns other models especially in 4d. Let us recall that the perturbative GFT expansion entails a sum over topologies, so that gaining control over it is a mathematically highly non-trivial issue with very important physical consequences.

- **How to relate to the other approaches to quantum gravity?** Even if one is optimistic and buys the picture of GFTs as a general framework for quantum gravity encompassing other approaches, or at least the main ingredients of other approaches, the links with these approaches need to be investigated in detail to start really believing the picture. For example, to obtain a clear link with simplicial approaches to quantum gravity, one needs first to construct a GFT that has Feynman amplitudes given by the exponential of the Regge action for the corresponding simplicial complex, probably building on the results of [26]. Then one would be left to investigate the properties of the measure in front of the exponential, to be compared with those used in Regge calculus, and to find a nice procedure for reducing the model to involve only equilateral triangulations and to admit a slicing structure, so to compare it with dynamical triangulations models. And this would be just a start.

- **What about doing some physics?** The ultimate aim is of course to have a consistent framework for describing quantum gravity effects and obtain predictions that can be compared to experiments. This may be seen as far-fetched at present, and maybe it is, but a consistent coupling of matter fields at the GFT level, a better understanding of its semiclassical states and of perturbations around them, and a better control over the continuum approximation of the GFT structures, all achievable targets for current and near future studies, may bring even this ultimate aim within our reach not too far from now.

Acknowledgements

It is my pleasure to thank K. Krasnov, E. Livine, K. Noui, A. Perez, C. Rovelli, J. Ryan and especially L Freidel for many discussions, clarifications and suggestions; I thank an anonymous referee for his/her helpful comments and criticisms; I also thank the organizers of the Blaubeuren Workshop on "Mathematical and Physical Aspects of Quantum Gravity" for their invitation to participate.

References

[1] D. Oriti, in the Proceedings of the 4th Meeting on Constrained Dynamics and Quantum Gravity, Cala Gonone, Italy (2005); gr-qc/0512048;

[2] L. Freidel, hep-th/0505016;

[3] D. Oriti, in *Approaches to Quantum Gravity - toward a new understanding of space, time and matter*, D. Oriti ed., Cambridge University Press (2006);

[4] J. Hartle, in Les Houches Sum.Sch.1992:0285-480, gr-qc/9304006;

[5] C. Isham, K. Kuchar, Annals Phys. 164, 316 (1985);

[6] G. Gibbons and S. Hawking, eds., *Euclidean Quantum Gravity*, World Scientific, Singapore (1993);

[7] F. Dowker, in *The future of theoretical physics and cosmology*, 436-452, Cambridge University Press (2002), gr-qc/0206020;

[8] G. Horowitz, Class. Quant. Grav. 8, 587-602 (1991);

[9] F. Dowker, R. Sorkin, Class. Quant. Grav. 15, 1153-1167 (1998), gr-qc/9609064;

[10] P. Anspinwall, B. Grene, D. Morrison, Nucl. Phys. B 416, 414-480 (1994), hep-th/9309097;

[11] T. Banks, Nucl. Phys. B 309, 493 (1988);

[12] S. Coleman, Nucl. Phys. B 310, 643 (1988);

[13] C. Isham, gr-qc/9510063;

[14] R. Sorkin, Int. J. Theor. Phys. 30, 923-948 (1991);

[15] H. Hamber, R. M. Williams, Nucl. Phys. B 415, 463-496 (1994), hep-th/9308099;

[16] C. Rovelli, *Quantum Gravity*, Cambridge University Press (2004);

[17] D. Oriti, Rept. Prog. Phys. 64, 1489 (2001), gr-qc/0106091;

[18] A. Perez, Class. Quant. Grav. 20, R43 (2003), gr-qc/0301113;

[19] F. Dowker, gr-qc/0508109;

[20] D. Oriti, *Spin foam models of a quantum spacetime*, PhD thesis, University of Cambridge (2003), gr-qc/0311066;

[21] S. B. Giddings, A. Strominger, Nucl. Phys. B 321, 481 (1989);

[22] M. McGuigan, Phys. Rev. D 38, 3031-3051 (1988);

[23] A. Morozov, hep-th/0502010;

[24] J. Ambjorn, J. Jurkiewicz, R. Loll, Phys. Rev. D 72, 064014 (2005), hep-th/0505154;

[25] M. Reisenberger, C. Rovelli, Class. Quant. Grav. 18, 121-140 (2001), gr-qc/0002095;

[26] D. Oriti, gr-qc/0512069;

[27] J. Baez, J. Barrett, Adv. Theor. Math. Phys. 3, 815-850 (1999), gr-qc/9903060;

[28] R. De Pietri, C. Petronio, J. Math. Phys. 41, 6671-6688 (2000), gr-qc/0004045;

[29] A. Mikovic, Class. Quant. Grav. 18, 2827-2850 (2001), gr-qc/0102110;

[30] A. Baratin, L. Freidel, E. Livine, in preparation;

[31] L. Freidel, D. Louapre, Class. Quant. Grav. 21, 5685 (2004), hep-th/0401076;

[32] D. Boulatov, Mod.Phys.Lett. A7 (1992) 1629-1646, hep-th/9202074;

[33] Laurent Freidel, David Louapre, Nucl. Phys. B662 (2003) 279-298;

[34] J. W. Barrett, L. Crane, Class.Quant.Grav. 17 (2000) 3101-3118, gr-qc/9904025;

[35] K. Krasnov, hep-th/0505174;

[36] L. Freidel, D. Oriti, J. Ryan, gr-qc/0506067;

[37] D. Oriti, J. Ryan, gr-qc/0602010;

[38] R. Oeckl, D. Oriti, J. Ryan, in preparation;

[39] L. Freidel, D. Louapre, Phys. Rev. D 68, 104004 (2003), hep-th/0211026.

Daniele Oriti
Department of Applied Mathematics and Theoretical Physics
Centre for Mathematical Sciences, University of Cambridge
Wilberforce Road, Cambridge CB3 0WA, England, EU
e-mail: d.oriti@damtp.cam.ac.uk

Quantum Gravity
B. Fauser, J. Tolksdorf and E. Zeidler, Eds., 127–150
© 2006 Birkhäuser Verlag Basel/Switzerland

An Essay on the Spectral Action and its Relation to Quantum Gravity

Mario Paschke

Abstract. We give a brief, critical account of Connes' spectral action principle, its physical motivation, interpretation and its possible relation to a quantum theory of the gravitational field coupled to matter. We then present some speculations concerning the quantization of the spectral action and the perspectives it might offer, most notably the speculation that the standard model, including the gauge groups and some of its free parameters, might be derived from first principles.

Mathematics Subject Classification (2000). Primary 58B34; Secondary 81R60.

Keywords. Noncommutative geometry, spectral action principle, standard model of particle physics, generally covariant quantum theory.

1. Introduction

The question is not whether god did create the world.
The question is whether he had any choice.
Albert Einstein

So there is the Beauty and there is the Beast:

In the eyes of most theoretical physicists, traditionally being addicted to symmetries, general relativity is a theory of exceptional beauty. If one assumes that the field equations are of second order, then – up to two free parameters – the theory is uniquely determined by the requirements of general covariance, the motion of test particles along geodesics and the assumption that the energy-momentum tensor of the matter fields is the sole source of the gravitational field.

As compared to this, the standard model of particle physics appears as a real beast. At first glance, one might say it wears the nice suit of being based on a Yang-Mills theory. But there are many possible gauge groups, and we should really wonder why nature has chosen this particular one. Furthermore one has to spontaneously break the symmetry by introducing the Higgs field in a rather ad hoc way, and add 48 Weyl Spinors in very specific representations of the gauge group.

Why 48 ? And why do they all sit in the trivial or the fundamental (respectively its conjugate) representation of the nonabelian gauge groups ? And why is parity broken *maximally*, which need not be the case? An ideal conception of a fundamental theory shouldn't have as many as 26 free parameters either. Only to mention the asthetic deficits of the classical theory. The consistency of the quantized theory – for example the necessary absence of anomalies – gives at least a few hints on the possible solution to these puzzles. But on the other hand it is actually not completely clear whether the theory exists in a strict, nonpertubative sense at the quantum level and if so whether it really describes all the observed phenomena like e.g. confinement correctly.

Yet, in the fairy tale the Beast has all the money, quite some power and endures all offences by its enemies completely unperturbed. The same can certainly be said about the pertubatively quantized standard model, whose phenomenological success is more than impressive. So maybe we should simply start to take it serious and ask ourselves whether it has some hidden (inner) beauty. Besides, we should never forget that the Beauty, as all beauties, is of a somewhat superficial nature: Not (yet) being quantized, it has a limited range of validity, and, in fact, even predicts its own breakdown in some regime, as is seen from the singularity theorems.

But if the Beauty and the Beast would turn out soul mates, they might be married, have children, so that general relativity finally becomes adult. It would then also be well conceivable that the appearance of the Beast is actually determined by its wife, as is the case in many marriages. Being romantic by nature, most theoretical physicists dream of such a wedlock ever since the invention of general relativity.

Noncommutative Geometry follows the above advice: It does take the standard model serious, but seeks after its hidden mathematical structure [16, 19, 14]. In the version we shall consider here it (almost completely) reformulates the standard model as part of the gravitational field on some noncommutative manifold [18]. The noncommutativity of the underlying spacetime is believed to appear via quantum effects of the gravitational field. However, at the present stage the conventional gravitational field is treated as classical, assuming that this is possible in the sense of an effective theory at presently accessible energy densities. Employing the *spectral action principle* one then obtains the *classical, ie. unquantized,* action of the standard model coupled to gravity in a way quite similar to the way one derives Einstein's equations from the assumptions given above.

Thus Noncommutative Geometry is not (yet) an approach to quantum gravity. It only takes up some heuristically derived speculations about the nature of spacetime in such a theory and adds some more support for them by revealing a plethora of unexpected and beautiful mathematical structures of the standard model. Since this program works out so smoothly there is some hope that it might now be used as a guideline on the way to some Noncommutative Geometry approach to quantum gravity.

As concerns the present status of the spectral action principle, Thomas Schücker compared it quite appropriately with Bohr's model of the hydrogen atom. The postulates of the latter aren't completely coherent, using some classical physics and combining it with some speculative ideas about the (at that time) sought for quantum theory. Yet it reproduced Rydbergs formula and paved the way to quantum mechanics. (In particular, Bohrs postulate of quantization of angular momentum might have led people to consider the matrix representaions of $SO(3)$.) Of course, it did not explain all the experimental data known at that time, e.g. it gave the wrong value for the ground state energy of the hydrogen atom. But also these open questions it uncovered played an important role for the later development of quantum mechanics.

In view of this analogy, it would be important to state the postulates of the noncommutative description of the standard model via the spectral action principle in the same way Bohr formulated his axioms for the hydrogen atom. I have tried to do so in this article, but to tell the truth I did not succeed. It seems that there are still some points to be clarified before the goal to shape such postulates as coherent as possible can be reached. Hence I decided to give only a preliminary version, but to work out all the weak spots, so that the reader can see where improvements might still be possible. Aspects that may lead to an experimental falsification of the whole program are mentioned as well. I much more dreamed of presenting here some interesting and important questions rather than well-established results.

In order not to draw the curtain of technical details over the main conceptual issues, the article is written in a very nontechnical style. But details can be devilish and readers interested to see that things really work out the way described here are referred to the excellent reviews [9, 12]. Still, the next two sections rather deliberately use some slang from operator theory. Readers who are uneasy about this might easily skip these two sections as the explanations and ideas given there do not represent essential prerequistes for the main part of the article, which starts in 4.2. In particular, Section 3 only gives a very intuitive picture of the noncommutative world and I'm well aware that this picture may turn out completely misleading. However, if spacetime is really noncommutative then we shall have to find a physical interpretation of such spaces and in that section it is at least pointed out that a picture guided by analogy to quantum mechanics is almost surely too naive.

This paper is intended to provide some food for thought, be it in the form of critical remarks or of highly speculative ideas. Since the spectral action is intended to provide a hint for a possible approach to quantum gravity, and as this is a workshop about such routes, I felt it appropriate to add some ideas on where this Tip of the TOE [1] that is maybe revealed by Noncommutative Geometry could finally lead to. They are gathered in the last two sections.

2. Classical spectral triples

The Gelfand-Naimark-Theorem establishes a complete equivalence of locally compact topological Hausdorff spaces $\mathcal{M}$ and commutative C^*-algebras $\mathcal{A}$: Any *commutative* C^*-algebra $\mathcal{A}$ is given as the C^*-algebra of continuous functions $C(\mathcal{M})$ (with the supremum norm) on the space $\mathcal{M}$ of *irreducible representations* of $\mathcal{A}$. For commutative algebras all irreducible representations are one-dimensional. Thus, labeling the normalized basis vector of such a representation space by $|p\rangle$ $p \in \mathcal{M}$, each algebra element $f \in \mathcal{A}$ can be viewed as a function on $\mathcal{M}$ by setting

$$f(p) := \langle p|f|p\rangle.$$

The nontrivial part of the Gelfand-Naimark-Theorem then shows that this space of irreducible representations inherits a suitable topology from the norm of the algebra. In fact, this is also true if the algebra is noncommutative, so that it makes sense to speak of noncommutative C^*-algebras as noncommutative topological spaces, identifying the points with the inequivalent irreducible representations, respectively with *equivalence classes of pure states* – which is the same, as is seen via the GNS-construction.

But there's more than topology to this. Almost all geometrical structures – like for instance vector bundles or differentiable and metrical structures – can equivalently be described in the language of commutative (pre-)C^*-algebras and, even more so, this description still makes sense for noncommutative algebras. However, the generalization to the noncommutative case is in general not unique. The noncommutative de Rham calculus, to give only one example, can be defined as the dual of the space of derivations on $\mathcal{A}$ [14], on Hopf algebras also as a bicovariant differential calculus [8], or via spectral triples. For commutative algebras all these definitions are equivalent. However, if the pre-C^*-algebra $\mathcal{A}$ is noncommutative they will in general provide different answers.

Spectral triples not only provide a differential calculus and a metric on the space of pure states (irreducible representations) over the algebra. They also give rise to an effective way of computing certain differential topological invariants via the local index formula of Connes and Moscovici [15, 17]. These invariants not only appear in quantum field theory, they do so precisely in the form of the local index formula [36]. In so far, spectral triples are quite natural candidates for geometries on which a quantum theory can reside. Moreover, the language of spectral triples is manifestly covariant under automorphisms of the algebra. i.e. the noncommutative "diffeomorphisms". Therefore this language is tailored to provide an alternative description of (quantum) general relativity.

A detailed account of spectral triples is given e.g. in [13]. Here I would only like to mention some essential axioms which turn out to provide phenomenological restrictions for the noncommutative description of the standard model. I therefore found it important to point out their geometrical significance. In the following $\mathcal{A} = C^\infty(\mathcal{M})$ will always be commutative. For simplicity we shall also assume that $\mathcal{M}$ is a compact, orientable, smooth manifold. This need not be the case: There do exist spectral triples for commutative spaces which describe geometries far beyond the realm of smooth manifolds, like for instance discrete spaces. I therefore prefer to call the manifold case "classical spectral triples".

A **real even spectral triple of dimension** d is given by the following data

$$(\mathcal{H}, D, \mathcal{A}, \gamma, J)$$

Here

- $\mathcal{H}$ is a Hilbert space.
- $\mathcal{A}$ is a pre-C^*-algebra represented on $\mathcal{H}$
- D is an unbounded, essentially selfadjoint operator on $\mathcal{H}$ such that $[D, f]$ is bounded for all $f \in \mathcal{A}$.

  Moreover we assume that the spectrum of D is discrete and that all the eigenvalues λ_n of D have only a finite degeneracy. The sum $\sum_{n=0}^{N} \lambda_n^{-d}$ is assumed to diverge logarithmically in N (the λ_n in increasing order). This defines the dimension d.
- $\gamma = \gamma^*$, $\gamma^2 = 1$ and $[\gamma, \mathcal{A}] = 0$. Moreover $D\gamma = (-1)^{d+1}\gamma D$.
- Finally, J is an antilinear isometry on $\mathcal{H}$ such that

$$[JfJ^{-1}, g] = 0 \qquad \qquad \forall f, g \in \mathcal{A}.$$

If $\mathcal{A}$ is commutative then it is additionally required that $JfJ^{-1} = \bar{f}$ for all $f \in \mathcal{A}$.

The data of a spectral triple are in particular required to obey the following **axioms:**

1. From the **"Order-One-Condition"**:

$$[JfJ^{-1}, [D, g]] = 0 \qquad \qquad \forall f, g \in \mathcal{A}$$

together with the above requirements on D one infers that D is a differential operator of first order. Thus *locally*, i.e. in each point $p \in \mathcal{M}$, D can be written as:

$$D = i\gamma^\mu \partial_\mu + \rho, \qquad \qquad [\gamma^\mu, f] = [\rho, f] = 0, \quad \forall f \in \mathcal{A}.$$

The *locally defined* matrices γ^μ are selfadjoint and bounded. It is not clear at this point, however, whether they exist globally as sections of some bundle of matrices over $\mathcal{M}$.

2. The regularity assumption $[\sqrt{D^2}, [D, f]] \in \mathcal{B}(\mathcal{H})$ then implies that there exist scalar functions $g^{\mu\nu}$ such that

$$\gamma^\mu\gamma^\nu + \gamma^\nu\gamma^\mu = 2g^{\mu\nu} \in \mathcal{A}.$$

For this conclusion one needs to employ the axiom of **Poincaré duality**.

3. From the axiom of **orientability** one then concludes that the $g^{\mu\nu}$ define a nondegenerate positiv definite matrix valued function on $\mathcal{M}$. Thus, the γ^μ span in each point of $\mathcal{M}$ the Clifford algebra of the correct dimension.

4. It now remains to show that the locally defined Clifford algebras (over each point $p \in \mathcal{M}$) glue together to define a (global) **spin structure over** $\mathcal{M}$. One does so with the help of a theorem of Plymen whose adaptability is (once more) ensured by **Poincaré duality**.

   Thus this axiom plays an important role in the reconstruction of the metric and the spin structure. Note that it is only shown that the $g^{\mu\nu}$ define a metrical tensor on $\mathcal{M}$ after the spin structure has been established. Without this step it would not be clear that these locally defined functions glue together appropriately and hence that they transform correctly under a change of coordinates.

5. Differential one-forms, like gauge potentials, are represented on the Hilbert space by $\pi(f dg) = f[D, g]$, and in particular $\pi(dx^{\mu}) = i\gamma^{\mu}$. This works because the matrices γ^{μ} are now shown to transform like the basis one-forms dx^{μ} under changes of coordinates.

6. The antilinear operator J is identified with **charge conjugation**. This is ensured by requiring relations

$$J^2 = \epsilon(d)\,\mathrm{id}, \qquad J\gamma = \epsilon'(d)\gamma J, \qquad DJ = \epsilon''(d)\,JD$$

where the signs $\epsilon(d), \epsilon'(d), \epsilon''(d) \in \{1, -1\}$ depend on the dimension according to the "spinorial chessboard". In particular, if the algebra is commutative then these relations imply that the Dirac-Operator D cannot contain a term $A = \gamma^{\mu}A_{\mu}$, i.e. an electromagnetic vector potential, as one would have $(D + A)J = \pm J(D - A)$ in such a case: Let's assume for simplicity that the vector potential is pure gauge $A = \bar{u}[D, u]$ with $u \in \mathcal{A}$ and $\bar{u} = u^{-1}$. Then, by the above condition on J we have $JuJ^{-1} = \bar{u}$ and thus

$$JAJ^{-1} = J(\bar{u}[D, u])J^{-1} = \epsilon''(d)\,u[D, \bar{u}] = -\epsilon''(d)\,u\bar{u}^2[D, u] = -\epsilon''(d)A.$$

(Similarly for generic selfadjoint one forms A.)

Remark 2.1. Note that essentially *all diffeomorphisms* φ of M are represented as unitary operators U_{φ} on $\mathcal{H}$, such that

$$U_{\varphi} f(p)\, U_{\varphi}^* = f(\varphi^{-1}(p)) \qquad \forall p \in \mathcal{M}, \quad f \in \mathcal{A}$$

and $U_{\varphi} X U_{\varphi}^* = X_{\varphi}$ where X is D, γ or J, and X_{φ} the respective image under the diffeomorphism φ.

Furthermore it is to be noted that not all unitaries on $\mathcal{H}$ correspond to such diffeomorphisms, even though that should be obvious.

As we have seen above, in the classical case $\mathcal{A} = C^{\infty}(\mathcal{M})$, the space of Dirac-Operators is essentially the same space as the space of all metrics over $\mathcal{M}$. The **spectral action** is defined as

$$S_{\Lambda}(D) = \mathrm{Tr}\left(\chi\left(\frac{D^2}{\Lambda^2}\right)\right),$$

where χ is some smooth cutoff function, cutting out the eigenvalues above 1, while $\Lambda \in \mathbb{R}$ is some scale. The function χ can be chosen such that $S_{\Lambda}(D)$ admits an asymptotic expansion in Λ. Note that $S_\Lambda(D)$ is **spectral invariant**, i.e. it is invariant under *all unitaries* on $\mathcal{H}$ and thus in particular under all diffeomorphisms. The same must be true for each term in the asymptotic expansion, and from dimensional reasons (D and Λ have the dimension of an energy) their repective order in the curvature is clear. Hence for the first two nontrivial orders it follows immediately that (as Λ tends to infinity):

$$S_\Lambda(D) \sim c_0 \Lambda^4 \int_{\mathcal{M}} \sqrt{g} \mathrm{d}^4 x + c_1 \Lambda^2 \int_{\mathcal{M}} R\sqrt{g} \mathrm{d}^4 x + \mathcal{O}(\Lambda^0).$$

The first two terms can be interpreted as the Einstein-Hilbert action with some cosmological constant. These are the only diffeomorphism invariant terms linear, respectively independent of the curvature. As it turns out the two constants c_0, c_1 can be chosen arbitrarily by adjusting χ appropriately. The same is true for higher orders in the curvature, which can therefore be suppressed (here). **Hence, the spectral action can be viewed as the analogue of the Einstein-Hilbert action of pure gravity**.

Remark 2.2. Honestly speaking this is not completely true: A physical action also requires the choice of initial values for the physical field – here the metric. The Einstein-Hilbert action is indeed defined as the integral of the scalar curvature over any spacetime region sandwiched by two Cauchy surfaces, with the (initial and final) values of the metric helded fixed on these Cauchy surfaces when applying the variational principle.

The spectral action, on the other hand, only reproduces the integral over all of $\mathcal{M}$. The resulting field equations therefore only state that $\mathcal{M}$ must be any Einstein manifold, but not more, as the initial values of the metric are not specified.

The reason for this unpleasant fact lies in the *spectral invariance* of the spectral action, which is much more than only diffeomorphism invariance. ("One cannot hear the shape of a drum.") The physical Einstein-Hilbert action is only diffeomorphism invariant, but it is not spectral invariant, and, in fact, that's quite fortunate: The only spectral invariant quantities one might construct out of the Dirac-Operator are its eigenvalues, which correspond to the masses of the elementary fermions. There is only a finite number of the latter. So we couldn't learn much about spacetime from the spectrum of D. But fortunately we can also measure scattering phases. [5]

We should stress however that it may be possible to reconstruct a Riemmanian space from the eigenvalues of the Dirac-Operator if the precise form of the volume element is given. To give an example, one may conclude that the underlying space is a circle if one knows that the spectrum of D is $\mathbb{Z}$, each eigenvalue occuring precisely once *and* that the volume form is given as $u^*[D, u]$, where u is some algebra element. (We need not know more about u.) A much more intriguing example is given in [35]. One may then speculate whether adding the volume form to the spectral action can lead to further progress. See the most recent [34], where this idea has been used to render the mass scale in D dynamical.

Remark 2.3. Related to the above remark, we should also stress that the spectral action only makes sense for *Riemannian* manifolds. On generic Lorentzian manifolds, when each of the eigenvalues of D is infinitely degenerate, such a cutoff trace would never exist. Even more so, the spectrum of D on generic Lorentzian manifolds is always the same, namely $\mathbb{R}$. But still we can infer most information about the geometry from the degeneracy of *a single eigenvalue of D*, i.e. the space of solutions given a fixed mass, and in addition the knowledge of the volume form [4]. We shall have to say more about this later.

3. On the meaning of noncommutativity

If $\mathcal{A}$ is commutative then all the different irreducible representations of $\mathcal{A}$ are obviously unitarily inequivalent, as they are one-dimensional:

Two representations π_1, π_2 on Hilbert spaces $\mathcal{H}_1, \mathcal{H}_2$ are said to be unitarily equivalent if there exists a unitary operator $U : \mathcal{H}_1 \to \mathcal{H}_2$ such that

$$U\,\pi_1(f)\,U^* = \pi_2(f) \qquad \forall f \in \mathcal{A}$$

and this implies $\pi_1 = \pi_2$ if $\dim(\mathcal{H}_1)= 1$.

For noncommutative algebras there do exist equivalent representations, respectively equivalent pure states. Thus, in identifying points with irreducible representations of the algebra, one has to understand the geometrical meaning of this equivalence.

First of all, note that for irreducible GNS representations such a unitary equivalence arises via **inner automorphisms of $\mathcal{A}$**, i.e. the corresponding unitaries U are elements of the algebra representations, $U \in \pi(\mathcal{A})$. A commutative algebra does not possess inner automorphisms. On the other hand, outer automorphisms, i.e. those which are not induced by algebra elements, do not give rise to equivalent *irreducible* representations. It should be stressed that this picture of noncommutative spaces is absolutely generic: A theorem of Dixmier [38] states that **a C^*-algebra is noncommutative if and only if there do exist unitarily equivalent irreducible representations**. Moreover one can formulate analogues of the Gelfand-Naimark theorem for noncommutative algebras as characterizing topological spaces together with some equivalence relation of points in many different ways (see [30] and references therein.)

Thus, as a first moral one might state that **there is a clear distinction between the inner and outer automorphisms of a noncommutative algebra: Inner automorphisms lead to equivalent irreducible representations.**

Readers may wonder how our intuition of the phase space of nonrelativistic quantum mechanics, i.e. the Heisenberg algebra $[x, p] = i$, fits into this picture. To be honest, I don't know a fully compelling answer to this question:

With a suitable regularity assumption there exists up to equivalence only one irreducible representation for this algebra – the one on $L^2(\mathbb{R})$ that we are all familiar with. Does this mean the phase space of quantum mechanics has only one point ? Certainly the above statement implies that, given any sharp value for x, all values for p are completely indistinguishable. That is in good agreement with our intuition. But it also implies that all values for x are equivalent as well, which is certainly not consistent with the fact that we can localize an electron within a region of radius of its Compton-wavelength. Well, only with respect to the coordinate x: In a special relativistic quantum field theory every inertial observer would agree that the particle is localized in such a region. But, in view of the Unruh effect, accelerated observers would not even agree that there is only one electron. And they would not necessarily see a localized state.

The point of view of Noncommutative Geometry is completely coordinate independent. I guess that this is behind the seemingly contradictionary intuitions of Noncommutative Geometry and quantum theory: the way we currently interpret it, quantum field theory is not generally covariant – but any dynamical theory of spacetime should be. Unfortunately we still lack an interpretation of quantum theories that is in accordance with Einstein's equivalence principle.

To be more concrete, let's consider the Moyal-plane,

$$[x, y] = i\theta.$$

The noncommutativity parameter θ is then, employing the above mentioned wrong analogy to $\hbar$ in quantum mechanics, often misinterpreted as a physical observable – even though that implies that the coordinates x, y obtain a special status. It would then, for instance, no longer be allowed to rescale these coordinates by $x \to \lambda x$, $y \to \lambda y$. In quantum mechanics, due to the presence of a symplectic form, the choice of the coordinates x, p of the phase space is really canonical, p being the canonical momentum associated to the chosen coordinate x. However, there should be no obstruction to the choice of coordinates of the configuration space of a physical system. An experimental physicist should, for instance, be allowed to choose his unit of length arbitrarily. Of course, a length is actually not described by the coordinates alone, but by a metric g^{ij} where the matrix indices refer to the chosen coordinates. Note that the determinant g of this metric transforms under coordinate transformations in the same way as θ^2 (in two dimensions) would do, if we allow such transformations. Thus, the combination

$$A_p := \frac{\theta}{\sqrt{g}}$$

is an invariant in this case and could really be viewed as a physical parameter. As it has the dimension of an area, it might be viewed as the minimal area that can be resolved on such a spacetime.

As for a physical interpretation of the picture of noncommutative spaces described above, let us consider the following example:

Suppose that the configuration space Bea of a system is given by two points, $Bea = \{uty, st\}$ say. The commutative algebra of functions on Bea is then isomorphic to the algebra of diagonal two-by-two matrices,

$$\mathcal{A} = C(Bea) = \left\{ \begin{pmatrix} f(uty) & 0 \\ 0 & f(st) \end{pmatrix} \;\middle|\; f(uty), f(st) \in \mathbb{C} \right\}.$$

We now want to describe a situation where the two points, i.e. the two inequivalent representations of $C(Bea)$ are equivalent. This is achieved by adding to $C(Bea)$ a further unitary element u that interchanges the two representations, i.e.

$$u = \begin{pmatrix} 0 & 1 \\ 1 & 0 \end{pmatrix}.$$

It is not hard to see that the matrix u together with the matrices $e_{uty} = \begin{pmatrix} 1 & 0 \\ 0 & 0 \end{pmatrix} \in$ $C(Bea)$ and $e_{st} = \begin{pmatrix} 0 & 0 \\ 0 & 1 \end{pmatrix} \in C(Bea)$ generate the full matrix algebra $M_2(\mathbb{C})$ of complex two-by-two matrices Up to equivalence $M_2(\mathbb{C})$ has only one irreducible representation. On the other hand, the set X of pure states of the algebra $M_2(\mathbb{C})$ is a two-sphere:

$$X = \left\{ \begin{pmatrix} \psi e^{i\sigma} \\ \sqrt{1 - \psi^2} \end{pmatrix} \ \middle| \ \psi \in [0, 1], \sigma \in [0, 2\pi] \right\}.$$

That is just the space of possible choices of bases in $\mathbb{C}^2$ up to an overall phase, i.e. the set of (mutually equivalent) representations of $M_2(\mathbb{C})$ on $\mathbb{C}^2$.

Thus, by identifying the two points we have blown up this two point set to obtain a bubble of foam. By doing so we have also introduced new degrees of freedom in our algebra $\mathcal{A}$ that we still interpret as functions on the configuration space. These new degrees of freedom, being given by off-diagonal two-by-two matrices correspond in a very precise sense (as we shall see in the next section) to **gauge degrees of freedom**. The inner automorphisms of the algebra, characterizing its noncommutativity, can, in fact, always be interpreted as gauge transformations.

4. The noncommutative description of the standard model and the physical intuition behind it

Let us now come to the main part of this article, namely the postulates governing the reformulation of the standard model of particle physics as part of the gravitational field on a certain noncommutative space. Before I describe these axioms, however, I shall first sketch its fundamental physical idea.

4.1. The intuitive idea: an effective picture of quantum spacetime at low energies

As has been shown in section 2, for classical spectral triples, $\mathcal{A} = C^\infty(\mathcal{M})$, it is not possible to add a one-form, i.e. a gauge field, to the Dirac-Operator, because of the reality condition $DJ = JD$. (We shall restrict to the case of 4 dimensions from now on, when $\epsilon''(4) = 1$.) The Dirac-Operator of real spectral triples therefore "only" describes the gravitational field.

If the algebra is noncommutative, however, one may add a term of the form $A + JAJ^{-1}$ to D, where $A = f[D, g]$ is a one-form: Now there exist A's with $JAJ^{-1} \neq -A$. This is so because there then exist some $f \in \mathcal{A}$ such that JfJ^{-1} is not in $\mathcal{A}$ since JAJ^{-1} belongs to the commutant of $\mathcal{A}$, which cannot contain all of the *noncommutative* algebra $\mathcal{A}$.

Consider an unitary $u \in \mathcal{A}$, i.e. the generator of an inner automorphism. We can associate to it the unitary operator U:

$$U\psi := u\, JuJ^{-1}\, \psi =: u\psi u^*, \qquad \psi \in \mathcal{H}.$$

This then represents the inner automorphism generated by u on $\mathcal{H}$ as $U f U^* = ufu^*$ for all $f \in \mathcal{A}$. The reason why this representation is chosen that way becomes clear if one considers its action on a Dirac-Operator $D_A = D + A + JAJ^{-1}$:

$$U\, D_A\, U^* = D_{A^u}, \qquad A^u := uAu^* + u[D, u^*].$$

Hence **adding $A + JAJ^{-1}$ to D gauges the inner automorphisms of** $\mathcal{A}$. Note that $JuJ^*\psi =: \psi u^*$ can consistently be interpreted as a right action of u^* on ψ because it commutes with the left action and because J is antilinear.

Of course, the terms $A + JAJ^{-1}$ are not one-forms in a strict sense. However, if one considers all Dirac-Operators for a given algebra – i.e. all metrics – then these degrees of freedom have to be taken into account. Readers familiar with Connes' distance formula will notice that these terms do infect the metric as they do not commute with algebra elements. Let us consider an **almost commutative** algebra like $\mathcal{A} = C^\infty(\mathcal{M}) \otimes M_2(\mathbb{C})$ represented on $\mathcal{H} = L^2(\mathcal{M}, S) \otimes M_2(\mathbb{C})$, the space of square integrable spinors on $\mathcal{M}$ with values in $M_2(\mathbb{C})$. The part $M_2(\mathbb{C})$ of $\mathcal{A}$ acts by matrix multiplication from the left on this space.

For D we can take the usual Dirac-Operator on $\mathcal{M}$ – for some metric g – acting trivially on $M_2(\mathbb{C})$. J is taken to interchange the left and the right action of matrices on matrices, i.e. for $f \in C^\infty(\mathcal{M})$, $\psi \in L^2(\mathcal{M}, S)$ and $\lambda, \sigma \in M_2(\mathbb{C})$ one defines

$$J\left(\bar{f} \otimes \lambda^*\right) J^{-1} \psi \otimes \sigma = (f\psi) \otimes (\sigma\lambda)$$

In such a case the terms $A + JAJ^{-1}$ can be interpreted as one-forms over $\mathcal{M}$ with values in the Lie-algebra $\mathbf{su(2)}$ – but of course not as one-forms over the noncommutative spacetime $\mathcal{A}$.

At first glance, the above example, where the part $A + JAJ^{-1}$ of D_A that gauges the inner automorphisms can be interpreted as gauge fields on the commutative part $\mathcal{M}$ of the underlying noncommutative spacetime might not seem to be relevant for a quantum theory of gravity: The typical arguments that the semi-classical states of such a theory should correspond to *non*commutative manifolds like e.g. [3, 24, 25] do not suggest such an almost commutative algebra but much more noncommutative ones.

However, as we have seen above the appearance of gauge degrees of freedom is an absolutely generic phenomenon on noncommutative spaces and there will be even more such degrees of freedom if the spacetime is more noncommutative. But since such a noncommutativity is expected to show up only at extremely high energy densities one might think that these degrees of freedom cannot be excited in experiments that are realistic to be performed in the near future. Thus we might expect these degrees of freedom to be frozen in contemporary experiments.

On the other hand we do see some nonabelian gauge fields already at energies of the order of the Z-mass. So maybe it is not true that all these degrees of freedom which reflect the noncommutativity can only be excited at the Planck energy. After all, the Planck energy only gives some scale where we expect that quantum gravity effects can no longer be neglected. It might well be possible that this is the case already at much lower energies and the above example then suggests to **interpret the strong and the weak interactions as the first shadow of the quantum corrections for the gravitational field.** In the present experiments of particle physics we do not have enough energy to resolve spacetime at scales much smaller than the Compton wavelength of the Z boson. But with this resolution we might already see a glow of the bubble of foam that replaces points: namely the two-sphere S^2 which is the set of pure states of the algebra $M_2(\mathbb{C})$.

This intuitive idea is made more precise in the following postulates for the noncommutative description of the standard model. Subsection 4.3. will then see their consequences. Yet, in order to show our colours rather than cumbersome matrix calculations, these two subsections are extremly rough and sketchy. Readers interested in the details are referred to [12, 9] As already stressed in the introduction, the postulates are not thought to be completely coherent, and many critical comments are in order. I tried to gather them in the next section.

4.2. The postulates

1. There exists an energy scale Λ up to which spacetime can effectively be described as an almost commutative geometry $C^\infty(\mathcal{M}) \otimes \mathcal{A}_F$. Here $\mathcal{M}$ denotes a four-dimensional (compact), orientable manifold, and $\mathcal{A}_F$ the finite dimensional *real C^*-algebra*

$$\mathcal{A}_F = \mathbb{C} \oplus \mathbb{H} \oplus M_3(\mathbb{C})$$

   where $\mathbb{H}$ denotes the algebra of quaternions.

   In experiments with energies below Λ we will not see a deviation from this picture of the topology of spacetime.

2. At energies below the energy scale Λ it is a very good approximation to neglect the backreaction of matter on the Riemannian metric on $\mathcal{M}$. Hence we can keep this metric as a *fixed, classical* background metric that therefore need not be quantized. Accordingly we only need to take into account the "inner fluctuations" of the metric on $\mathcal{A}$, which correspond to the dynamics of the gauge degrees of freedom for the inner automorphisms of $\mathcal{A}$.

3. The spectral triple $(\mathcal{A}, \mathcal{H}, D_A, \gamma, J)$ describing the space of metrics for this situation is given by:
   - $\mathcal{H} = L^2(\mathcal{M}, S) \otimes \mathbb{C}^{90}$ where the space $\mathbb{C}^{90}$ is explained as follows:
     If one does not (yet) add right handed neutrinos to the standard model, then there do exist 45 elementary Weyl Spinors. Here we consider the respective antiparticles separately, thus there are 90 Weyl Spinors. (We comment on this later)
     The representation of $\mathcal{A}_F$ on $\mathbb{C}^{90}$ is chosen according to the action of

the gauge group – i.e. the unitaries in $\mathcal{A}_F$ – on the elementary Weyl Spinors.

- The Dirac-Operator D is given schematically as:

$$D_A = D_{\mathbb{M}} + \mathbf{M}\gamma + A + JAJ^{-1},$$

where $D_{\mathbb{M}}$ denotes the Dirac-Operator on $\mathbb{M}$ corresponding the fixed background metric. $\mathbf{M}$ is the fermionic mass matrix. It contains all the masses and Kobayashi-Maskawa mixing angles. Finally A denotes an arbitrary one-form built with $D_0 = D_{\mathbb{M}} + \mathbf{M}\gamma$. Thus A parametrizes the aforementioned gauge degrees of freedom which are dynamical – unlike D_0.

- γ has eigenvalue $+1$ on the right handed fermions and -1 on the left handed ones.

- $J = C \otimes J_F$ where C denotes the charge conjugation on $\mathbb{M}$, while J_F, acting on $\mathbb{C}^{90}$ interchanges the particles with the antiparticles (as they both appear separately in $\mathbb{C}^{90}$.)

4. Only fermion fields $\psi \in \mathcal{H}$ obeying the Majorana condition $J\psi = \psi$ are physical. (Hence the double counting of particles and antiparticles is removed. It is however needed to get the quantum numbers correct.)

   The action for these fermions is given as $S(\psi) = \langle \psi, D\psi \rangle$.

5. At the energy scale Λ, the action of the dynamical degrees of freedom A of D_A is invariant under all unitaries on $\mathcal{H}$.

   Thus, there exists a cutoff function χ (smooth, with a Laplace-transform of rapid decay) at this scale such that

$$S_\Lambda(A) = \mathrm{Tr}\left(\chi\left(\frac{D_A^2}{\Lambda^2}\right)\right).$$

6. The effective action at energy scales below Λ is then obtained via the renormalization group flow of the pertubatively quantized action $S_\Lambda(A) + S(\psi)$ from Λ to the scale under consideration.

4.3. How such a noncommutative spacetime would appear to us

1. Under the above assumptions, the most general "inner fluctuation A of the metric" is given as

$$A = i\gamma^\mu (A_\mu + W_\mu + G_\mu) + \gamma\phi\mathbf{M}$$

where A_μ, W_μ, G_μ denote the gauge potentials for the electro-weak and strong interactions, while ϕ is the *scalar Higgs-doublet*.

   The appearance of the gauge bosons is, of course, put in by hand, as we have chosen the algebra appropriately – considering $\mathcal{A}$ as a phenomenological input. But the appearance of the Higgs, which comes automatically if the mass matrix $\mathbf{M}$ is nontrivial is quite surprising and really a result.

2. $10^{13} GeV \leq \Lambda \leq 10^{17} GeV$. Note that this means that there is a "big desert" in which no new physics is to be expected.

3. The spectral action $S_\Lambda(A)$ possesses an asymptotic expansion and is given at the Z-mass as

$$S_\Lambda(A) \sim \int_{\mathcal{M}} \sqrt{g} d^4 x \left\{ \mathcal{L}_{EH}(g) + \mathcal{L}_{st.mod}(A) + \frac{9\alpha}{64\pi^2} C^{\mu\nu\rho\sigma} C_{\mu\nu\rho\sigma} + \frac{\beta}{12} |\phi|^2 R + \ldots \right\}.$$

Here $\mathcal{L}_{EH}(g)$ denotes the Einstein-Hilbert Lagrangian with a cosmological constant term that is to be fixed by experiments. Note that this term is viewed as a constant, since the background metric g is fixed. The same applies for the Weyl-Tensor $C^{\mu\nu\rho\sigma}$ and the scalar curvature R for g.

$\mathcal{L}_{st.mod}(A)$ denotes the complete bosonic Lagrangian of the standard model of particle physics. In particular it always contains the Higgs potential that spontaneously breaks the electroweak energy. There is no freedom to eliminate this potential if the mass matrix $\mathbf{M}$ is nonvanishing.

The dots in the above formula indicate possible higher order terms in the curvature, that may be suppressed however. Note that this is not so for the terms $\frac{9\alpha}{64\pi^2} C^{\mu\nu\rho\sigma} C_{\mu\nu\rho\sigma}$ and $\frac{\beta}{12} |\phi|^2 R$ as the freedom in the choice of χ corresponding to second order terms in curvature is used to adjust the coupling constants of the standard model Lagrangian. (The precise value of the numerical factors α, β can be found in [9].)

4. At the scale Λ the coupling constants of the three interactions of the standard model obey the relation $g_{weak} = g_{strong} = \sqrt{\frac{5}{3}} g_{hypercharge}$. However under the flow of the renormalization group of the pertubatively quantized standard model one approximately obtains the experimentally measured values at the Z-mass. For that reason Λ has to lie in the range indicated above. We should remark, however, that the so obtained values are not completely consistent with the experimental values. But there are more severe problems of the spectral action anyway, as we shall see.

5. The Higgs mass m_H is constrained as $m_H = 182 \pm 20 \ GeV$.

5. Remarks and open questions

5.1. Remarks

General Relativity relies on the empirical fact that the inertial and the gravitational mass are identical. Only for this reason is it possible to describe the motion of test particles as free motion along geodesics. In this subsection I shall describe the empirical facts which are necessary for the above description of the standard model as part of the gravitational field on some noncommutative space to work – well, as good or bad as it does. General relativity does not explain the equality of the gravitational and the inertial mass. Neither does the Noncommutative Geometry interpretation of the standard model explain any of the following facts on which it relies:

- The elementary fermions of the standard model only appear in the trivial, the fundamental (respectively its conjugate) representations of the gauge group. In principle it would also be possible that some fermions sit in the adjoint representation. But no other representation can occur, since the gauge groups are induced by the unitaries in some algebra.
- It is highly nontrivial that the electrical charges are assigned correctly. At first sight this would not be possible for models based on almost commutative spectral triples, as then only the $U(1)$ charges $0, \pm 1$ seem to be allowed. However the algebra $\mathcal{A}_F$ contains two $U(1)$ factors – one in $\mathbb{C}$, one in $M_3(\mathbb{C})$ – a certain combination of which drops out due to a mechanism deeply rooted in the geometrical interpretation of spectral triples [33]. The remaining $U(1)$ factor then asigns the correct weak hypercharges to the fermions, so that – after the weak symmetry is spontaneously broken – the experimentally observed electrical charges show up.
- Parity is broken maximally. In the standard model the weak interaction could couple to the left handed fermions with some strength $1 - \epsilon$ and to the right handed ones with strength ϵ. In models based on noncommutative geometry this would be impossible. Fortunately experiments tell us $\epsilon = 0$.
- Fermion masses arise in such models only via the Higgs effect. So for this description to be true, the Higgs must exist – which it hopefully (?!) does.
- The fact that the coupling constants of the standard model under the flow of the renormalization group almost meet in one point is essential: Else their derivation via the spectral action would be too bad an approximation to provide any hope that a modification of this picture of spacetime at higher energies could make contact with the precise experimental values.
- The "Order-One Condition" excludes the possibility that color symmetry is broken (if one additionally requires "S^0-reality", see [7] however). Thus, if the gluons would turn out massive, this picture of spacetime would break down.
- Finally it has to be said that there is a minor problem with the observed neutrino masses. Due to the axiom of Poincaré duality one can only add two massive neutrinos to the model. One neutrino has to remain massless. Unfortunately, it is probably impossible to detect whether all or only two neutrinos are massive, since oscillation experiments are only sensitive to mass differences, while, as it seems, only the mass of the electron neutrino could in principle be measured directly. However, on the side of noncommutative geometry it is possible to replace the axiom of Poincaré duality by other axioms. But how could we find out empirically whether we need to do so ?

5.2. Open problems, perspectives, more speculations

- It is certainly an aesthetic deficit that one has to first count particle and antiparticles as different, but then impose the Majorana condition to identify the ones with the charge conjugates of the other – even though there is something appealing about Majorana fermions.

- We should stress that spectral invariance, which is the underlying principle to construct the action $S_\Lambda(A)$ is not preserved under the renormalization group flow: It implies the constraints on the coupling constants given above and these are obviously not fullfilled at the Z-mass.

  But this breaking of spectral invariance is of course to be expected since special relativistic quantum field theories – like the standard model whose renormalization group flow is used here – are not compatible with the equivalence principle. One may hope that in a full theory of quantum gravity such a problem will not show up.

- It is not very encouraging that Λ comes out so large, implying the "big desert" [9].

- It is definitely a major drawback of this approach that the gravitational field – i.e. the metric on $\mathcal{M}$ – has to be kept fixed. The same is in fact true for the mass matrix $\mathbf{M}$ which would also be a freedom in the choice of Dirac-Operators on the noncommutative spacetime $\mathcal{A}_F$. For the latter I don't see any good physical reason why it's dynamics – if there are any – should not be visible at the Z-mass. As the former is concerned, one might argue that the gravitational constant is too small, so the backreaction of matter on g can be ignored at the considered energy scales. But one has to do so, as one needs to take the renormalization group flow of the quantized theory into account, and there's as yet no way to include the gravitational field into this game.

  In the long term perspective, one has to consider the full dynamics of the Dirac-Operators D. Only then the philosophy of Noncommutative Geometry would be fully implemented.

  Note that then the fermionic mass matrices could be rendered dynamical as well. One may speculate that in this way a mechanism that dynamically generates the funny pattern of the fermion masses could be revealed [5, 6].

- At the present stage, the model still keeps many of the aesthetic defects of the standard model. In particular the gauge group has to be put in by hand. However, as shown in the series of papers ([11, 6], [31] and references given therein) with very mild irreducibility assumptions it turns out that the number of possible almost noncommutative geometries is fairly restricted. If this program succeeds, then it will be possible to infer the gauge group of the standard model from the assumption that it can be described as the inner fluctuations of the metric on an almost commutative geometry – i.e. without having to assume the algebra from the start.

- Of course, a theory of quantum gravity should be valid also at higher energies as Λ. In that case we would probably have to modify $\mathcal{A}$ however, as we expect that then the full noncommutativity of spacetime is revealed. So there is the question which algebra might replace the algebra of the standard model, but lead to the same physics at energies less than Λ. Many people have tried deformations of the Lorentz group to provide candidates for such algebras. Another rather interesting candidate for this algebra is given in [2]. I will briefly describe this idea in the next subsection. But of course, the answer to this question can only be given to us by physics, rather than by mathematical speculations:

- In [32] the author proposed a novel model for an almost commutative geometry which extends the algebra $\mathcal{A}_F$ and might exhibit interesting phenomenological consequences with respect to dark matter. This is certainly a promising route to take as it may lead to definite predictions for future experiments.

- If Noncommutative Geometry could really be valorized to an approach to quantum gravity, then of course we would also hope to understand the dynamical mechanism that leads to the noncommutative geometry of spacetime. Note that such a spacetime as the one used for the above description has a larger diffeomorphism group as commutative manifolds. In the standard model example there are for instance also the transformations of the different families of fermions into each other among the symmetries of the spectral action. If we could understand where this enlarged symmetry stems from, then we might understand why there are three families. [5]

- The spectral action is only well defined over compact *Riemmanian* manifolds. For me this has always been its major drawback. On *Lorentzian* spectral geometries one probably has to take another route to a generalization of Einsteins equations. I will come to this problem in the next section, where a proposal for a Noncommutative Geometry approach to quantum gravity will be presented.

  It is sometimes said that this problem is not important as one may always "Wick-rotate" from the Riemannian to the Lorentzian case. As concerns gravity that's just wrong: Wick rotation requires the existence of a timelike Killing vector to be unambiguos. So it is only possible if we severely restrict the allowed space of metrics, which is certainly not in the spirit of the equivalence principle.

- The prediction of the Higgs mass might turn out a spectacular virtue of the theory of course. However, it is not unlikely that the Higgs turns out to be lighter than $180 GeV$. I wouldn't consider this a real problem, though. For me this prediction has always had a similar flavor as the false prediction of the ground state energy of the hydrogen atom by the Bohr model. In fact, as we have seen above, the real lessons about the standard model we can hope to learn from Noncommutative Geometry concern its particle content rather than the spectral action, which is not defined for the physically realistic Lorentzian manifolds anyway.

5.3. Comparision of the intuitive picture with other approaches to Quantum Gravity

One could actually keep this story very short: As yet the effective picture of spacetime at the Z-mass that Noncommutative Geometry assumes is not shown to be in contradiction to the expectations indicated by any of the prominent approaches to quantum gravity. Mainly because most other approaches start at the Planck-energy and don't have definite predictions at the Z-mass yet, of course. Noncommutative Geometry, on the other hand, follows a bottom-up strategy and, as we have pointed out above, will now have to refine its picture by taking corrections into account that will become relevant at energies higher than Λ. So it could be viewed as a complementary approach that may well turn out to be equivalent to one (or all) of the other approaches:

- String theory leads quite naturally to an effective description via noncommutative gauge theories at lower energies [23, 10]. Moreover String theory predicts that there is an infinite tower of elementary particles, thus there are many degrees of freedom which are frozen at the Z-mass. This is very much analogous to the picture that I described in section 3: If we believe that spacetime is not only almost commutative but "really noncommutative" then we would expect many more degrees of freedom than only the gauge bosons of the standard model to show up.

  As a side remark – not meant too seriously – one could add that String theory replaces four dimensional spacetime $\mathcal{M}$ by $\mathcal{M} \times CY$ where CY is an appropriately chosen compact six-dimensional Calabi-Yau manifold. Noncommutative Geometry replaces $C^\infty(\mathcal{M})$ by $C^\infty(\mathcal{M}) \otimes \mathcal{A}_F$. But $\mathcal{A}_F$ does of course not correspond to extra dimensions. We only observe it via the gauge bosons of the standard model and hence we need not invent any mechanism to hide it. This is so, because the geometrical concepts behind the spectral action are formulated in a completely background independent way. Unfortunately, in order to make contact with the real world, the spectral invariance has to be sacrified at the present stage.

  Hence, both theories, String theory and Noncommutative Geometry, still have to come up with a truely background independent formulation, before one can compare them more sensibly with each other, and, much more importantly, with reality.

- Besides the heuristic arguments in [3], quite recently many different approaches to quantum gravity have found clues for a noncommutativity of the spacetimes corresponding to semi-classical states of these theories, see [24, 25] for nice examples. In particular the nontrivial dimension spectrum seen independently in [26, 27] is a generic feature of noncommutative spectral triples [15, 17]

- As concerns Loop Quantum Gravity, there will be many remarks in the next chapter. See however [20] for an alternative approach to combine Loop Quantum Gravity with Noncommutative Geometry.

- Finally I would like to mention the noncommutative approach to quantum gravity advocated in [2]: Here the basic idea is to consider a noncommutative algebra that is constructed via the frame-bundle E over $\mathcal{M}$ and its structure group, i.e. the local Lorentz-transformations. To me this seems to provide a very natural candidate for a noncommutativity of spacetime, not only because it is obviously related to the algebra found in [24, 25]: If noncommutativity is interpreted as appearance of equivalence relations among points of spacetime, and if the dynamical coupling of quantum matter to geometry leads to such a noncommutativity, then one would expect that these equivalence relations are related to the fundamental principles underlying this coupling. Thus, it is well conceivable that the noncommutativity of spacetime is related to local changes of frames, i.e. the local Lorentz transformations.

6. Towards a quantum equivalence principle

The noncommutative description of the standard model certainly has many very appealing features. However, the appropriateness of the spectral action appears somewhat problematic: It is only available over (noncommutative) Riemannian manifolds and it is not compatible with initial conditions for field equations. Moreover its guiding principle – spectral invariance – is not fullfilled in the standard model alone. That problem could of course be overcome once the quantization of the gravitational field is achieved. In that case the metric might be considered as a dynamical degree of freedom, which is not really the case at the present stage.

Last not least, one would actually not want to put in by hand the noncommutativity of spacetime. Rather one would like to infer the dynamical mechanism that leads to this noncommutativity in a theory of quantum matter coupled to geometry. Concrete proposals for such mechanisms have been made in [3, 24, 25]. However, while the first one of these arguments is not yet in accordance with the equivalence principle, the other two are still restricted to low dimensional models, when gravity only possesses topological degrees of freedom. What would really be required first is a background independent formulation of quantum theories in the spirit of General Relativity.

6.1. Globally hyperbolic spectral triples

The approach for such a formulation proposed in [21, 22] requires not only a notion of *Lorentzian* spectral triples but also of causal structures and a generalization of the Cauchy problem for the Dirac equation to such noncommutative spacetimes. In this subsection the logic behind this notion shall be briefly sketched.

A **Lorentzian spectral triple (Lost) over** $\mathcal{A}$ has the data

$$L = (\mathcal{A}, \mathcal{H}, D, \beta, \gamma, J).$$

There is one new ingredient: the fundamental symmetry $\beta = -\beta^*$, $\beta^2 = -1$, which can be viewed as a timelike one-form. In fact it is required that there exist algebra elements $f^{(i)}, g^{(i)}$ such that $\beta = \sum_i f^{(i)}[D, g^{(i)}]$ (time-orientability). Most of the axioms of spectral triples remain unchanged except that the Lorentzian Dirac-Operator D is, of course, no longer selfadjoint but β-symmetric,

$$D^* = \beta D \beta,$$

on the common domain of D and $\beta D\beta$ which is required to be dense in $\mathcal{H}$. Another difference is that we now allow for nonunital algebras $\mathcal{A}$, corresponding to *non-compact* manifolds. If $\mathcal{A}$ is the algebra of smooth functions of compact support on some smooth manifold $\mathcal{M}$ then the Losts over $\mathcal{A}$ correspond to Lorentzian metrics and spin structures over $\mathcal{M}$.

An important remark is in order here: Given the other data, the choice of β (and thus D^*) is not uniquely determined by the axioms. That should not be the case anyway, as β involves the choice of a timelike direction in the commutative case and there is no preferred such direction in general. Indeed it can be shown that two Losts which differ only by the choice of β are unitarily equivalent. We shall denote the set of all admissable β given the remaining data of $\mathbf{L}$ by $\mathbf{B_L}$.

To any such β we associate the symmetric operator

$$\partial_\beta = \frac{1}{2}\{D,\beta\} = \frac{1}{2}\beta(D - D^*).$$

In the classical case $\mathcal{A} = C_c^\infty(\mathcal{M})$ one can prove that ∂_β is a derivation on the algebra, i.e $[\partial_\beta, f] \in \mathcal{A}$ for all $f \in \mathcal{A}$.

We call a Lost *timelike foliated* if ∂_β is essentially selfadjoint for all choices of β and if there exists an unitary algebra element u such that $\beta = u^*[D,u]$. Moreover it is required that $u^\kappa \in \mathcal{A}$, $\forall \kappa \in \mathbb{R}$.

Needless to say that these conditions ensure in the classical case that one may reconstruct a foliation of $\mathcal{M} = \mathbb{R} \times \Sigma$ along the timelike direction specified by β. But due to the required essential selfadjointness the ∂_β will – even in the noncommutative case – give rise to one-parameter groups $U_\tau = e^{i\partial_\beta\tau}$ of unitatries on $\mathcal{H}$. We denote by

$$\mathcal{T}_\mathrm{L} := \{e^{i\partial_\beta\tau} \,|\, \beta \in \mathbf{B_L}\}$$

the set of all these "time-flows". As a matter of fact, the set $\mathcal{T}_\mathrm{L}$ can now be used for all Losts to

- Define the noncommutative analogue of the space $\mathcal{H}_c^\infty$ of smooth spinors of compact support.
- Given two distributions $\xi, \eta \in (\mathcal{H}_c^\infty)'$ one can characterize whether the support of η lies in the causal future/past of the support of ξ with the help of $\mathcal{T}_\mathrm{L}$.
- One may even reformulate the wavefront sets of such distributions by employing the elements of $\mathcal{T}_\mathrm{L}$.

A **globally hyperbolic spectral triple (ghyst)** is now simply defined as a timelike foliated Lost for which there exist **uniquely determined advanced and retarded propagators** $E_\pm : \mathcal{H}_c^\infty \to (\mathcal{H}_c^\infty)'$.

Here a propagator E is defined as a map $E : \mathcal{H}_c^\infty \to (\mathcal{H}_c^\infty)'$ such that $D\,E(\psi) = \psi$ in a distributional sense:

$$\langle E(\psi), D\varphi \rangle = \langle \psi, \varphi \rangle \qquad \forall \psi, \varphi \in \mathcal{H}_c^\infty.$$

Such propagators are called advanced (respectively retarded) if the support of $E(\psi)$ is contained in the causal future (respectively past) of that of ψ. Given an advanced propagator E_+ and a retarded one E_- one can construct all the solutions for the Cauchy-problem for D with the help of $G = E_+ - E_-$.

Remark 6.1. We still have to work out the precise conditions under which a time-like foliated Lost is a ghyst. An even more urgent and important open problem is to generalize the concept of **isometric embeddings** of globally hyperbolic space-times into each other to the noncommutative case. In the classical case this can be done by replacing the unitaries representing the diffeomorphisms on $\mathcal{H}$ by appropriate partial isometries. However this notion turns out to restrictive if $\mathcal{A}$ is noncommutative, as is seen in noncommutative examples for ghysts.

6.2. Generally covariant quantum theories over spectral geometries

An important virtue of Noncommutative Geometry that has not been mentioned so far is that it provides a mathematical language ideally tailored to reconstruct (noncommutative) spacetimes which approximate the semiclassical states of a background independent quantum theory. We do not yet have such a theory, of course, and accordingly the definition of "semiclassical" states for such a theory is far from clear. But however such a theory may look like: it should provide observables to reconstruct spacetime, and hence there should exist a map from a suitable subalgebra of observables to globally hyperbolic spectral triples (ghysts) or some variant therof. Moreover this map should be covariant with respect to diffeomorphisms and respect **locality**, i.e. the fact that observables localized at spacelike separated regions commute.

In view of arguments like the one presented in [3], it has to be expected that the image of this map does not contain classical, i.e. commutative, manifolds. But there are probably also many noncommutative ghysts which are not realized in a quantum theory of gravity coupled to matter.

Now suppoose we can invert the above map. The inverse would map ghysts to algebras of observables. It then follows that classical ghysts cannot lie in the domain of this map, as else the map would produce a theory that cannot exist according to [3]– namely a theory of quantized gravity coupled to quantum matter for which there does exist a semiclassical state corresponding to a commutative manifold on which the matter fields reside. If one attempts to show that a full quantum theory of gravity coupled to matter implies a noncommutativity of spacetime one may therefore adopt the following strategy [21]:

1. **Reformulate local quantum theory as a generally covariant map** *Obs* **from the category of ghysts to the category of algebras**. Thus, *Obs* is required to be a covariant tensor functor between these categories. Note that the above requirements on *Obs* imply that $Obs(\mathcal{N})$ is a subalgebra of $Obs(\mathcal{M})$ whenever $\mathcal{N}$ can be embedded as a globally hyperbolic submanifold of $\mathcal{M}$. In particular every diffeomorphism φ of manifolds induces an algebra homomorphism α_φ on the image of *Obs*.
2. **Require as a further constraint on** *Obs* **that there exists a causal dynamical law**. This is achieved by demanding that for any submanifold $\mathcal{N}$ of $\mathcal{M}$ which

contains a *full Cauchy surface* of $\mathcal{M}$ it follows that $Obs(\mathcal{N}) = Obs(\mathcal{M})$, i.e. it is sufficient to know the restriction of observables to one Cauchy surface to know them everywhere on $\mathcal{M}$.

3. **Demand now that geometry and matter are dynamically coupled by requiring the existence of a diffeomorphism invariant state for** *Obs*. Diffeomorphism invariance of states ω here means that, with the above notation

$$\omega \circ \alpha_\varphi = \omega \qquad \forall \varphi.$$

4. **Show that the domain of a map** *Obs* **meeting all the above requirements contains only certain noncommutative ghysts.**

This is of course a very (over- ?)ambitious program, that I could sketch only very roughly here. A complete definition of generally covariant quantum theories and some remarks concerning its applicability to quantum gravity can be found in [37] and the contribution of Romeo Brunetti Klaus Fredenhagen to the present volume. I should stress that it is far from clear whether the last two points of the above program are well-defined. In particular we have not yet shown that there exists any nontrivial functor *Obs* for which there exists a diffeomorphism invariant state. However for Loop Quantum Gravity there does exist a uniqely defined diffeomorphism invariant state [28, 29] and we hope to show that this provides an example for such a functor. Even more so, we have not shown that the existence of such a state really implies diffeomorphism invariance – we only have some indications for this so far. Moreover, as mentioned above, there are still many technical and conceptual difficulties with the definition of the category of ghysts that we have to overcome in the zeroth step of the program.

Nevertheless, once these problems are overcome, we hope that we may uncover the dynamical mechanism that leads to the noncommutativity of spacetime. At least it is to be expected that the required existence of a diffeomorphism invariant state is such a strong demand that it excludes many (commutative) ghysts from the domain of the corresponding functor. Thus in order to clarify this question it will be essential to understand in which way the morphisms of the category of ghysts $\mathcal{M}$ act on the states over the algebras $Obs(\mathcal{M})$ associated to it.

Note that, if it succeeds, the above program can really be viewed as a quantum version of the postulates that lead to Einstein's equations: Namely diffeomorphism invariance, the motion of test particles along geodesics and the existence of second order field equations.

This then leads us back to the spectral action. It is not the idea of the above program to produce a diffeomorphism invariant classical action and then to quantize it, but rather to directly infer the "Quantum Einstein equations" from first principles. Only in that way one may also hope that the dynamical mechanism leading to the noncommutativity of spacetime can be uncovered – if there exists such a mechanism. If we are able to reveal the precise nature of this noncommutativity it will turn out whether it can really be approximated at the Z-mass by the almost commutative geometry underlying the standard model.

References

[1] P.P.Marvol, *A Tip of the TOE*, [physics/0309055]

[2] M.Heller, L.Pysiak and W. Sasin, *Noncommutative Unification of General Relativity and Quantum Mechanics*, Gen.Rel.Grav. **36** (2004), 111-126

[3] S.Doplicher, K.Fredenhagen and J.Roberts, *The quantum structure of spacetime at the Planck scale and quantum fields* Comm.Math.Phys. **172** (1995), 187-220

[4] T.Kopf, M.Paschke, *A spectral quadruple for de Sitter space*, J.Math.Phys. **43** (2002), 818–846

[5] M.Paschke, *Von Nichtkommutativen Geometrien, ihren Symmetrien und etwas Hochenergiephysik*, Ph.D. thesis (2001), Mainz

[6] M.Paschke, A.Sitarz, *Discrete spectral triples and their symmetries* J.Math.Phys. **39** (1998), 6191

[7] M.Paschke, F.Scheck, A.Sitarz *Can noncommutative geometry accommodate leptoquarks?* Phys. Rev. D **59** (1999) 035003

[8] S.L.Woronowicz, *Differential calculus on compact matrix pseudogroups* Comm. Math.Phys. **122** (1989) 125

[9] T.Schücker, *Forces from Connes' geometry* hep-th/0111236

[10] M.R.Douglas, *Two Lectures on D-Geometry and Noncommutative Geometry* hep-th/9901146

[11] T.Krajewski, *Classification of finite spectral triples* J.Geom.Phys. **28** (1998) 1

[12] K.Elsner, *Elektroschwaches Modell und Standardmodell in der Nichtkommutativen Geometrie* Diploma thesis, Marburg 1999

[13] H.Figueroa, J.Gracia-Bondia, J.Varilly, *Elements of Noncommutative Geometry* Birkhäuser, 2000

[14] M.Dubois-Violette, R.Kerner, J.Madore, *Gauge bosons in a noncommutative geometry* Phys.Lett. B **217** (1989) 485

[15] A.Connes, H.Moscovici *The local index formula in noncommutative geometry* Geom.Func.Anal. **5** (1995) 174-243

[16] A.Connes, J.Lott, *Particle models and noncommutative geometry* Nucl.Phys.B (Proc.Suppl.) **18** (1990) 29-47

[17] A.Connes, *Noncommutative Geometry* Acad. Press, San Diego 1994

[18] A.Chamseddine, A.Connes *The spectral action principle* Comm.Math.Phys. **186** (1997) 731-750

[19] R.Coquereaux, G.Esposito-Farèse, F.Scheck, *Noncommutative Geometry and graded algebras in electroweak interactions* J.Mod.Phys.A **7** (1992) 6555

[20] J. Aastrup and J. M. Grimstrup, *Spectral triples of holonomy loops*, arXiv:hep-th/0503246.

[21] M. Paschke, R. Verch, *Local covariant quantum field theory over spectral geometries*, Class.Quantum Grav. **21** (2004), 5299–5316

[22] M. Paschke, R. Verch, *Globally hyperbolic noncommutative geometries*, in preparation

[23] R.Szabo *Quantum Field Theory on Noncommutative Spaces*, Phys.Rept. **378** (2003) 207-299

[24] H.-J. Matschull and M.Welling *Quantum Mechanics of a Point Particle in 2+1 Dimensional Gravity*, Class.Quant.Grav. **15** (1998) 2981-3030

[25] L.Freidel, E.R. Livine, *Effective 3d Quantum Gravity and Non-Commutative Quantum Field Theory*, hep-th/0512113

[26] J.Ambjorn, J.Jurkewicz, R.Loll, *Emergence of a 4D World from Causal Quantum Gravity*, Phys.Rev.Lett. **93** (2004) 131301

[27] O.Lauscher, M.Reuter, *Fractal Spacetime Structure in Asymptotically Safe Gravity*, JHEP **0510** (2005) 050

[28] J.Lewandowski, A.Okolow, H.Sahlmann, T.Thiemann, *Uniqueness of diffeomorphism invariant states on holonomy-flux algebras*, gr-qc/0504147

[29] C.Fleischhack, *Representations of the Weyl Algebra in Quantum Geometry*, math-ph/0407006

[30] M. Patel, *Noncommmutative theorems: Gelfand Duality, Spectral Invariant Subspace, and Pontryagin Duality*, math.OA/0503127

[31] J.-H. Jureit, T. Schücker, C.Stephan, *On a Classification of Irreducible Almost Commutative Geometries III*, J.Math.Phys. **46** (2005), 072303

[32] C.Stephan, *Almost-Commutative Geometries Beyond the Standard Model*, hep-th/0509213

[33] S.Lazzarini, T. Schücker, *A Farewell To Unimodularity*, Phys.Lett. B **510** (2001) 277-284

[34] A. Chamseddine, A.Connes, *Scale Invariance in the spectral action*, hep-th/0512169

[35] G.Landi, A.Connes, *Noncommutative Manifolds, The Instanton Algebra and Isospectral Deformations*, Comm.Math.Phys **221** (2001), 141-159

[36] D.Perrot, *Chern character, Hopf algebras, and BRS cohomology* math-ph/0210043

[37] R.Brunetti, K.Fredenhagen, R.Verch *The generally covariant locality principle – A new paradigm for local quantum physics* , Comm.Math.Phys. **237** (2003) 31-68

[38] J.Dixmier, C^*-algebras, North-Holland, 1977

Mario Paschke
Max-Planck Institute for Mathematics in the Sciences
Inselstrasse 22
04103 Leipzig
Germany
e-mail: paschke@mis.mpg.de

Quantum Gravity
B. Fauser, J. Tolksdorf and E. Zeidler, Eds., 151–159

Towards a Background Independent Formulation of Perturbative Quantum Gravity

Romeo Brunetti and Klaus Fredenhagen

Abstract. The recent formulation of locally covariant quantum field theory may open the way towards a background independent perturbative formulation of Quantum Gravity.

Mathematics Subject Classification (2000). Primary 81T20; Secondary 83C45; 81T05.

Keywords. Perturbative quantum gravity, locally covariant quantum field theory, background independence, quantum field theory as a functor.

1. Problems of perturbative Quantum Gravity

In quantum field theory the fields are defined as operator-valued distributions on a given spacetime, and many of their properties, in particular the commutativity at spacelike separated points, depend in a crucial way on properties of the background. In a perturbative approach to quantum gravity, one decomposes the metric $g_{\mu\nu}$ into a background metric $\eta_{\mu\nu}$ and a quantum field $h_{\mu\nu}$, which is treated according to standard methods in perturbation theory. The (up to now observed) effects of this quantum field are very small, hence a perturbative approach seems to be appropriate. There are, however, several obstructions which raise doubts on the validity of the perturbative approach:

1. The arising quantum field theory is nonrenormalizable [10]. Hence, infinitely many counterterms occur in the process of renormalization, and it is unclear how the arising ambiguities can be fixed [9].
2. The perturbatively constructed theory depends on the choice of the background. It is unlikely that a perturbative formulation can describe a drastic change of the background.

The two main approaches to quantum gravity try to cope with these difficulties in different ways. String theory, in a first attempt, accepts the choice of a fixed background, and aims at a more general theory where the perturbation series is finite in every order. Loop quantum gravity, on the other hand, uses a background free formulation where the degrees of freedom of gravity are directly quantized. Problems with renormalizability do not occur, but it seems to be difficult to check whether such theories describe the world as we see it.

Instead of following these routes one may take a more conservative approach and study first the influence of classical gravitational fields on quantum fields. Because of the weakness of gravitational forces this approximation is expected to have a huge range of validity. Surprisingly, this seemingly modest approach leads to many conceptual insight, and it may even lead to a new approach to quantum gravity itself [8].

In this paper we want to review the recently developped new formulation of quantum field theory on curved spacetimes, which satisfies the conditions of general covariance [3, 11, 12, 16]. We will show that the arising structure has great similarities with Segal's concept of topological quantum field theories [15] and its generalization to Riemannian spaces. It may be considered as a Lorentzian version of thisapproach. It is gratifying that the axiom of local commutativity is implied in this framework by the tensorial structure of the theory, while the time slice axiom (i.e., a form of dynamics) is related to cobordisms.

It is remarkable that the new structures emerged from the (finally successful) attempt to construct interacting theories in the sense of renormalized perturbation theory.

Up to now the complete proofs apply only to scalar field theories. The extension to gauge theories requires the control of BRST invariance. Preliminary steps in this direction have been performed [4, 5, 6], and no obstruction is visible.

More or less, the same construction then should apply to quantum gravity, treated in a background formulation. The leading idea is that background independence can be reached from a background dependent formulation provided the change of background amounts to a symmetry of the theory.

2. Locally covariant quantum field theory

We adopt the point of view [3] of algebraic quantum field theory and identify physical systems with $*$-algebras with unit (if possible, C^*-algebras) and subsystems with subalgebras sharing the same unit. In quantum field theory the subsystems can be associated to spacetime regions. Every such region may be considered as a spacetime in its own right, in particular it may be embedded into different spacetimes. It is crucial that the algebra of the region does not depend on the way it is embedded into a larger spacetime. For instance, in a Schwartzschild spacetime the physics outside the horizon should not depend on a possible extension to a Kruskal spacetime.

We formulate our requirements in form of five axioms:

1. Systems: To each time oriented globally hyperbolic spacetime M we associate a unital $*$-algebra $\mathscr{A}(M)$.

2. Subsystems: Let $\chi : M \to N$ be an isometric causality preserving embedding of globally hyperbolic spacetimes. Then, there exists a uniquely defined (injective) $*$-homomorphism $\alpha_\chi : \mathscr{A}(M) \to \mathscr{A}(N)$.
3. Covariance: If $\chi : M_1 \to M_2$ and $\chi' : M_2 \to M_3$ are embeddings as above, then $\alpha_{\chi\chi'} = \alpha_\chi \alpha_{\chi'}$.
4. Causality: If $\chi_1 : M_1 \to M$ and $\chi_2 : M_2 \to M$ are embeddings as above, such that $\chi_1(M_1)$ and $\chi_2(M_2)$ cannot be connected by a causal curve in M, then

$$\alpha_{\chi_1}(\mathscr{A}(M_1)) \vee \alpha_{\chi_2}(\mathscr{A}(M_2)) \simeq \alpha_{\chi_1}(\mathscr{A}(M_1)) \otimes \alpha_{\chi_2}(\mathscr{A}(M_2))$$

where $\vee$ indicates the generated subalgebra of $\mathscr{A}(M)$.
5. Dynamics: Let $\chi : M \to N$ be an embedding as above such that $\chi(M)$ contains a Cauchy surface of N. Then $\alpha_\chi(\mathscr{A}(M)) = \mathscr{A}(N)$.

The axioms above describe a functor $\mathscr{A}$ from the category **Loc** (the localization category) whose objects are time-oriented globally hyperbolic spacetimes and whose arrows are the causal isometric embeddings, to the category **Obs** (the observables category) whose objects are unital $*$-algebras and whose arrows are (injective) $*$-homomorphisms.

Axiom 1 is similar to the usual axiom in local quantum theories on a fixed background, where the arrow has specific (un)bounded regions on that background as a domain. Here, it is imperative to quantize *simultaneously* on all globally hyperbolic spacetimes (of the given type).

Axiom 2 may be pictured in the form

$$
\begin{array}{ccc}
M & \xrightarrow{\ \chi\ } & N \\
{\scriptstyle\mathscr{A}}\Big\downarrow & & \Big\downarrow{\scriptstyle\mathscr{A}} \\
\mathscr{A}(M) & \xrightarrow{\ \alpha_\chi\ } & \mathscr{A}(N)
\end{array}
$$

where $\alpha_\chi \doteq \mathscr{A}\chi$.[1]

Axiom 3 says that the functor $\mathscr{A}$ is covariant.

Axiom 4 may be reformulated in terms of a tensor structure. Namely, require for disjoint unions,

$$\mathscr{A}(M_1 \amalg M_2) = \mathscr{A}(M_1) \otimes \mathscr{A}(M_2) , \qquad \mathscr{A}(\emptyset) = \mathbb{C} ,$$

with $\chi_i : M_i \to M, i = 1, 2$ for which $\alpha_{\chi_1 \amalg \chi_2} = \alpha_{\chi_1} \otimes \alpha_{\chi_2}$. Let χ be a causal embedding of $M_1 \amalg M_2$ into M. Then $\chi(M_1)$ and $\chi(M_2)$ are spacelike separated, hence with $i_k : M_k \to M_1 \amalg M_2$, and with $\chi_k = \chi \circ i_k$ (see fig.1), we see that $\alpha_\chi(\mathscr{A}(M_1) \otimes \mathscr{A}(M_2))$ is equal to the algebra generated by $\alpha_{\chi_1}(\mathscr{A}(M_1))$ and $\alpha_{\chi_2}(\mathscr{A}(M_2))$, hence the causality axiom is satisfied. In short, the functor $\mathscr{A}$ is promoted to a *tensor* functor. This is very reminiscent of G. Segal's approach [15].

---

[1] Annotation by the Eds.: The diagram shown resembles not a commutative diagram. It displays the definition of the functor $\mathscr{A}$ on morphisms, it is common usage to express this by the $\doteq$ notation.

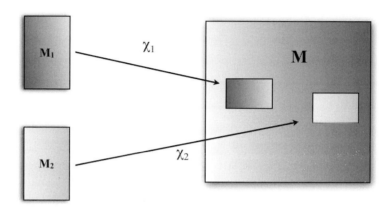

FIGURE 1. Causal Embedding

Axiom 5 may be interpreted as a description of motion of a system from one Cauchy surface to another. Namely, let N_+ and N_- be two spacetimes that embed into two other spacetimes M_1 and M_2 around Cauchy surfaces, via causal embeddings given by $\chi_{k,\pm}, k = 1, 2$. Figure 2 gives a hint.

Then $\beta = \alpha_{\chi_{1+}} \alpha_{\chi_{2+}}^{-1} \alpha_{\chi_{2-}} \alpha_{\chi_{1-}}^{-1}$ is an automorphism of $\mathscr{A}(M_1)$. One may say that in case M_1 and M_2 are equal as topological manifold but their metrics differ by a (compactly supported) symmetric tensor $h_{\mu\nu}$ with supp $h \cap J^+(N_+) \cap J^-(N_-) = \emptyset$, the automorphism depends only on the spacetime between the two Cauchy surfaces, hence in particular, on the tensor h. It can then be shown that

$$\Theta_{\mu,\nu}(x) \doteq \frac{\delta \beta_h}{\delta h_{\mu,\nu}(x)}\big|_{h=0}$$

is a derivation valued distribution which is covariantly conserved, an effect of the diffeomorphism invariance of the automorphism β_h, and may be interpreted as the commutator with the energy-momentum tensor. Indeed, in the theory of the free scalar field this has been explicitly verified [3], and it remains true in perturbatively constructed interacting theories [14].

The structure described may also be understood as a version of cobordism. Namely, one may associate to a Cauchy surface Σ of the globally hyperbolic spacetime M, the algebra

$$\mathscr{A}(\Sigma) \doteq \varprojlim_{N \supset \Sigma} \mathscr{A}(N) ,$$

where the inverse limit extends over the globally hyperbolic neighborhoods N of Σ. Clearly, $\mathscr{A}(\Sigma)$ depends only on the germ of Σ as a submanifold of M. The elements of $\mathscr{A}(\Sigma)$ are germs of families $(A_N)_{N \supset \Sigma}$ with the coherence condition $\alpha_{N_1 N_2}(A_{N_2}) = A_{N_1}$, where $\alpha_{N_1 N_2}$ is the homomorphism associated to the inclusion $N_2 \subset N_1$.

We then define a homomorphism

$$\alpha_{M\Sigma} : \mathscr{A}(\Sigma) \to \mathscr{A}(M)$$

by $\alpha_{M\Sigma}(A) \doteq \alpha_{MN}(A_N)$, $N \supset \Sigma$, where the r.h.s. is independent of the chioce of the neighborhood N. By the time slice axiom, $\alpha_{M\Sigma}$ is invertible, hence for a choice of two Cauchy hypersurfaces Σ_1, Σ_2 of M we find a homomorphism

$$\alpha^M_{\Sigma_1 \Sigma_2} : \mathscr{A}(\Sigma_2) \to \mathscr{A}(\Sigma_1) \ ,$$

with $\alpha^M_{\Sigma_1 \Sigma_2} \doteq \alpha^{-1}_{M\Sigma_1} \alpha_{M\Sigma_2}$. We may interepret Σ_1 and Σ_2 as past and future boundaries, respectively, of M and obtain for any spacetime M connecting Σ_1 and Σ_2 a homomorphism of the corresponding algebras.

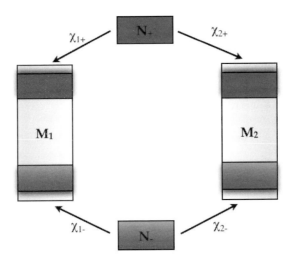

FIGURE 2. Evolution

3. Locally covariant fields

One problem with a theory on a generic spacetime is that it is not clear what it means to do the same experiment at different spacetime points. In quantum theory we are however forced to repeat the experiments in order to obtain a probability distribution. In a spacetime with a large symmetry group one may use these symmetries to compare measurements on different spacetime points. In the framework of locally covariant quantum field theory, as described above, quantum fields can serve as means for comparison of observables at different points. Namely, we define a locally covariant quantum field A as a family of algebra valued distributions (A_M) indexed by the objects $M \in \mathbf{Loc}$ which satisfy the following covariance condition

$$\alpha_\chi(A_M(x)) = A_N(\chi(x)) , \qquad \chi : M \to N .$$

More formally, one may define a locally covariant quantum field as a natural transformation from the functor $\mathscr{D}$, that associates to each manifold its test-function space, to the functor $\mathscr{A}$. Actually, we pinpoint that fields, differently from the traditional point of view, are now objects as fundamental as observables. Not only, they might even be more fundamental in cases like that of quantum gravity where *local* observables would be difficult to find (if they exist at all).

Let us look at an example. We define the theory of a real Klein-Gordon field in terms of the algebras $\mathscr{A}_0(M)$ which are generated by elements $\varphi_M(f)$, $f \in \mathscr{D}(M)$, satisfying the relations

(i) $f \to \varphi_M(f)$ is linear;
(ii) $\varphi_M(f)^* = \varphi_M(\bar{f})$;
(iii) $\varphi_M((\Box_M + m^2)f) = 0$;
(iv) $[\varphi_M(f), \varphi_M(g)] = i(f, \Delta_M g)\mathbf{1}$

where $\Delta_M = \Delta_M^{\mathrm{ret}} - \Delta_M^{\mathrm{adv}}$ with the retarded and advanced propagators of the Klein-Gordon operator, respectively. The homomorphism α_χ, $\chi : M \to N$, is induced by

$$\alpha_\chi(\varphi_M(f)) = \varphi_N(\chi_*(f))$$

where $\chi_* f$ is the push-forward of the test function f. We see immediately that $\varphi = (\varphi_M)$ is a locally covariant quantum field associated to the functor $\mathscr{A}_0$.

To find also other locally covariant fields it is convenient to enlarge the previuosly constructed algebra by neglecting the field equation (iii). Notice that the new functor $\mathscr{A}_{00}$ does no longer satisfy the time slice axiom. We then introduce localized polynomial functionals $\mathscr{F}(M)$ on the space of classical field configurations $\phi \in C^\infty(M)$,

$$\mathscr{F}(M) \ni F(\phi) = \sum_{n=0}^{\mathrm{ord}(F)} \langle f_n, \phi^{\otimes n} \rangle , \qquad \phi \in C^\infty(M)$$

where $f_n \in \mathscr{E}'_{\Gamma_n}(M^n)$, i.e., f_n is a distribution of compact support whose wave front set $\mathrm{WF}(f_n) \subset \Gamma_n$ and $\Gamma_n \cap \{(x, k) \in T^*M^n, k \in \overline{V}_+^n \cup \overline{V}_-^n\} = \emptyset$. We choose a decomposition of Δ_M, $\Delta_M(x, y) = H(x, y) - H(y, x)$, such that $\mathrm{WF}(H) = \{(x, k) \in \mathrm{WF}(\Delta_M), k \in \overline{V}_+ \times \overline{V}_-\}$. (This is a microlocal version of the decomposition into positive and negative frequencies, for explanations see, for instance, [2].) Then, we define a product on $\mathscr{F}(M)$ by

$$F *_H G = \sum_n \frac{1}{n!} \left\langle \frac{\delta^n F}{\delta\phi^n} \otimes \frac{\delta^n G}{\delta\phi^n}, H^{\otimes n} \right\rangle ,$$

which makes $\mathscr{F}(M)$ to an associative algebra $(\mathscr{F}(M), *_H)$.

But H is not unique. If we change H to $H' = H + w$, $w \in C^\infty_{\text{symm}}(M^2)$ (since the difference between two H's is always a smooth symmetric function), we find

$$\gamma_w(F *_H G) = \gamma_w(F) *_{H'} \gamma_w(G)$$

with

$$\gamma_w(F) = \sum_n \frac{1}{2^n n!} \left\langle \frac{\delta^{2n} F}{\delta\phi^{2n}}, w^{\otimes n} \right\rangle .$$

The algebra $\mathscr{A}_{00}(M)$ may be embedded into $(\mathscr{F}(M), *_H)$ by

$$\alpha_H(\varphi_M(f)) = \langle f, \phi \rangle ,$$

hence, in particular

$$\alpha_H(\varphi_M(f)\varphi_M(g)) = \langle f \otimes g, \phi^{\otimes 2} \rangle + \langle f, Hg \rangle .$$

Then $\alpha_H(\mathscr{A}_{00}(M))$ is a subalgebra of $(\mathscr{F}(M), *_H)$ with coefficients $f_n \in \mathscr{D}(M^n)$. Since the space $\mathscr{F}(M)$ is uniquely determined by the coefficients, we may equip it with the inductive topology of the direct sum of the spaces $\mathscr{E}'_{\Gamma_n}(M^n)$. Since $\mathscr{D}(M^n)$ is dense in $\mathscr{E}'_{\Gamma_n}(M^n)$, $\alpha_H(\mathscr{A}_{00}(M))$ is dense in $\mathscr{F}(M)$. We may now equip the algebra $\mathscr{A}_{00}(M)$ with the initial topologies of α_H. Since γ_w is a homeomorphism, it turns out that all induced topologies coincide, and the completion is identified with our seeked algebra $\mathscr{A}(M)$.

This algebra contains, besides the usual fields, also their normal ordered products.

Now, for the sake of constructing interacting quantum fields one may use the following steps:

1. Construction of locally covariant Wick polynomials; this turns out to the solution of a cohomological problem [3] (see also [12]) which can be solved in terms of an explicit Hadamard parametrix of the Klein Gordon equation.
2. Construction of locally covariant retarded and time ordered products; this requires the generalization of the Epstein-Glaser renormalization scheme to curved spacetime [1] and again a solution of a cohomological problem in order to be able to impose the same renormalization conditions on every point of a given spacetime and even on different spacetimes.
3. Construction of the algebras of interacting fields, together with a family of locally covariant fields [1, 5, 11, 12].

We refrain from giving details of these steps and refer to the original publications [1, 3, 5, 11, 12, 13, 14].

4. Quantization of the background

As usual, we view the metric as a background plus a fluctuation, namely,

$$g_{\mu\nu} = \eta_{\mu\nu} + h_{\mu\nu}$$

and we look at the fluctuation as a quantum field. Note that differently from other approaches the background metric does not need to be Minkowskian, we only restrict to backgrounds complying with the requirements of the previous sections.

So we have a quantum field h that propagates via the linearized Einstein equations on a fixed background η.

One may now proceed by the general strategy for constructing gauge theories by the BRST method. It is crucial that this method can be adapted to localized interactions, as was done in [4, 6]. Furthermore, the freedom in renormalization can be used to arrive at quantized metric and curvature fields satisfying Einstein's equation.

The condition of background independence may be formulated as the condition that the automorphism β_κ describing the relative Cauchy evolution corresponding to a change of the background between two Cauchy surfaces must be trivial. In perturbation theory, it is sufficient to check the infinitesimal version of this condition. This amounts to the equation

$$\frac{\delta\beta_\kappa}{\delta\kappa_{\mu\nu}(x)} = 0 \ .$$

In contrast to the situation for an unquantized background metric the left hand side involves in addition to the energy momentum tensor also the Einstein tensor. Hence the validity of Einstein's equation for the quantized fields should imply background independence.

There remain, of course, several open questions. First of all, the details of the proposal above have to be elaborated, and in particular the question of BRST invariance has to be checked. A possible obstruction could be that locally the cohomology of the BRST operator is trivial, corresponding to the absense of local observables in quantum gravity. Another problem is the fact that the theory is not renormalizable by power counting. Thus the theory will have the status of an effective theory. Nevertheless, due to the expected smallness of higher order counter terms the theory should still have predictive power.

References

[1] Brunetti, R., Fredenhagen, K., Microlocal analysis and interacting quantum field theories: Renormalization on physical backgrounds, *Commun. Math. Phys.* **208** (2000) 623.

[2] Brunetti, R., Fredenhagen, K., Köhler, M., The microlocal spectrum condition and Wick polynomials of free fields on curved spacetimes," *Commun. Math. Phys.* **180** (1996) 633.

[3] Brunetti, R., Fredenhagen, K., Verch, R., The generally covariant locality principle – A new paradigm for local quantum physics, *Commun. Math.Phys.* **237** (2003) 31.

[4] Dütsch, M., Fredenhagen, K., A local (perturbative) construction of observables in gauge theories: The example of QED, *Commun. Math. Phys.* **203** (1999) 71.

[5] Dütsch, M., Fredenhagen, K., Causal perturbation theory in terms of retarded products, and a proof of the Action Ward Identity, *Rev. Math. Phys.* **16** (2004) 239.

[6] Dütsch, M., Fredenhagen, K., Perturbative renormalization and BRST, preprint hep-th/0411196.

[7] Epstein, H., Glaser, V., "The role of locality in perturbation theory, *Ann. Inst. H. Poincaré A* **19** (1973) 211.

[8] Fredenhagen, K., Locally Covariant Quantum Field Theory," preprint hep-th/0403007. Proceedings of the XIVth International Congress on Mathematical Physics, Lisbon 2003.

[9] Gomis, J., Weinberg, S., Are nonrenormalizable gauge theories renormalizable?," *Nucl. Phys.* **B 469** (1996) 473.

[10] Goroff, M., Sagnotti, A., The ultraviolet behaviour of Einstein gravity, *Nucl.Phys.* **B266** (1986) 709.

[11] Hollands, S., Wald, R. M., Local Wick Polynomials and Time-Ordered-Products of Quantum Fields in Curved Spacetime, *Commun. Math. Phys.* **223** (2001) 289.

[12] Hollands, S., Wald, R. M., Existence of Local Covariant Time-Ordered-Products of Quantum Fields in Curved Spacetime, *Commun. Math. Phys.* **231** (2002) 309.

[13] Hollands, S., Wald, R. M., On the Renormalization Group in Curved Spacetime, *Commun. Math. Phys.* **237** (2003) 123.

[14] Hollands, S., Wald, R. M., Conservation of the stress tensor in interacting quantum field theory in curved spacetimes, *Rev. Math. Phys.* **17** (2005) 227.

[15] Segal, G., Two dimensional conformal field theory and modular functors, in IXth International Congress on Mathematical Physics, Simon, Truman, Davies (eds.), Adam Hilger, Bristol 1989: and talk given at the XIVth Oporto Meeting on Geometry, Topology and Physics, held in Oporto (Portugal) in 21-24 july 2005.

[16] Verch, R., A spin-statistics theorem for quantum fields on curved spacetime manifolds in a generally covariant framework, *Commun. Math. Phys.* **223** (2001) 261.

Romeo Brunetti and Klaus Fredenhagen
II Institute für Theoretische Physik
Universität Hamburg
Luruper Chaussee 149, D-22761 Hamburg
Germany
e-mail: `romeo.brunetti@desy.de`
          `klaus.fredenhagen@desy.de`

Quantum Gravity
B. Fauser, J. Tolksdorf and E. Zeidler, Eds., 161–201
© 2006 Birkhäuser Verlag Basel/Switzerland

Mapping-Class Groups of 3-Manifolds in Canonical Quantum Gravity

Domenico Giulini

This contribution is dedicated to Rafael Sorkin on the occasion of his 60th birthday

Abstract. Mapping-class groups of 3-manifolds feature as symmetry groups in canonical quantum gravity. They are an obvious source through which topological information could be transmitted into the quantum theory. If treated as gauge symmetries, their inequivalent unitary irreducible representations should give rise to a complex superselection structure. This highlights certain aspects of spatial diffeomorphism invariance that to some degree seem physically meaningful and which persist in all approaches based on smooth 3-manifolds, like geometrodynamics and loop quantum gravity. We also attempt to give a flavor of the mathematical ideas involved.

Mathematics Subject Classification (2000). Primary 83C45; Secondary 57N10.

Keywords. Canonical quantum gravity, 3-manifolds, mapping classes, diffeomorphism invariance, residual finiteness, super selection sectors, spinorial primes, geons.

1. Some facts about Hamiltonian general relativity

1.1. Introduction

As is well known, Einstein's field equations for General Relativity can be cast into the form of a constrained Hamiltonian system. The unreduced phase space is then given by the cotangent bundle over the space $\text{Riem}(\Sigma)$ of Riemannian metrics on a 3-manifold Σ. This phase space is coordinatized by (q, p), where q is a Riemannian metric on Σ and p is a section in the bundle of symmetric contravariant tensors of rank two and density-weight one over Σ.

I sincerely thank Bertfried Fauser and Jürgen Tolksdorf for organizing Blaubeuren II and inviting me to it.

The relation of these objects to the description of a solution to Einstein's equations in terms of a four-dimensional globally hyperbolic Lorentzian manifold (g, M) is as follows: let Σ be a Cauchy surface in M and $q := g|_{T(M)}$ its induced metric (first fundamental form). Let n be the normal field to Σ, where $g(n, n) = -1$, i.e. we use the 'mostly plus' signature convention where timelike vectors have a negative g-square. Let D and ∇ be the Levi-Civita covariant derivatives on (q, Σ) and (g, M) respectively. Then for any vector X tangent to Σ and any vector field $\tilde{Y}$ on M whose restriction Y to Σ is tangential to Σ, we have

$$\nabla_X \tilde{Y} = D_X Y + n\, K(X, Y)\,, \tag{1.1}$$

were $n\, K(X, Y)$ represents the normal component of $\nabla_X \tilde{Y}$. It is easy to see that K is a symmetric covariant tensor of rank 2 on Σ, also called the extrinsic curvature (or second fundamental form) of Σ in M. The canonical momentum field p can now be expressed in terms of these data:

$$p = \sqrt{\det(q)} \left(K - q\, \mathrm{Tr}_q(K) \right)^{\sharp}\,. \tag{1.2}$$

Here $\sharp$ denotes the isomorphism $T^*(M) \to T(M)$ induced by q ('index raising'), extended to tensors of all ranks.

Einstein's equations in Hamiltonian form now decompose into two evolution equations (of six independent component equations each):

$$\dot{q} = F_1(q, p; \alpha, \beta; \phi)\,, \tag{1.3a}$$

$$\dot{p} = F_2(q, p; \alpha, \beta; \phi)\,, \tag{1.3b}$$

and two equations without time derivatives in q and p (of one and three independent component equations respectively), thereby implying constraints on the initial data:

$$C_s(q, p; \phi) := G_q(p, p) - \sqrt{\det(q)} \left(S(q) - 2\Lambda \right) + 2\rho_m(\phi, q) = 0\,, \tag{1.4a}$$

$$C_v(q, p; \phi) := -2\mathrm{div}_q p + j_m(\phi, q) = 0\,. \tag{1.4b}$$

These are referred to as the scalar and the vector (or diffeomorphism) constraints respectively.

The meanings of the symbols in (1.3) and (1.4) are as follows: $F_{1,2}$ are local functionals whose explicit forms need not interest us at this moment. α and β are a scalar function and a vector field on Σ respectively (the 'lapse' and 'shift' function) which are not determined by the equations of motion but which one needs to specify by hand. They represent the four free functions out of the ten component functions $g_{\mu\nu}$ which are *not* determined by the equations of motion due to spacetime-diffeomorphism invariance. $S(q)$ is the scalar curvature (Ricci scalar) of (q, Σ) and div_q denotes the covariant divergence with respect to the Levi Civita derivative on (q, Σ), i.e. in components $(\mathrm{div}_q p)^b = D_a p^{ab}$. The symbol ϕ collectively represents possible matter fields. ρ_m and j_m are respectively the scalar and vector densities of weight one for the energy and momentum density of the matter. As usual, Λ is the cosmological constant. Finally, G_q is a bilinear form, the so-called DeWitt metric, that maps a pair of symmetric contravariant 2nd-rank tensors of density weight one to a scalar density of weight one. In components one has

$$G_q(p_1, p_2) = \left[\det(q_{ab})\right]^{-1/2} \tfrac{1}{2}\left(q_{ac}q_{bd} + q_{ad}q_{bc} - q_{ab}q_{cd}\right)p_1^{ab}p_2^{cd}$$
$$= \left[\det(q_{ab})\right]^{-1/2}\left(p_1^{ab}p_{2\,ab} - \tfrac{1}{2}p_1^{a}{}_{a}p_2^{b}{}_{b}\right). \tag{1.5}$$

Pointwise (on Σ) it defines a Lorentzian metric of signature $(1,5)$ (the 'negative direction' being the trace mode) on the six dimensional space of symmetric 2nd-rank tensor densities, a discussion of which may be found in [10]. Some relevant aspects of the infinite-dimensional geometry that is obtained by integrating $G_q(p_1, p_2)$ over Σ (the so-called 'Wheeler-DeWitt metric') are discussed in [27].

1.2. Topologically closed Cauchy surfaces

If Σ is closed (compact without boundary) the constraints (1.4) actually generate all of the dynamical evolution (1.3). That is, we can write

$$F_1 = \frac{\delta H}{\delta p}, \tag{1.6a}$$

$$F_2 = -\frac{\delta H}{\delta q}, \tag{1.6b}$$

where

$$H[q, p; \alpha, \beta; \phi] = \int_\Sigma \alpha C_s(q, p; \phi) + \int_\Sigma \beta \cdot C_v(q, p; \phi). \tag{1.7}$$

Here $\beta \cdot C_v$ denote the natural pairing between a vector (β) and a one-form of density weight one (C_v). This means that the entire dynamical evolution is generated by constraints. These constraints form a first-class system, which means that the Poisson bracket of two of them is again a linear combination (generally with phase-space dependent coefficients) of the constraints. Writing

$$C_s(\alpha) := \int_\Sigma \alpha C_s, \quad \text{and} \quad C_v(\beta) := \int_\Sigma \beta \cdot C_v \tag{1.8}$$

we have

$$\{C_s(\alpha), C_s(\alpha')\} = C_v\left(\alpha(d\alpha')^\sharp - \alpha'(d\alpha)^\sharp\right), \tag{1.9a}$$

$$\{C_v(\beta), C_s(\alpha)\} = C_s\left(\beta \cdot d\alpha\right), \tag{1.9b}$$

$$\{C_v(\beta), C_v(\beta')\} = C_v\left([\beta, \beta']\right). \tag{1.9c}$$

In passing we remark that (1.9c) says that the vector constraints form a Lie-subalgebra which, however, according to (1.9b), is *not* an ideal. This means that the flows generated by the scalar constraints are not tangential to the constraint-hypersurface that is determined by the vanishing of the vector constraints, except for the points where the constraint-hypersurfaces for the scalar and vector constraints intersect. This means that generally one cannot reduce the constraints in steps: *first* the vector constraints and *then* the scalar constraints, simply because the scalar constraints do not act on the solution space of the vector constraints.

This difficulty clearly persists in any implementation of (1.9) as operator constraints in canonical quantum gravity.

According to the orthodox theory of constrained systems [11, 47], all motions generated by the *first class* constraints should be interpreted as gauge transformations, i.e. be physically unobservable.[1] In other words, states connected by a motion that is generated by first-class constraints are to be considered as *physically identical*.

The conceptual question of how one should interpret the fact that all evolution is pure gauge is know as the *problem of time* in classical and also in quantum general relativity. It is basically connected with the constraints $C_s(\alpha)$, since their Hamiltonian flow represents a change on the canonical variables that corresponds to the motion of the hypersurface Σ in M in normal direction. For a detailed discussion see Section 5.2 in [52].

In contrast, the meaning of the flow generated by the constraints $C_v(\beta)$ is easy to understand: it just corresponds to an infinitesimal diffeomorphism within Σ. Accordingly, its action on a local[2] phase-space function $F[q,p](x)$ is just given by its Lie derivative:

$$\{F, C_v(\beta)\} = L_\beta F. \tag{1.10}$$

Hence the gauge group generated by the vector constraints is the identity component $\text{Diff}^0(\Sigma)$ of the diffeomorphism group $\text{Diff}(\Sigma)$ of Σ. Note that this is true despite the fact that $\text{Diff}(\Sigma)$ is only a Fréchet Lie group and that, accordingly, the exponential map is not surjective on any neighborhood of the identity (cf. [17]). The point being that $\text{Diff}^0(\Sigma)$ is simple (cf. [63]) and that the subgroup generated[3] by the image of the exponential map is clearly a non-trivial normal subgroup of, and hence equivalent to, $\text{Diff}^0(\Sigma)$.

What about those transformations in $\text{Diff}(\Sigma)$ which are not in the identity component (i.e. the so called 'large' diffeomorphisms)? Are they, too, to be looked at as pure gauge transformations, or are they physically meaningful (observable) symmetries? Suppose we succeeded in constructing the reduced phase space with

---

[1]In [11] Dirac proposed this in the form of a conjecture. It has become the orthodox view that is also adopted in [47]. There are simple and well known—though rather pathological— counterexamples (e.g. [47] § 1.2.2 and § 1.6.3). The conjecture can be proven under the hypothesis that there are no *ineffective* constraints, i.e. whose Hamiltonian vector fields vanish on the constraint hypersurface (e.g. [8]; also [47] § 3.3.2). For further discussion of this rather subtle point see [33] and [34]. Note however that these issues are only concerned with the algebraic form in which the constraints are delivered by the formalism. For example, if $\phi(q,p) = 0$ is effective (i.e $d\phi|_{\phi^{-1}(0)} \neq 0$) then $\phi^2(q,p) = 0$ is clearly ineffective, even though it defines the *same* constraint subset in phase space. In a proper geometric formulation the algebra of observables just depends on this constraint subset: define the 'gauge Poisson-algebra', Gau, by the set of all smooth functions that vanish on this set (it is clearly an ideal with respect to pointwise multiplication, but not a Lie-ideal). Then take as algebra of physical observables the quotient of the Lie idealizer of Gau (in the Poisson algebra of, say, smooth functions on unconstrained phase space) with respect to Gau. See e.g. [30] for more details.

[2]'Local' meaning that the real-valued function F on Σ depends on $x \in \Sigma$ through the values of q and p as well their derivatives up to *finite* order at x.

[3]The subgroup 'generated' by a set is the subgroup of all finite products of elements in this set.

respect to $\mathrm{Diff}^0(\Sigma)$, we would then still have a residual non-trivial action of the discrete group

$$\mathcal{G}(\Sigma) := \mathrm{Diff}(\Sigma)/\mathrm{Diff}^0(\Sigma) =: \pi_0\big(\mathrm{Diff}(\Sigma)\big). \tag{1.11}$$

Would we then address as physical *observables* only those functions on phase space which are invariant under $\mathcal{G}(\Sigma)$? The answer to this questions may well depend on the specific context at hand. But since $\mathcal{G}(\Sigma)$ is generically a non-abelian and infinite group, the different answers will have significant effect on the size and structure of the space of physical states and observables. A 2+1 dimensional model where this has been studied in some detain is presented in [24].

Reducing the configuration space $\mathrm{Riem}(\Sigma)$ by the action of $\mathrm{Diff}(\Sigma)$ leads to what is called 'superspace'[4]:

$$\mathcal{S} = \mathrm{Riem}(\Sigma)/\mathrm{Diff}(\Sigma). \tag{1.12}$$

It can be given the structure of a stratified manifold [13], where the nested singular sets are labeled by the isotropy groups of $\mathrm{Diff}(\Sigma)$ (i.e. the singular sets are the geometries with non-trivial isotropy group and nested according to the dimensionality of the latter.)

There is a natural way to resolve the singularities of $\mathcal{S}$ [14], which can be described as follows: pick a point $\infty \in \Sigma$ (we shall explain below why the point is given that name) and consider the following subgroups of $\mathrm{Diff}(\Sigma)$ that fix ∞ and frames at ∞ respectively:

$$\mathrm{Diff}_\infty(\Sigma) := \big\{\phi \in \mathrm{Diff}(\Sigma) : \phi(\infty) = \infty\big\}, \tag{1.13a}$$

$$\mathrm{Diff}_\mathrm{F}(\Sigma) := \big\{\phi \in \mathrm{Diff}_\infty(\Sigma) : \phi_*|_\infty = \mathrm{id}|_{T_\infty(\Sigma)}\big\}. \tag{1.13b}$$

The resolved Superspace, $\mathcal{S}_\mathrm{F}$, is then isomorphic to[5]

$$\mathcal{S}_\mathrm{F} := \mathrm{Riem}(\Sigma)/\mathrm{Diff}_\mathrm{F}(\Sigma). \tag{1.14}$$

The point is that $\mathrm{Diff}_\mathrm{F}(\Sigma)$ acts freely on $\mathrm{Riem}(\Sigma)$ due to the fact that diffeomorphisms that fix the frames at one point cannot contain non-trivial isometries.[6] This, as well as the appropriate slicing theorems for the surjection $\mathrm{Riem}(\Sigma) \to \mathcal{S}_\mathrm{F}$ (which already holds for the action of $\mathrm{Diff}(\Sigma)$, see [13] and references therein) then establish a manifold structure of $\mathcal{S}_\mathrm{F}$. In fact, we have a principle bundle

$$\mathrm{Diff}_\mathrm{F}(\Sigma) \overset{i}{\rightarrowtail} \mathrm{Riem}(\Sigma) \overset{p}{\twoheadrightarrow} \mathcal{S}_\mathrm{F}. \tag{1.15}$$

The contractibility of $\mathrm{Riem}(\Sigma)$ (which is an open convex cone in a topological vector space) implies that $\mathrm{Riem}(\Sigma)$ is a universal classifying bundle and $\mathcal{S}_\mathrm{F}$ a universal classifying space for the group $\mathrm{Diff}_\mathrm{F}(\Sigma)$. It also implies, via the long exact homotopy sequence for (1.15), that

$$\pi_n\big(\mathrm{Diff}_\mathrm{F}(\Sigma)\big) \cong \pi_{n+1}\big(\mathcal{S}_\mathrm{F}\big) \qquad \text{for} \quad n \geq 0. \tag{1.16}$$

---

[4]This notion is independent to similarly sounding ones in the theory of supersymmetric field theories

[5]The isomorphism is non canonical since we had to select a point $\infty \in \Sigma$.

[6]To see this, assume $\phi \in \mathrm{Diff}_\mathrm{F}(\Sigma)$ is an isometry of (q, Σ), where Σ is connected. The set of fixed points is clearly closed. It is also open, as one readily sees by using the exponential map.

Recall that π_0 of a topological group G is a group (this is not true for arbitrary topological spaces) which is defined by G/G^0, where G^0 is the identity component. Hence we have

$$\mathcal{G}_{\mathrm{F}}(\Sigma) := \mathrm{Diff}_{\mathrm{F}}(\Sigma)/\mathrm{Diff}_{\mathrm{F}}^0(\Sigma) =: \pi_0\big(\mathrm{Diff}_{\mathrm{F}}(\Sigma)\big) \cong \pi_1\big(\mathcal{S}_{\mathrm{F}}\big). \qquad (1.17)$$

In this way we recognize the mapping-class group for frame-fixing diffeomorphisms of Σ as the fundamental group of the singularity-resolved $\mathrm{Diff}(\Sigma)$–reduced configuration space of canonical gravity. Next to (1.11) and (1.17) we also introduce the analogous mapping class groups for point-fixing diffeomorphisms:

$$\mathcal{G}_{\infty}(\Sigma) := \mathrm{Diff}_{\infty}(\Sigma)/\mathrm{Diff}_{\infty}^0(\Sigma) =: \pi_0\big(\mathrm{Diff}_{\infty}(\Sigma)\big). \qquad (1.18)$$

1.3. Topologically open Cauchy surfaces

So far we assumed Σ to be closed. This is the case of interest in cosmology. However, in order to model isolated systems, one is interested in 3-manifolds Σ' with at least one asymptotically flat end, where here we restrict to the case of one end only. The topological implication behind 'asymptotical flatness' is simply the requirement that the one-point compactification $\Sigma = \Sigma' \cup \infty$ (here ∞ is the point added) be a manifold. This is equivalent to the existence of a compact set $K \subset \Sigma'$ such that $\Sigma' - K$ is homeomorphic to $S^2 \times \mathbb{R}$ (i.e. $\mathbb{R}^3 -$ ball).

The analytic expressions given in Section 1.1 made no reference to whether Σ is open or closed. In particular, the constraints are still given by (1.4). However, in the open case it is not true anymore that the dynamical evolution is entirely driven by the constraints, as in (1.6) and (1.7). Rather, we still have (1.6) but must change (1.7) to[7]

$$H[q, p; \alpha, \beta; \phi] = \lim_{R \to \infty} \left\{ \int_{B_R} \alpha C_s(q, p; \phi) + \int_{B_R} \beta \cdot C_v(q, p; \phi) \right.$$
$$\left. + \int_{S_R} \mathcal{E}(\alpha; q, p) \quad + \int_{S_R} \mathcal{M}(\beta; q, p) \right\}. \qquad (1.19)$$

Here B_R is a sequence of compact sets, labeled by their 'radius' R, so that $R' > R$ implies $B_{R'} \supset B_R$ and $\lim_{R \to \infty} B_R = \Sigma'$. S_R is equal to the boundary ∂B_R, which we assume to be an at least piecewise differentiable embedded 2-manifold in Σ'. $\mathcal{E}$ and $\mathcal{M}$ are the fluxes for energy and linear momentum if asymptotically for large R the lapse function α assumes the constant value 1 and β approaches a translational Killing vector. Correspondingly, if β approaches a rotational Killing vector, we obtain the flux for angular momentum (see [6] for the analytic expressions in case of pure gravity). Since the constraints (1.9) must still be satisfied as part of Einstein's equations, we see that 'on shell' the Hamiltonian (1.19) is a sum of surface integrals. Note also that even though the surface integrals do not explicitly depend on the matter variables ϕ, as indicated in (1.19), there is an implicit dependence through the requirement that (q, p, ϕ) satisfy the constraints (1.9). This must be so since

---

[7]Here we neglect other surface integrals that arise in the presence of gauge symmetries other than diffeomorphism invariance whenever globally charged states are considered.

these surface integrals represent the total energy and momentum of the system, including the contributions from the matter.

Let us consider the surface integral associated with the spatial vector field β. It is given by

$$P(q, p; \beta) := 2 \lim_{R \to \infty} \left\{ \int_{S_R} p(n^b, \beta^b) \right\} \tag{1.20}$$

where n is the outward pointing normal of S_R in Σ' and $n^b := q(n, \cdot)$ etc. It is precisely minus the surface integral that emerges by an integration by parts from the second integral on the right hand side of (1.7) and which obstructs functional differentiability with respect to p. The addition of (1.20) to (1.7) just leads to a cancellation of both surface integrals thereby restoring functional differentiability for non-decaying β. This is precisely what was done in (1.19). Conversely, this shows that the constraint $C_v(\beta)$ (cf. (1.8)) only defines a Hamiltonian vector field if β tends to zero at infinity. Hence the constraints C_v only generate asymptotically trivial diffeomorphisms. The rate of this fall-off is of crucial importance for detailed analytical considerations, but is totally unimportant for the topological ideas we are going to present. For our discussion it is sufficient to work with $\Sigma = \Sigma' \cup \infty$. In particular, the group of spatial diffeomorphisms generated by the constraints may again be identified with $\mathrm{Diff}_F^0(\Sigma)$. This is true since we are only interested in homotopy invariants of the diffeomorphism group[8] and the group of diffeomorphisms generated by the constraints is homotopy equivalent to $\mathrm{Diff}_F^0(\Sigma)$, whatever the precise fall-off conditions for the fields on Σ' are. Moreover, the full group of diffeomorphisms, $\mathrm{Diff}(\Sigma')$, is homotopy equivalent to $\mathrm{Diff}_\infty(\Sigma)$.[9]

To sum up, the configuration space topology in Hamiltonian General Relativity is determined by the topology of $\mathrm{Diff}_F(\Sigma)$, where Σ is a closed 3-manifold. This is true in case the Cauchy surface is Σ and also if the Cauchy surface is open with one regular end, in which case Σ is its one-point compactification. In particular, the fundamental group of configuration space is isomorphic to the mapping-class groups (1.17). This is the object we shall now focus attention on. It has an obvious interest for the quantization program: for example, it is well known from elementary Quantum Mechanics that the inequivalent irreducible unitary representations of the fundamental group of the classical configuration space (the domain of the Schrödinger function) label inequivalent quantum sectors; see e.g. [26] and references therein. Even though in field theory it is not true that the classical configuration space is the proper functional-analytic domain of the Schrödinger state-functional, it remains true that its fundamental group—$\mathcal{G}_F(\Sigma)$ in our case—acts as group of (gauge) symmetries on the space of quantum states. Hence one is naturally interested in the structure and the representations of such groups. Some applications of these concepts in quantum cosmology and 2+1 quantum gravity

---

[8] We already alert to the fact that homotopy invariants of the groups of homeomorphisms or diffeomorphisms of a 3-manifold Σ are topological invariants of Σ but *not* necessarily also homotopy invariants of Σ. We will come back to this below.

[9]To see this one needs, in particular, to know that $\mathrm{Diff}(S^2)$ is homotopy equivalent to $SO(3)$ [75].

may be found in [31] [24] respectively. The whole discussion on θ-sectors[10] in quantum gravity started in 1980 with the seminal paper [18] by John Friedman and Rafael Sorkin on the possibility of spin-1/2 states in gravity. We will discuss this below. Further discussion of mapping-class groups as symmetry groups in canonical quantum gravity and their physical relevance are given in [50, 76, 78, 2].

In analogy to standard gauge theories of Yang-Mills type, one may speculate that the higher homotopy groups of $\mathrm{Diff}_\mathrm{F}(\Sigma)$ are also of physical significance, e.g. concerning the question of various types of anomalies [1, 68]. Such groups may be calculated for large classes of prime 3-manifolds [25] (the concept of a prime manifold is explained below), but not much seems to be known in the general reducible case.[11]

As already pointed out (cf. footnote 8), the homotopy invariants of $\mathrm{Diff}_\mathrm{F}(\Sigma)$ are topological but not necessarily also homotopy invariants of Σ (cf. [60]). For example, if Σ is a spherical space form, that is $\Sigma \cong S^3/G$ with finite $G \subset SO(4)$, the mapping class group $\mathcal{G}_\mathrm{F}(\Sigma)$ often fully characterizes Σ and can even sometimes distinguish two non-homeomorphic Σ's which are homotopy equivalent. The latter happens in case of lens spaces, $L(p,q)$, where generally p and q denote any pair of coprime integers. Their mapping-class groups[12] are as follows: 1.) $\mathcal{G}_\mathrm{F}(\Sigma) \cong \mathbb{Z}_2 \times \mathbb{Z}_2$ if $q^2 = 1 \,(\mathrm{mod}\,p)$ and $q \neq \pm 1 \,(\mathrm{mod}\,p)$, 2.) $\mathcal{G}_\mathrm{F}(\Sigma) \cong \mathbb{Z}_4$ if $q^2 = -1 \,(\mathrm{mod}\,p)$, and 3.) $\mathcal{G}_\mathrm{F}(\Sigma) \cong \mathbb{Z}_2$ in the remaining cases; see Table II. p. 581 in [85]. On the other hand, it is known that two lens spaces $L(p,q)$ and $L(p,q')$ are homeomorphic iff[13] $q' = \pm q \,(\mathrm{mod}\,p)$ or $qq' = \pm 1 \,(\mathrm{mod}\,p)$ [70] and homotopy equivalent iff qq' or $-qq'$ is a quadratic residue mod p, i.e. iff $qq' = \pm n^2 \,(\mathrm{mod}\,p)$ for some integer n [84]. For example, this implies that $L(15,1)$ and $L(15,4)$ are homotopy equivalent but not homeomorphic and that the mapping class group of $L(15,1)$ is $\mathbb{Z}_2$ whereas that of $L(15,4)$ is $\mathbb{Z}_2 \times \mathbb{Z}_2$. Further distinctions can be made using the fundamental group of $\mathrm{Diff}_\mathrm{F}(\Sigma)$. See the table on p. 922 in [23] for more information.

---

[10]θ symbolically stands for the parameters that label the equivalence classes of irreducible unitary representations. This terminology is borrowed from QCD, where the analog of $\mathrm{Diff}_\mathrm{F}(\Sigma)$ is the group of asymptotically trivial $SU(3)$ gauge-transformations, whose associated group of connected components—the analog of $\mathrm{Diff}_\mathrm{F}(\Sigma)/\mathrm{Diff}_\mathrm{F}^0(\Sigma)$—is isomorphic to $\pi_3(SU(3)) \cong \mathbb{Z}$. The circle-valued parameter θ then just labels the equivalence classes of irreducible unitary representations of that $\mathbb{Z}$.

[11]For example, for connected sums of three or more prime manifolds, the fundamental group of the group of diffeomorphisms is not finitely generated [51].

[12]A special feature of lens spaces is that $\mathcal{G}(\Sigma) \cong \mathcal{G}_\infty(\Sigma) \cong \mathcal{G}_\mathrm{F}(\Sigma)$, where $\Sigma = L(p,q)$. These groups are also isomorphic to $\mathrm{Isom}(\Sigma)/\mathrm{Isom}^0(\Sigma)$, where Isom denotes the group of isometries with respect to the metric of constant positive curvature. The property $\mathcal{G}(\Sigma) \cong \mathrm{Isom}(\Sigma)/\mathrm{Isom}^0(\Sigma)$ is known to hold for many of the spherical space forms; an overview is given in [23] (see the table on p. 922). It is a weakened form of the *Hatcher Conjecture* [38], which states that the inclusion of $\mathrm{Isom}(\Sigma)$ into $\mathrm{Diff}(\Sigma)$ is a homotopy equivalence for all spherical space forms Σ. The Hatcher conjecture generalizes the Smale conjecture [75] (proven by Hatcher in [41]), to which it reduces for $\Sigma = S^3$.

[13]Throughout we use 'iff' for 'if and only if'.

2. 3-Manifolds

It is well known (e.g. [86]) that Einstein's equations (i.e. the constraints) pose no topological obstruction to Σ. Hence our Σ can be any closed 3-manifold. For simplicity (and no other reason) we shall exclude non-orientable manifolds and shall from now on simply say '3-manifold' if we mean *closed oriented* 3-manifold.

The main idea of understanding a general 3-manifold is to decompose it into simpler pieces by cutting it along embedded surfaces. Of most interest for us is the case where one cuts along 2-spheres, which results in the so-called *prime decomposition*. The inverse process, where two 3-manifolds are glued together by removing an embedded 3-disc in each of them and then identifying the remaining 2-sphere boundaries in an orientation reversing (with respect to their induced orientations) way is called *connected sum*. This is a well defined operation in the sense that the result is independent (up to homeomorphisms) of 1.) how the embedded 3-discs where chosen and 2.) what (orientation reversing) homeomorphism between 2-spheres is used for boundary identification (this is nicely discussed in § 10 of [7]). We write $\Sigma_1 \uplus \Sigma_2$ to denote the connected sum of Σ_1 with Σ_2. The connected sum of a 3-manifold Σ with a 3-sphere, S^3, is clearly homeomorphic to Σ.

Let us now introduce some important facts and notation. The classic source is [43], but we also wish to mention the beautiful presentation in [39]. Σ is called *prime* if $\Sigma = \Sigma_1 \uplus \Sigma_2$ implies that Σ_1 or Σ_2 is S^3. Σ is called *irreducible* if every embedded 2-sphere bounds a 3-disk. Irreducibility implies primeness and the converse is almost true, the only exception being the *handle*, $S^1 \times S^2$, which is prime but clearly not irreducible (no $p \times S^2$ bounds a 3-disc). Irreducible prime manifolds have vanishing second fundamental group.[14] The converse is true if every embedded 2-sphere that bounds a homotopy 3-disk also bounds a proper 3-disk; in other words, if fake 3-disks do not exist, which is equivalent to the Poincaré conjecture. So, if the Poincaré conjecture holds, a (closed orientable) 3-manifold P is prime iff either $\pi_2(P) = 0$ or $P = S^1 \times S^2$.

Many examples of irreducible 3-manifolds are provided by space forms, that is, manifolds which carry a metric of constant sectional curvature. These manifolds are covered by either S^3 or $\mathbb{R}^3$ and hence have trivial π_2.

- Space forms of positive curvature (also called 'spherical space forms') are of the form S^3/G, where G is a finite freely acting subgroup of $SO(4)$. Next to the cyclic groups $\mathbb{Z}_p$ these G e.g. include the $SU(2)$ double covers of the symmetry groups of n-prisms, the tetrahedron, the octahedron, and the icosahedron, as well as direct products of those with cyclic groups

---

[14]Note that this is *not* obvious from the definition of irreducibility, since non-zero elements of π_2 need not be representable as *embedded* 2-spheres. However, the *sphere theorem* for 3-manifolds (see Thm. 4.3 in [43]) implies that at least *some* non-zero element in π_2 can be so represented if π_2 is non-trivial.

of relatively prime (coprime) order.[15] For example, the *lens spaces* $L(p,q)$, where q is coprime to p, are obtained by letting the generator of $\mathbb{Z}_p$ act on $S^3 = \{(z_1, z_2) \in \mathbb{C}^2 : |z_1|^2 + |z_2|^2 = 1\}$ by $(z_1, z_2) \mapsto (rz_1, r^q z_2)$ with $r = \exp(2\pi i/p)$. See e.g. [85] for explicit presentations of the groups $G \subset SO(4)$.

- The flat space forms are of the form $\mathbb{R}^3/G$ where $\mathbb{R}^3$ carries the Euclidean metric and G is a freely, properly-discontinuously acting subgroups of the group $E_3 = \mathbb{R}^3 \rtimes O(3)$ of Euclidean motions. There are six such groups leading to orientable compact quotients (see [87], Thm. 3.5.5 and Cor. 3.5.10): the lattice $\mathbb{Z}^3 \subset \mathbb{R}^3$ of translations and five finite downward extensions[16] of it by $\mathbb{Z}_2$, $\mathbb{Z}_3$, $\mathbb{Z}_4$, $\mathbb{Z}_6$, and $\mathbb{Z}_2 \times \mathbb{Z}_2$. These gives rise to the 3-torus T^3 and five spaces regularly covered by it.

- Space forms of negative curvature (also called 'hyperbolic space forms') are given by H^3/G, where $H^3 = \{(t, \vec{x}) \in \mathbb{R}^{(1,3)} : t = \sqrt{\vec{x}^2 + 1}\}$ is the hyperbola in Minkowski space and G is a freely, properly-discontinuously acting subgroup of the Lorentz group $O(1,3)$ leading to orientable compact quotients. They are much harder to characterize explicitly.

Flat and hyperbolic space forms are covered by $\mathbb{R}^3$ so that all their homotopy groups higher than the first are trivial. The class of topological spaces for which $\pi_k = \{0\}$ for all $k > 1$ is generally called $K(\pi, 1)$ (Eilenberg MacLane spaces of type $(\pi, 1)$). In a sense, most primes are $K(\pi, 1)$ and much remains to be understood about them in general. Considerably more is known about a special subclass, the so called *sufficiently large* $K(\pi, 1)$ 3-manifolds or *Haken manifolds*. They are characterized by the property that they contain an embedded *incompressible* Riemann surface R_g, i.e. that if $e : R_g \to \Pi$ is the embedding then $e_* : \pi_1(R_g) \to \pi_1(\Pi)$ is injective.[17] Simple examples are provided by the products $S^1 \times R_g$. An important conjecture in 3-manifold theory states that every irreducible 3-manifold with infinite fundamental group is virtually Haken, that is, finitely covered by a Haken manifold. If this is the case, any prime with infinite fundamental group allows for an immersion $e : R_g \to \Pi$ such that $e_* : \pi_1(R_g) \to \pi_1(\Pi)$ is injective.

---

[15]As far as I am aware, it is still an open conjecture that spherical space forms comprise all primes with finite fundamental group, even given the validity of the Poincaré conjecture. In other words, it is only conjectured that 3-manifolds covered by S^3 are of the form S^3/G where $G \subset SO(4)$ acting in the standard linear fashion. In [64] Milnor classified all finite groups that satisfied some necessary condition for having a free action on S^3. The validity of the Smale conjecture [41] (which states that the embedding of $O(4)$ into $\mathrm{Diff}(S^3)$ is a homotopy equivalence) eliminates those groups from the list which are not subgroups of $SO(4)$ [38]. What remains to be shown is that these groups do not admit inequivalent (equivalence being conjugation by some diffeomorphism) actions. The undecided cases are some cyclic groups of odd order; see [81].

[16] Let G be a group with normal subgroup N and quotient $G/N = Q$. We call G either an *upward extension* of N by Q or a *downward extension* of Q by N; see [9], p. XX.

[17]In other words, every loop on $e(R_g) \subset \Pi$ that bounds a 2-disc in Π (and is hence contractible in Π) also bounds a 2-disc on $i(R_g)$ (an is hence contractible in $e(R_g)$).

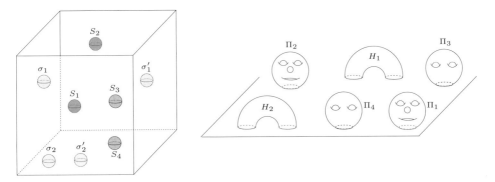

FIGURE 1. The connected sum of two handles $H_{1,2}$ and four irreducible primes $\Pi_1, \cdots \Pi_4$, with Π_1 diffeomorphic to Π_2 and Π_3
diffeomorphic to Π_4. For later application concerning the map-
ping class groups it is advisable to represent a handle as a cylinder
$[0,1] \times S^3$ with both ends, $0 \times S^2$ and $1 \times S^2$, separately attached
by a connecting 2-sphere. The connecting spheres are denoted by
σ_i, σ_i' for H_i $(i = 1, 2)$ and S_i for Π_i $(i = 1, \cdots, 4)$. The left picture
gives an *internal view,* in which only the connecting spheres are
seen (and not what is behind), the right picture gives an *external view* from three dimensions onto a two-dimensional analogous
situation that also reveals the topological structures behind the
connecting spheres.

Now, given a connected sum $(N = n + m)$

$$\Sigma = \biguplus_{i=1}^{N} P_i = \left\{ \biguplus_{i=1}^{n} \Pi_i \right\} \uplus \left\{ \biguplus_{i=1}^{m} S^1 \times S^2 \right\}, \qquad (2.1)$$

where notationally we distinguish between unspecified primes, denoted by P_i, and
irreducible primes (i.e. those different from $S^1 \times S^2$), denoted by Π_i, so that $P_i =
\Pi_i$ for $1 \leq i \leq n$ and $P_i = S^1 \times S^2$ for $n < i \leq N$. A simple application of
van Kampen's rule gives that the fundamental group of the connected sum in
3 (and higher) dimensions is isomorphism to the free product (denoted by $*$) of
the fundamental groups of the factors (since the connecting spheres are simply
connected):

$$\pi_1(\Sigma) = \pi_1(P_1) * \cdots * \pi_1(P_N). \qquad (2.2)$$

The converse is also true: a full decomposition of the group $\pi_1(\Sigma)$ into free products
corresponds to a decomposition of Σ into the connected sum of primes (known as
Kneser's conjecture).

The existence of prime decompositions was first shown by Kneser [53], the
uniqueness (up to permutations of factors) by Milnor [65]. Regarding the latter
we need to recall that we consider all manifolds to be oriented. Many orientable
primes do not allow for orientation reversing self diffeomorphism; they are called

chiral. The table on p. 922 of [23] lists which spherical and flat space forms are chiral; most of them are. Chiral primes with opposite orientation must therefore be considered as different prime manifolds.

Irreducible primes can be further decomposed by cutting them along 2-tori, which is the second major decomposition device in Thurston's *Geometrization Program* of 3-manifolds. Here we will not enter into this.

3. Mapping class groups

Mapping class groups can be studied through their action on the fundamental group. Consider the fundamental group of Σ based at $\infty \in \Sigma$. We write $\pi_1(\Sigma, \infty)$ or sometimes just π_1 for short. There are homomorphisms

$$h_{\mathrm{F}} \; : \; \mathcal{G}_{\mathrm{F}}(\Sigma) \; \to \; \mathrm{Aut}(\pi_1) \,, \tag{3.1}$$

$$h_{\infty} \; : \; \mathcal{G}_{\infty}(\Sigma) \; \to \; \mathrm{Aut}(\pi_1) \,, \tag{3.2}$$

$$h \;\;\; : \; \mathcal{G}(\Sigma) \;\;\; \to \; \mathrm{Out}(\pi_1) \,, \tag{3.3}$$

where the first two are given by

$$[\phi] \mapsto \left([\gamma] \mapsto [\phi \circ \gamma]\right). \tag{3.4}$$

Here $[\phi]$ denotes the class of $\phi \in \mathrm{Diff}_{\mathrm{F}}(\Sigma)$ in $\mathrm{Diff}_{\mathrm{F}}(\Sigma)/\mathrm{Diff}_{\mathrm{F}}^0(\Sigma)$ (or of $\phi \in \mathrm{Diff}_{\infty}(\Sigma)$ in $\mathrm{Diff}_{\infty}(\Sigma)/\mathrm{Diff}_{\infty}^0(\Sigma)$) and the other two square brackets the homotopy classes of the curves γ and $\phi \circ \gamma$. As regards (3.3), it is not difficult to see that in $\mathrm{Diff}(\Sigma)$ any inner automorphism of $\pi_1(\Sigma, \infty)$ can be generated by a diffeomorphism that is connected to the identity (in $\mathrm{Diff}(\Sigma)$, not in $\mathrm{Diff}_{\infty}(\Sigma)$ or $\mathrm{Diff}_{\mathrm{F}}(\Sigma)$). Hence we have to factor out the inner automorphisms for the map h to account for the possibility to move the basepoint ∞. More precise arguments are given in Section 3 of [28].

The images of h_{∞} and h_{F} coincide but their domains may differ, i.e. the groups $\mathcal{G}_{\mathrm{F}}(\Sigma)$ and $\mathcal{G}_{\infty}(\Sigma)$ are not necessarily isomorphic. Let us explain this a little further. Consider the fibration

$$\mathrm{Diff}_{\mathrm{F}}(\Sigma) \;\rightarrowtail^{\;i\;}\; \mathrm{Diff}_{\infty}(\Sigma) \;\xrightarrow{\;p\;}\!\!\!\!\twoheadrightarrow\; \mathrm{GL}^+(3,\mathbb{R}) \,, \tag{3.5}$$

where $p(\phi) := \phi_*|_{\infty}$ (here we identify $GL^+(3,\mathbb{R})$ with the orientation preserving linear isomorphisms of $T_{\infty}(\Sigma)$). Associated with this fibration is a long exact sequence of homotopy groups, which ends with

$$1 \longrightarrow \pi_1(\mathrm{Diff}_{\mathrm{F}}(\Sigma)) \longrightarrow \pi_1(\mathrm{Diff}_{\infty}(\Sigma)) \xrightarrow{\;p_*\;} \mathbb{Z}_2 \xrightarrow{\;\partial_*\;} \mathcal{G}_{\mathrm{F}}(\Sigma) \xrightarrow{\;i_*\;} \mathcal{G}_{\infty}(\Sigma) \longrightarrow 1 \tag{3.6}$$

where the leftmost zero comes from $0 = \pi_2(\mathrm{GL}^+(3,\mathbb{R}))$ and the $\mathbb{Z}_2$ in the middle is $\pi_1(\mathrm{GL}^+(3,\mathbb{R}))$. Now, there are only two possibilities as regards the image of p_*:

1. $\mathrm{Image}(p_*) = \mathbb{Z}_2 = \mathrm{kernel}(\partial_*) \Rightarrow \mathcal{G}_{\mathrm{F}}(\Sigma) \cong \mathcal{G}_{\infty}(\Sigma)$.
2. $\mathrm{Image}(p_*) = \{0\} = \mathrm{kernel}(\partial_*) \Rightarrow \mathcal{G}_{\mathrm{F}}(\Sigma)$ is a downward extension (recall footnote 16) of $\mathcal{G}_{\infty}(\Sigma)$ by $\mathbb{Z}_2$.

Let us focus on the second possibility. We first note that image(∂_*) lies in the kernel of h_F. This is true because the images of h_F and h_∞ coincide, so that $h_F \circ \partial_* = h_\infty \circ i_* \circ \partial_*$, which is the trivial map onto the identity in $\text{Aut}(\pi_1)$, since by exactness $i_* \circ \partial_*$ is the trivial map. The diffeomorphism that represents the non-trivial image of ∂_* can be represented by a rotation parallel to two concentric small spheres centered at ∞; see the left picture in Fig. 2. From this picture it also becomes clear that the diffeomorphism representing the 2π-rotation can be chosen of disjoint support from those diffeomorphisms representing other elements of $\mathcal{G}_F(\Sigma)$. Hence, if $\mathcal{G}_F(\Sigma)$ is a $\mathbb{Z}_2$ extension of $\mathcal{G}_\infty(\Sigma)$, it is *central*.

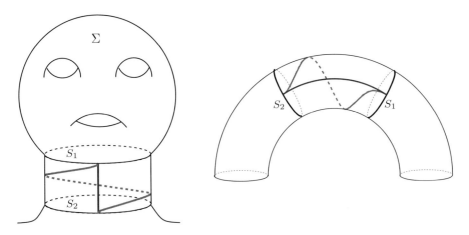

FIGURE 2. Both pictures show rotations parallel to spheres S_1 and S_2. On the left, a rotation of the manifold Σ parallel to spheres both centered at ∞, or, if $\Sigma = \Pi$ is a prime manifold in a connected sum, parallel to the connecting 2-sphere. On the right is a rotation parallel to two meridian spheres in a handle. The support of the diffeomorphism is on the cylinder bound by S_1 and S_2. In either case its effect is depicted by the two curves connecting the two spheres: the diffeomorphism maps the straight to the curved line. This 2-dimensional representation is deceptive insofar as here the two straight and the curved lines are not homotopic (due to $\pi_1(S^1) = \mathbb{Z}$), whereas in 3 dimensions they are (due to the triviality of $\pi_1(S^2)$).

Manifolds for which $\mathcal{G}_F(\Sigma)$ is a (downward) central $\mathbb{Z}_2$ extension of $\mathcal{G}_\infty(\Sigma)$ are called *spinorial*. If they are used to model isolated systems (by removing ∞ and making the end asymptotically flat) their asymptotic symmetry group is not the Poincaré group (as discussed in [6]) but its double cover (i.e. the identity component of the homogeneous symmetry group is $SL(2,\mathbb{C})$ rather than the proper orthochronous Lorentz group). The origin of this is purely topological and has nothing to do with quantum theory, though the possible implications for quantum gravity are particularly striking, as was first pointed out in a beautiful paper by Friedman and Sorkin [18]: it could open up the possibility to have half-integer spin states in pure gravity.[18]

A connected sum is spinorial iff it contains at least one spinorial prime [45]. Except for the lens spaces and handles all primes are spinorial, hence a 3-manifold is non-spinorial iff it is the connected sum of lens spaces and handles. Let us digress a little to explain this in somewhat more detail.

3.1. A small digression on spinoriality

That lens spaces and handles are non spinorial is easily visualized. Just represent them in the usual fashion by embedding a lens or the cylinder $S^2 \times [0,1]$ in $\mathbb{R}^3$, with the boundary identifications understood. Place the base point, ∞, on the vertical symmetry axis and observe that the rotation around this axis is compatible with the boundary identifications and therefore defines a diffeomorphism of the quotient space. A rotation parallel to two small spheres centered at ∞ can be continuously undone by rotating the body in $\mathbb{R}^3$ and keeping a neighborhood of ∞ fixed. This visualization also works for arbitrary connected sums of lens spaces and handles.

Spinoriality is much harder to prove. The following theorem has been shown by Hendriks ([45], Thm. 1 in § 4.3), and later in a more constructive fashion by Plotnick ([69], Thm. 7.4):

Theorem 3.1. *Let Σ be a closed (possibly non-orientable) 3-manifold and $\Sigma' := \Sigma - B_3$, where B_3 is an open 3-disc. A 2π-rotation in Σ' parallel to the boundary 2-sphere $\partial\Sigma'$ is homotopic to $\mathrm{id}_{\Sigma'}$ rel. $\partial\Sigma'$ (i.e. fixing the boundary throughout) iff every prime summands of Σ is taken from the following list:*

1.) S_h^3/G, *where S_h^3 is a homotopy sphere and G a finite freely acting group all Sylow subgroups of which are cyclic,*
2.) *the handle $S^1 \times S^2$,*
3.) *the (unique) non-orientable handle $S^1 \tilde{\times} S^2$,*
4.) $S^1 \times \mathbb{RP}^3$, *where $\mathbb{RP}^3$ denotes 3-dimensional real projective space.*

Since here we excluded non-orientable manifolds from our discussion, we are not interested in 3.) and 4.). Clearly, S^3 is the only homotopy 3-sphere if the Poincaré conjecture holds. Of the remaining spherical space forms S^3/G the following have cyclic Sylow subgroups[19]:

a.) $G = \mathbb{Z}_p$ (giving rise to the lens spaces),

---

[18]Topologically speaking, this is somewhat analogous to the similar mechanism in the Skyrme model [74], where loops in configuration space generated by 2π-rotations are non-contractible iff the skyrmion's winding number (its baryon number) is odd [22]). The analogy to the mechanism by which half-integer spin states can arise in gauge theories of integer spin fields [37, 32] is less close, as they need composite objects, e.g. from magnetic monopoles and electric charges.

[19]Each of the other groups contains as subgroup the 'quaternion group' $D_8^* := \{\pm 1, \pm i, \pm j, \pm k\}$, which is non abelian and of order $8 = 2^3$. Hence their 2-Sylow subgroups are not cyclic.

b.) $G = D^*_{4m} \times \mathbb{Z}_p$ for $m = $ odd and $4m$ coprime to $p \geq 0$. Here D^*_{4m} is the $SU(2)$ double cover of $D_{2m} \subset SO(3)$, the order $2m$ symmetry group of the m–prism.

c.) $G = D'_{2^k m} \times \mathbb{Z}_p$ for $m = $ odd, $k > 3$, and $2^k m$ coprime to $p \geq 0$. Here $D'_{2^k m}$ is a (downward) central extension of D^*_{4m} by $\mathbb{Z}_{2^{k-2}}$.[20]

Now there is a subtle point to be taken care of: that a diffeomorphism is *homotopic* to the identity means that there is a one-parameter family of *continuous* maps connecting it to the identity. This does not imply that it is *isotopic* to the identity, which means that there is a one-parameter family of *diffeomorphisms* connecting it to the identity. In case of the lens spaces it is easy to 'see' the isotopy, as briefly explained above. However, in the cases b.) and c.) it was proven by Friedman & Witt in [19] that the homotopy ensured by Thm. 3.1 does not generalize to an isotopy, so that these spaces again are spinorial. Taken together with Thm 3.1 this completes the proof of the statement that the only non-spinorial 3-manifolds are lens spaces, handles, and connected sums between them.

Note that this result also implies the existence of diffeomorphisms in $\mathrm{Diff}(\Sigma)$ which are homotopic but not isotopic to the identity. For example, take the connected sum $\Sigma = \Pi_1 \uplus \Pi_2$ of two primes listed under b.) or c.). The 2π-rotation parallel to the connecting 2-sphere will now be an element of $\mathrm{Diff}(\Sigma)$ that is homotopic but not isotopic to the identity [19]; see Fig. 3. This provides the first known example of such a behavior in 3-dimensions (in two dimensions it is known not to occur), though no example is known where this happens for a prime 3-manifold. In fact, that homotopy implies isotopy has been proven for a very large class of primes, including all spherical space forms, the handle $S^1 \times S^2$, Haken manifolds and many non-Haken $K(\pi, 1)$ (those which are Seifert fibered). See e.g. Thm. A1 of [25] for a list of references.

3.2. General Diffeomorphisms

It can be shown that if all primes in a connected sum satisfy the homotopy-implies-isotopy property, the kernel of h_F is isomorphic to $\mathbb{Z}_2^{m+n_s}$, where m is the number of handles and n_s the number of spinorial primes.[21] This group is generated by the diffeomorphisms depicted in Fig. 2, one neck-twist (left picture) for each spinorial primes and one handle-twist (right picture) for each handle. It can also be shown that the mapping class groups of each prime injects into the mapping class groups of the connected sum in which it occurs [46]. This means that a diffeomorphism

---

[20]We have $D_{2m} = \langle \alpha, \beta : \alpha\beta = \beta\alpha^{-1}, \alpha^m = \beta^2 = 1 \rangle$, where α is a $2\pi/m$–rotation of the m-prism (vertical axis) and β is a π rotation about a horizontal axis. Then $D^*_{4m} = \langle \alpha, \beta : \alpha\beta = \beta\alpha^{-1}, \alpha^m = \beta^2 \rangle = \langle a, b : ab = ba^{-1}, a^m = b^4 = 1 \rangle$, where $a := \alpha\beta^2$ and $b := \beta$ (to show equivalence of these two presentations one needs that m is odd), and $D'_{2^k m} = \langle A, B : AB = BA^{-1}, A^m = B^{2^k} = 1 \rangle$. The center of $D'_{2^k m}$ is generated by B^2 and isomorphic to $\mathbb{Z}_{2^{k-1}}$. B^4 generates a central subgroup $\langle B^4 \rangle$ isomorphic to $\mathbb{Z}_{2^{k-2}}$ and $D'_{2^k m}/\langle B^4 \rangle \cong D^*_{4m}$.

[21]This follows from Thm. 1.5 in [61] together with the fact that for a manifold Π with vanishing π_2 two self-diffeomorphisms $\phi_{1,2}$ are homotopic if their associated maps $h_\infty : [\phi_{1,2}] \mapsto \mathrm{Aut}(\pi_1(\Pi, \infty))$ coincide.

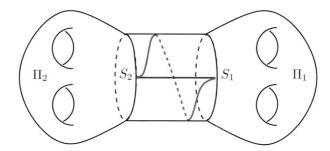

FIGURE 3. The connected sum of two irreducible primes Π_1 and Π_2. The relative 2π-rotation is a diffeomorphism with support inside the cylinder bounded by the 2-spheres S_1 and S_2. It transforms the straight line connecting S_1 and S_2 into the curved line. If $\Pi_i = S^3/G_i$ $(i = 1, 2)$, where $G_{1,2} \in SO(4)$ are taken from the families $D^*_{4m} \times \mathbb{Z}_p$ or $D'_{2^k m} \times \mathbb{Z}_p$ mentioned under b.) and c.) in the text, this diffeomorphism is homotopic but not isotopic to the identity.

that has support in a prime factor P (we call such diffeomorphisms *internal*) and is not isotopic to the identity within in the space of all diffeomorphisms fixing the connecting 2-sphere is still not isotopic to the identity in $\mathrm{Diff_F}(\Sigma)$. This statement would be false if $\mathrm{Diff_F}(\Sigma)$ were replaced by $\mathrm{Diff}(\Sigma)$. We will briefly come back to this point at the end of this section.

In analogy to (3.1), for each prime P, there is a map $h_F : \mathcal{G}_F(P) \to \mathrm{Aut}(\pi_1(P))$ which is (almost) surjective in many cases. For example, if Π is Haken, h_F maps onto $\mathrm{Aut}^+(\pi_1(\Pi))$ [40], the subgroup of orientation preserving automorphisms. If Π is not chiral, i.e. allows for orientation reversing self-diffeomorphisms, $\mathrm{Aut}^+(\pi_1(\Pi)) \subset \mathrm{Aut}(\pi_1(\Pi))$ is a subgroup of index two (hence normal). However, if Π is chiral, we have $\mathrm{Aut}^+(\pi_1(\Pi)) = \mathrm{Aut}(\pi_1(\Pi))$ and hence surjectivity. Since Haken manifolds are all spinorial, we can now say that their mapping class group is a central $\mathbb{Z}_2$ extension of $\mathrm{Aut}^+(\pi_1(\Pi))$.

For spherical space forms the mapping class groups have all been determined in [85]. For $S^1 \times S^2$ it is $\mathbb{Z}_2 \times \mathbb{Z}_2$, where, say, the first $\mathbb{Z}_2$ is generated by the twist as depicted on the right in Fig. 2. The second $\mathbb{Z}_2$ corresponds to $\mathrm{Aut}(\pi_1(\Pi)) = \mathrm{Aut}(\mathbb{Z})$ and is generated by a reflection in the circle S^1. If one thinks of a handle in a prime decomposition as a cylinder being attached with both ends, as depicted in Fig. 1, the latter diffeomorphism corresponds to exchanging the two cylinder ends (in an orientation preserving fashion), which is sometimes called a *spin* of a handle.

Suppose now that we are given a general connected sum (2.1) and that we know the mapping class group of each prime in terms of a finite presentation (finitely many generators and relations). We can then determine a finite presentation of $\mathcal{G}_F(\Sigma)$ by means of the so-called Fouxe-Rabinovitch presentation for the automorphism group of a free product of groups developed in [15, 16]; see also [62] and [21]. Let generally

$$G = \underbrace{G_{(1)} * \cdots * G_{(n)}}_{\text{each} \not\cong \mathbb{Z}} * \underbrace{G_{(n+1)} * \cdots * G_{(n+m)}}_{\text{each} \cong \mathbb{Z}} \tag{3.7}$$

be a free product of groups corresponding to the decomposition (2.1). Let a set $\{g_{(i)1}, \cdots, g_{(i)n_i}\}$ of generators for each $G_{(i)}$ be chosen. Clearly, for $n < i \leq n+m$ we have $n_i = 1$ so that we also write $g_{(i)1} =: g_{(i)}$. The generators of $\text{Aut}(G)$ can now be characterized by their action on these generators as follows (only non-trivial actions are listed):

1.) The generators of each $\text{Aut}(G_{(i)})$ for $1 \leq i \leq n$. As mapping-class generator these are called *internal*.

2.) The m generators σ_i $(1 \leq i \leq m)$ whose effect is $\sigma_i(g_{(n+i)}) = g_{(n+i)}^{-1}$, i.e. generating $\text{Aut}(\mathbb{Z}) \cong \mathbb{Z}_2$. As mapping-class generator σ_i is called a *spin of the i-th handle*.

3.) One generator $\omega_{(i)(k)}$ for each pair of distinct but isomorphic groups $G_{(i)}, G_{(k)}$ $(1 \leq i,k \leq n+m)$, whose effect it is to slotwise exchange the sets $\{g_{(i)1}, \cdots, g_{(i)n_i}\}$ and $\{g_{(k)1}, \cdots, g_{(k)n_k}\}$. (Here, for $1 \leq i,k \leq n$, we assume the generators of isomorphic groups to be chosen such that they correspond under a fiducial isomorphism, in particular $n_i = n_k$). As mapping-class generator $\omega_{(i)(k)}$ is called the *exchange of prime i with prime k*.

4.) One generator $\mu_{(i)j,(k)}$ for each $1 \leq i \leq n+m$, $1 \leq j \leq n_i$, and $1 \leq k \leq n$, whose effect is to map each generator $g_{(k)l}$ $(1 \leq l \leq n_k)$ to $g_{(i)j}^{-1} \cdot g_{(k)l} \cdot g_{(i)j}$. As mapping-class generator $\mu_{(i)j,(k)}$ is called a *slide of the (irreducible) prime k through prime i along $g_{(i)j}$*.

5.) One pair of generators, $\lambda_{(i)j,(k)}$ and $\rho_{(i)j,(k)}$ for each $1 \leq i \leq n+m$, $1 \leq j \leq n_i$, and $n \leq k \leq n+m$. The effect of $\lambda_{(i)j,(k)}$ is to map each generator $g_{(k)}$ to $g_{(i)j}^{-1} \cdot g_{(k)}$ (i.e. left multiplication) and that of $\rho_{(i)j,(k)}$ to map each $g_{(k)}$ to $g_{(k)} \cdot g_{(i)j}$ (i.e. right multiplication). As mapping-class generator $\lambda_{(i)j,(k)}$ is called the *slide of the left end of handle k through prime i along $g_{(i)j}$* and $\rho_{(i)j,(k)}$ is called the *slide of the right end of handle k through prime i along $g_{(i)j}$*.

Of these generators the mapping class group realizes all those listed in 2.)-5.), but might leave out some in 1.) in case $h_{\text{F}} : \mathcal{G}_{\text{F}}(\Pi_i) \rightarrow \text{Aut}(\pi_1(\Pi_i))$ is not surjective for some $i \leq n$. In that case just replace 1. by the generators of the image of h_{F} (which might be a larger set than the generators of $\text{Aut}(\pi_1(\Pi_i))$). Finally, we have to add the generators of the kernel of h_{F}. As already stated, this kernel is given by the direct product of $n_s + m$ copies of $\mathbb{Z}_2$, where n_s is the number of spinorial primes, if we assume the 'homotopy-implies-isotopy'-property for all primes. In that case we have found all generators after adjoining these additional $n_s + m$ generators. A complete list of relations can then be found from the Fouxe-Rabinovitch relations for $\text{Aut}(G) = G_1 * \cdots * G_{n+m}$ (see Chapter 5.1 of [62])[22] and some added relations which the $n_s + m$ added generators have to satisfy. The latter are not difficult to find due to the simple geometric interpretation of the diffeomorphisms of Fig. 2 that represent the added generators.[23]

The procedure just outlined reduces the problem of finding a presentation of $\mathcal{G}_F(\Sigma)$ to that of finding presentations $\mathcal{G}_F(\Pi_i)$ for each irreducible prime. As already stated, they are explicitly known for all spherical space forms. We also mentioned that for Haken manifolds $\mathcal{G}_F(\Pi_i)$ is a $\mathbb{Z}_2$ extension of $\text{Aut}^+(\pi_1)$, which in simple cases allows to find an explicit presentation. For example, for the 3-torus we have $\text{Aut}^+(\pi_1 = \mathbb{Z}^3) = SL(3,\mathbb{Z})$ and the appropriate central $\mathbb{Z}_2$ extension can be shown to be given by the Steinberg group $St(3,\mathbb{Z})$, a presentation of which may e.g. be found in § 10 of [66].

All the generators listed in 2.-5. can be realized by appropriate diffeomorphisms. This is not difficult to see for 2. and 3., as diffeomorphisms that 'spin' a handle or 'exchange' two diffeomorphic primes are easily visualized. A visualization of the slide transformations 4. and 5. is attempted in Fig. 4: The general idea—and this is where the name 'slide' derives from—is similar to that of rotation of parallel (i.e. concentric) spheres explained in Fig. 2. But now we take two 'parallel' (i.e. coaxial) tori, T_1 and T_2, and consider a diffeomorphism whose support is confined to the region between them. This toroidal region is of topology $[1,2] \times S^1 \times S^1$ (i.e. does not contain prime summands) and hence is foliated by a one parameter (r) family of parallel (coaxial) tori $T_r = r \times S^1 \times S^1$, where $r \in [1,2]$. Each of these tori we think of as being coordinatized in the standard fashion by two angles, θ and φ with range $[0,2\pi]$ each, where θ labels the latitude and φ the longitude (the circles of constant φ are the small ones that become contractible in the solid torus). The slide now corresponds to a diffeomorphism which is the identity outside the toroidal region and which inside $(1 \leq r \leq 2)$ is given by

$$(r, \theta, \varphi) \mapsto (r, \theta, \varphi + \beta(2 - r)), \tag{3.8}$$

where β is a C^∞ step-function $\beta : [0,1] \to [0,2\pi]$ with $\beta(0) = 0$ and $\beta(1) = 2\pi$. In Fig. 4, the loop γ generates a non-trivial element $[\gamma] \in \pi_1(\Pi, p)$ (the non-triviality is indicated by the little knot inside Π for lack of better representation). This loop γ, after having been acted on by the slide, will first follow ℓ and go through the handle, then travel trough Π as before, and finally travel the handle in a reversed sense. That is, the slide conjugates $[\gamma]$ with $[\ell]$.

---

[22]The original papers by Fouxe-Rabinovitch [16, 15] contained some errors in the relations which were corrected in [62]; see also [21].

[23]The added generators are internal transformations (i.e. have support within the prime factors) and hence behave naturally under exchanges. They commute with all other internal diffeomorphisms and slides *of* the prime in question, since their supports may be chosen to be disjoint from the diffeomorphisms representing the other internal diffeomorphisms and slides. They also commute with slides of other primes *through* the one in question, since their action on such slides (by conjugation) is a slide along a curve isotopic to the original one, which defines the same mapping class. Finally, the conjugation of a handle's 'twist' (right picture in Fig. 2) with a spin of that handle is isotopic to the original twist.

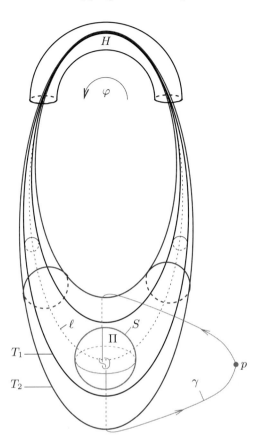

FIGURE 4. The slide of a prime as seen form the 'inside view' (compare the left picture in Fig. 1) through a handle H in the background. The prime to be slid, Π, hides behind its connecting 2-sphere S. The loop ℓ, representing the generator of $\pi_1(H)$ along which the prime is slid, is thickened to two coaxial tori, T_1 and T_2, such that the prime is contained in the inner torus T_1. The diffeomorphism in the toroidal region is given by (3.8), where the angle φ measures the axial direction. The slide acts on $[\gamma] \in \pi_1(\Pi, p) \subset \pi_1(\Sigma, p)$ (the non-triviality of which being indicated by the little knot inside Π) by conjugation with $[\ell]$, when the latter is appropriately considered as element of $\pi_1(\Sigma, p)$.

The slides of irreducible primes described in 4.), or the slides of ends of handles described in 5.), are then obtained by choosing the tori such that the only connecting sphere contained inside the inner torus is that of the prime, or handle-end, to be slid. The common axis of the tori trace out a non-contractible loop through another prime, the prime through which the first one is slid. All this is depicted in Fig. 4.

Of crucial importance is the different behavior of slides in 4.) on one hand, and slides in 5.) on the other. Algebraically this has to do with the different behavior, as regards the automorphism group of the free product, of the free factors $\mathbb{Z}$ on one hand and the non-free factors $G_{(i)}$ on the other. Whereas left or right multiplication of all elements in one $\mathbb{Z}$ factor with any element from the complementary free product defines an automorphism of G, this is not true for the non-free factors. Here only conjugation defines an automorphism. Geometrically this means that we have to consider slides of both ends of a handle separately in order to be able to generate the automorphism group. It is for this reason that we pictured the handles in Fig. 1 as being attached to the base manifold with two rather than just one connecting sphere.

In passing we remark that certain important interior diffeomorphisms (i.e. falling under case 1.) above) have an interpretation in terms of certain *internal* slides. Imagine Fig. 4 as an *inside* view from some prime Π, i.e. everything seen on Fig. 4 is inside Π. The handle H, too, is now interpreted as some topological structure of Π itself that gives rise to non-contractible loops within Π. S is again the connecting sphere of Π, now seen from the inside, beyond which the part of the manifold Σ outside Π lies. The diffeomorphism represented by Fig. 4 then slides the connecting sphere S once around the loop ℓ in Π. Its effect is to conjugate each element of $\pi_1(\Pi)$ by $[\ell] \in \pi_1(\Pi)$. Hence we see that such internal slides generate all *internal* automorphisms for each irreducible factor Π. In case of handles, there are no non-trivial inner automorphisms, and the only non-trivial outer automorphism ($\mathbb{Z} \mapsto -\mathbb{Z}$) is realized by spinning the handle, as already mentioned.

Having said that, we will from now on always understand by 'slides' *external* transformations as depicted in 4, unless explicitly stated otherwise (cf. last remark at the end of this section). But let us for the moment forget about slides altogether and focus attention only on those mapping classes listed in 1.)-3.), i.e. internal transformations and exchanges. In doing this we think of a spin of a handle as internal, which we may do as long as no slides are considered. It is tempting to think of the manifold Σ as being composed of $N = n + m$ 'particles' from d species, each with its own characteristic internal symmetry group G_r, $1 \leq r \leq d$. In this analogy diffeomorphic primes correspond to particles of one species and the symmetry groups G_r correspond to $\mathcal{G}_F(P_r)$. Let there be n_r primes in the r-th diffeomorphism class, so that $\sum_{r=1}^{d} n_r = N$. In this 'particle picture' the symmetry group would be a semi-direct product of the internal symmetry group, G^I, with an external symmetry group, G^E, both respectively given by

$$G^I := \overset{n_1}{\prod} G_1 \times \cdots \times \overset{n_d}{\prod} G_d, \tag{3.9a}$$

$$G^E := S_{n_1} \times \cdots \times S_{n_d}, \tag{3.9b}$$

where here S_n denotes the order $n!$ permutation group of n objects. The semi-direct product is characterized through the homomorphism $\theta : G^E \to \mathrm{Aut}(G^I)$, where $\theta = \theta_1 \times \cdots \times \theta_d$ and

$$\theta_i : S_{n_i} \to \mathrm{Aut}\left(\prod^{n_i} G_i\right) \tag{3.10}$$

$$\sigma \mapsto \theta_i(\sigma) : (g_1, \cdots, g_{n_i}) \mapsto \left(g_{\sigma(1)}, \cdots, g_{\sigma(n_i)}\right).$$

The semi-direct product $G^I \rtimes G^E$ with respect to θ is now defined by the following multiplication law: let $\gamma_i \in \prod^{n_i} G_i$, $i = 1, \cdots, d$, then

$$\begin{aligned}
&\left(\gamma_1', \cdots, \gamma_d'; \sigma_1', \cdots, \sigma_d'\right)\left(\gamma_1, \cdots, \gamma_d; \sigma_1, \cdots, \sigma_d\right) \\
&= \left(\gamma_1'\,\theta_1(\sigma_1')(\gamma_1), \cdots, \gamma_d'\,\theta_d(\sigma_d')(\gamma_d); \sigma_1'\sigma_1, \cdots, \sigma_d'\sigma_d\right).
\end{aligned} \tag{3.11}$$

We call $G^P = G^I \rtimes G^E$ the *particle group*. From the discussion above it is clear that this group is a subgroup of the mapping class group. But we also had to consider slides which were neither internal nor exchange diffeomorphisms and which are not compatible with this simple particle picture, since they mix internal and external points of the manifold. How much do the slides upset the particle picture? For example, consider the normal closure, G^S, of slides (i.e. the smallest normal subgroup in the group of mapping classes that contains all slides). Does it have a non-trivial intersection with G^P, i.e. is $G^P \cap G^S \neq \{1\}$? If this is the case $\mathcal{G}_{\mathrm{F}}(\Sigma)/G^S$ will be a non-trivial factor of G^P. Representations whose kernels contain G^S will then not be able to display all particle symmetries. This would only be the case if $G^P \cap G^S = \{1\}$.

Questions of this type have been addressed and partly answered in [28] (see also [79]). Here are some typical results:

Proposition 3.2. $G^P \cap G^S = \{1\}$ *and* $\mathcal{G}_{\mathrm{F}}(\Sigma) = G^S \rtimes G^P$ *if* Σ *contains no handle in its prime decomposition.*

Proposition 3.3. G^S *is perfect if* Σ *contains at least 3 handles in its prime decomposition.*

The last proposition implies that slides cannot be seen in abelian representations of mapping class groups of manifolds with at least three handles. In [28] an explicit presentation with four generators of the mapping class group of the connected sum of $n \geq 3$ handles was given and the following result was shown:

Proposition 3.4. *Let* $\Sigma = \overset{n}{\uplus} S^1 \times S^2$ *where* $n \geq 3$. *Then* $\mathcal{G}_{\mathrm{F}}(\Sigma)/G^S \cong \mathbb{Z}_2 \times \mathbb{Z}_2$ *where one* $\mathbb{Z}_2$ *is generated by the twist (right picture in Fig. 3) of, say, the first handle and the other* $\mathbb{Z}_2$ *by either the exchange of, say, the first and second handle or the spin of, say, the first handle. Hence we have a strict correlation between spins and exchanges of handles. Generally, given a representation* ρ *of* $\mathcal{G}_{\mathrm{F}}(\Sigma)$, *the following statements are equivalent:*

   a.) *ρ is abelian,*
   b.) *slides are in the kernel of ρ,*
   c.) *ρ strictly correlates exchanges and spins (i.e. $\rho(\mathrm{spin}) = \rho(\mathrm{exchange})$),*
   d.) *slides and exchanges commute under ρ.*

Reference [28] also deals with connected sums of arbitrarily many real projective spaces $\mathbb{R}P^3$. A presentation in terms of three generators was written down and various features studied. Since projective spaces are the most simple lens spaces ($\mathbb{R}P^3 = L(2,1)$) they are not spinorial, as one easily visualizes. Since the automorphism group of $\pi_1(\mathbb{R}P^3) = \mathbb{Z}_2$ is trivial there are no non-trivial mapping classes from internal diffeomorphisms ($\mathbb{R}P^3$ satisfies the homotopy-implies-isotopy property). Therefore, the particle group G^P is just the permutation group S_n and the mapping class group is the semi-direct product $G^S \rtimes S_n$, according to Prop. 3.2. A a very interesting systematic study of representaions of this group was started in [79] using Mackey theory (theory of induced representations). For illustrative purposes we consider in some detail the case of the connected sum of just two projective spaces in the next section, also considered in [3] and [79].

Even though we here restrict attention to frame-fixing diffeomorphisms, which is physically well motivated, we nevertheless wish to end this section with a few remarks that give an idea of the essential changes that result if we relaxed from $\mathrm{Diff}_F(\Sigma)$ (or $\mathrm{Diff}_\infty(\Sigma)$) to $\mathrm{Diff}(\Sigma)$. In the frame-fixing context, non-trivial mapping classes of prime factors (i.e. generated by internal diffeomorphisms) are non-trivial mapping classes in the total manifold Σ. In other words, there is an injection $\mathcal{G}_F(P) \to \mathcal{G}_F(\Sigma)$ [46]. As already remarked above, this is not true if $\mathcal{G}_F(\Sigma)$ is replaced by $\mathcal{G}(\Sigma)$. In fact, it follows from the above that any diffeomorphism in $\mathrm{Diff}_F(\Sigma)$ whose image under h_F in $\mathrm{Aut}(\pi_1(\Sigma))$ is an *inner* automorphism is isotopic in $\mathrm{Diff}(\Sigma)$ to transformations of the type depicted in Fig. 2. Hence there are generally many inner diffeomorphisms of prime factors P which represent non-trivial elements of $\mathrm{Diff}_F(P)$ but trivial elements in $\mathrm{Diff}(\Sigma)$. Also, the distinction between inner and non-inner (exchanges and slides) diffeomorphisms ceases to be meaningful in $\mathrm{Diff}(\Sigma)$. A trivial example is the diffeomorphism depicted in Fig. 3, i.e. the 2π-rotation parallel to the common connecting sphere of two irreducible prime manifolds. Up to isotopy in $\mathrm{Diff}(\Sigma)$ it can clearly be considered as inner diffeomorphism of *either* prime. A less trivial example is the following: consider $\Sigma = \Pi_1 \uplus \Pi_2$, where $\{g_{(1)1}, \cdots, g_{(1)n_1}\}$ and $\{g_{(2)1}, \cdots, g_{(2)n_2}\}$ are the generators of $\pi_1(\Pi_1)$ and $\pi_1(\Pi_2)$ respectively. As explained above, a slide of Π_1 through Π_2 along, say, $g_{(2)1}^{-1}$ acts on these sets of generators by conjugating each in the first set with $g_{(2)1}^{-1}$. On the other hand, each inner automorphisms of $\pi_1(\Sigma)$ can be produced by a diffeomorphism that is isotopic to the identity in $\mathrm{Diff}(\Sigma)$.[24] Taking for that the diffeomorphism that conjugates $\pi_1(\Sigma)$ by $g_{(2)1}$ we see that the slide just considered is isotopic (in $\mathrm{Diff}(\Sigma)$, not in $\mathrm{Diff}_F(\Sigma)$ of $\mathrm{Diff}_\infty(\Sigma)$) to a diffeomorphism that leaves the $g_{(1)i}$ untouched and conjugates the $g_{(2)i}$ by $g_{(2)1}$. This, in turn, can

---

[24]That there is a diffeomorphism generating any inner automorphism of $\pi_1(\Sigma)$ is shown exactly as in our discussion of 'internal slides' above. See Fig. 4, where now everything is in Σ and the inside of T_1 is taken to be a solid torus, i.e. there is *no* prime factor Π inside T_1. Then it is obvious that the slide depicted is isotopic to the identity in $\mathrm{Diff}(\Sigma)$ by a diffeomorphism whose support is inside T_2 but extends into the inside of T_1: just take the isotopy $[0,1] \ni s \mapsto \phi_s : (r, \theta, \varphi) \mapsto (r, \theta, s\beta(2-r))$, where $\beta : [0,1] \to [0, 2\pi]$ is the step function as in (3.8), continued by the constant value 2π to $[1,2]$.

be represented by an inner diffeomorphisms of Π_2 (an internal slide, possibly with a 2π-rotation). This shows that the original slide of Π_1 trough Π_2 is isotopic in Diff(Σ) (not in Diff$_F(\Sigma)$ or Diff$_\infty(\Sigma)$) to an internal diffeomorphism of Π_2.[25] For example, if Π_2 is a lens space, which is non-spinorial and has abelian fundamental group, the internal diffeomorphism just considered is clearly isotopic to the identity. Hence the original slide of the first prime through the lens space is isotopic to the identity in Diff(Σ), but not in Diff$_\infty(\Sigma)$ or Diff$_F(\Sigma)$. All this shows that the division of mapping-class generators into internal diffeomorphisms, exchanges, and slides only makes sense in Diff$_\infty(\Sigma)$ and Diff$_F(\Sigma)$, but not in Diff(Σ).

4. A simple yet non-trivial example

In this section we wish to discuss in detail the mapping class group $\mathcal{G}_F(\Sigma)$ for $\Sigma = \mathbb{R}\mathrm{P}^3 \uplus \mathbb{R}\mathrm{P}^3$. Before doing this, let us say a few words on how the single $\mathbb{R}\mathrm{P}^3$ manifold can arise in an exact black-hole solution in General Relativity.

4.1. The $\mathbb{R}\mathrm{P}^3$ geon

Recall that we limited attention to asymptotically flat manifolds with a single end (no 'internal infinity'). Is this not too severe a restriction? After all, we know that the (maximally extended) manifold with one (uncharged, non-rotating) black hole is the Kruskal manifold [54] (see also Chapter 5.5. in [42]), in which space has *two* ends. Figure 5 shows the familiar conformal diagram, where the asymptotically flat ends lie in regions I and III. In Kruskal coordinates[26] (T, X, θ, φ), where T and X each range in $(-\infty, \infty)$ obeying $T^2 - X^2 < 1$, the Kruskal metric reads (as usual, we write $d\Omega^2$ for $d\theta^2 + \sin^2 \theta \, d\varphi^2$):

$$g = \frac{32m^2}{r} \exp(-r/2m) \left(-dT^2 + dX^2\right) + r^2 d\Omega^2 \,, \tag{4.1}$$

where r is a function of T and X, implicitly defined by

$$\left((r/2m) - 1\right) \exp(r/2m) = X^2 - T^2 \tag{4.2}$$

and where $m > 0$ represents the mass of the hole in geometric units. The metric is spherically symmetric and allows for the additional Killing field[27]

$$K = \tfrac{1}{4m}\left(X\partial_T + T\partial_X\right), \tag{4.3}$$

which is timelike for $|X| > |T|$ and spacelike for $|X| < |T|$.

---

[25]This point is not correctly taken care of in [12], where it is argued that a slide of one prime through the other is *always* either isotopic to the identity or the relative 2π-rotation of both primes ([12], p. 1162). But this is only true if the loop along which is slid generates a central element of $\pi_1(\Pi_2)$. In particular, the isotopy claimed in the last sentence of the footnote on p. 1162 of [12] cannot exist in general. However, the application in [12] is eventually restricted to primes with abelian fundamental group so that this difficulty does not affect the specific conclusions drawn in [12]. I thank Fay Dowker and Bob Gompf for discussions of that point.
[26]Kruskal [54] uses (v, u) Hawking Ellis [42] (t', x') for what we call (T, X).
[27]That K is Killing is immediate, since r depends only on the combination $X^2 - T^2$ which is annihilated by K.

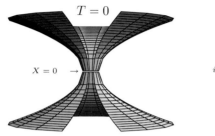
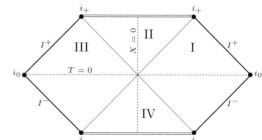

FIGURE 5. To the right is the conformal (Penrose) diagram of Kruskal spacetime in which each point of this 2-dimensional representation corresponds to a 2-sphere (an orbit of the symmetry group of spatial rotations). The asymptotic regions are i_0 (spacelike infinity), $I^\pm$ (future/past lightlike infinity), and $i^\pm$ (future/past timelike infinity). The diamond and triangular shaped regions I and II correspond to the exterior ($r > 2m$) and interior ($0 < r < 2m$) Schwarzschild spacetime respectively, the interior being the black hole. The triangular region IV is the time reverse of II, a white hole. Region III is another asymptotically flat end isometric to the exterior Schwarzschild region I. The double horizontal lines on top an bottom represent the singularities ($r = 0$) of the black and white hole respectively. The left picture shows an embedding diagram of the hypersurface $T = 0$ (central horizontal line in the conformal diagram) that serves to visualize its geometry. Its minimal 2-sphere at the throat corresponds to the intersection of the hyperplanes $T = 0$ and $X = 0$ (bifurcate Killing Horizon).

The familiar exterior Schwarzschild solution is given by region I in Fig. 5, where the transformation from Schwarzschild coordinates (t, r, θ, φ), where $2m < r < \infty$ and $-\infty < t < \infty$, to Kruskal coordinates is given by

$$T = \sqrt{(r/2m) - 1} \, \exp(r/4m) \, \sinh(t/4m) \,, \tag{4.4a}$$

$$X = \sqrt{(r/2m) - 1} \, \exp(r/4m) \, \cosh(t/4m) \,. \tag{4.4b}$$

This obviously just covers region I: $X > |T|$. In Schwarzschild coordinates the Killing field (4.3) just becomes $K = \partial_t$.

Now consider the following discrete isometry of the Kruskal manifold:

$$J : (T, X, \theta, \varphi) \mapsto (T, -X, \pi - \theta, \varphi + \pi) \,. \tag{4.5}$$

It generates a freely acting group $\mathbb{Z}_2$ of smooth isometries which preserve space- as well as time-orientation. Hence the quotient is a smooth space- and time-orientable manifold, the $\mathbb{R}P^3 geon.$[28] Its conformal diagram is just given by cutting away the $X < 0$ part (everything to the left of the vertical $X = 0$ line) in Fig. 5 and taking into account that each point on the remaining edge, $X = 0$, now corresponds to a 2-sphere with antipodal identification, i.e. a $\mathbb{R}P^2$ which is non-orientable. The spacelike hypersurface $T = 0$ has now the topology of the once punctured $\mathbb{R}P^3$. In the left picture of Fig. 5 this corresponds to cutting away the lower half and eliminating the inner boundary 2-sphere $X = 0$ by identifying antipodal points. The latter then becomes a minimal one-sided non-orientable surface in the orientable space-section of topology $\mathbb{R}P^3 - \{\text{point}\}$. The $\mathbb{R}P^3$ geon isometrically contains the exterior Schwarzschild spacetime (region I) with timelike Killing field K. But K ceases to exits globally on the geon spacetime since it reverses direction under (4.5).

Even though the Kruskal spacetime and its quotient are, geometrically speaking, locally indistinguishable, their physical properties are different. In particular, the thermodynamic properties of quantum fields are different. For details we refer to [55] and references therein. We also remark that the mapping-class group $\mathcal{G}_F(\mathbb{R}P^3)$ is trivial [85], as are the higher homotopy groups of $\mathcal{S}_F(\mathbb{R}P^3)$ [25]. Equation (1.16) then shows that the configuration space $\mathcal{S}_F(\mathbb{R}P^3)$ is (weakly) homotopically contractible. In fact, the three-sphere S^3 and the real projective 3-space $\mathbb{R}P^3$ are the only 3-manifolds for which this is true; see [23] (table on p. 922).

4.2. The connected sum $\mathbb{R}P^3 \uplus \mathbb{R}P^3$

Asymptotically flat initial data on the once punctured manifold $\mathbb{R}P^3 \uplus \mathbb{R}P^3 - \{\text{point}\}$ can be explicitly constructed. For this one considers time-symmetric conformally-flat initial data. The constraints (1.4) then simply reduce to the Laplace equation for a positive function Φ, where Φ^4 is the conformal factor. The 'method of images' known from electrostatics can then be employed to construct special solutions with reflection symmetries about two 2-spheres. The topology of the initial data surface is that of $\mathbb{R}^3$ with two disjoint open 3-discs excised. This excision leaves two inner boundaries of S^2 topology on each of which antipodal points are identified. The metric is constructed in such a way that it projects in a smooth fashion to the resulting quotient manifold whose topology is that of $\mathbb{R}P^3 \uplus \mathbb{R}P^3 - \{\text{point}\}$. Details and analytic expressions are given in [29]. These data describe two black holes momentarily at rest. The spacetime they involve into (via (1.3)) is not known analytically, but since the analytical form of the data is very similar indeed to the form of the Misner-wormhole data (cf. [29]), which were often employed in numerical studies, I would not expect the numerical evolution

---

[28]The $\mathbb{R}P^3$ geon is different from the two mutually different 'elliptic interpretations' of the Kruskal spacetime discussed in the literature by Rindler, Gibbons, and others. In [71] the identification map considered is $J' : (T, X, \theta, \varphi) \mapsto (-T, -X, \theta, \varphi)$, which gives rise to singularities on the set of fixed-points (a two-sphere) $T = X = 0$. Gibbons [20] takes $J'' : (T, X, \theta, \varphi) \mapsto (-T, -X, \pi - \theta, \varphi + \pi)$, which is fixed-point free, preserves the Killing field (4.3) (which our map J does not), but does not preserve time-orientation. J'' was already considered in 1957 by Misner & Wheeler (Section 4.2 in [67]), albeit in so-called 'isotropic Schwarzschild coordinates', which only cover the exterior regions I and III of the Kruskal manifold.

to pose any additional difficulties. All this is just to show that the once punctured manifold $\mathbb{RP}^3 \uplus \mathbb{RP}^3$ is not as far fetched in General Relativity as it might seem at first: it is as good, and no more complicated than, the Misner wormhole which is the standard black-hole data set in numerical studies of head-on collisions of equal-mass black holes.[29]

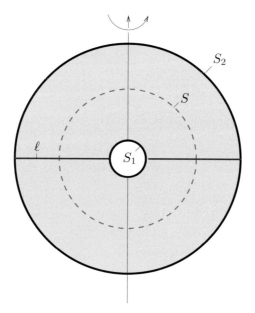

FIGURE 6. Visualization of $\mathbb{RP}^3 \uplus \mathbb{RP}^3$, which corresponds to the shaded region between the 2-spheres S_1 and S_2, on each of which antipodal points are identified. As indicated, the whole picture is to be thought of as rotating about the vertical axis, except for the solid vertical line ℓ, which, like any other radial line, corresponds to a closed loop, showing that $\mathbb{RP}^3 \uplus \mathbb{RP}^3$ is fibered by circles over $\mathbb{RP}^2$. Each fiber intersects the connecting 2-sphere, S, in two distinct points.

We wish to study and visualize the mapping class group $\mathcal{G}_F(\Sigma)$ where $\Sigma = \mathbb{RP}^3 \uplus \mathbb{RP}^3$. We represent Σ by the annular region depicted in Fig. 6, which one should think of as representing a 3-dimensional spherical shell inbetween the outer boundary 2-sphere S_2 and the inner boundary 2-sphere S_1. In addition, on each boundary 2-sphere we identify antipodal points. The result is the connected sum $\mathbb{RP}^3 \uplus \mathbb{RP}^3$ where we might take S for the connecting 2-sphere that lies 'half way'

---

[29]The $\mathbb{RP}^3$ data even have certain advantages: they generalize to data where the masses of the black holes are not equal (for the wormhole identification the masses need to be equal) and even to data for any number of holes with arbitrary masses (in which case the holes may not be 'too close').

between S_1 and S_2 in Fig. 6. The radial lines, like ℓ, fiber the space in loops showing that $\mathbb{RP}^3 \uplus \mathbb{RP}^3$ is an S^1 bundle over $\mathbb{RP}^2$.[30] Interestingly, it is doubly covered by the prime manifold $S^1 \times S^2$, the corresponding deck transformation of the latter being $(\psi, \theta, \varphi) \mapsto (2\pi - \psi, \pi - \theta, \varphi + \pi)$, where $\psi \in [0, 2\pi]$ coordinatizes S^1 and (θ, φ) are the standard spherical polar coordinates on S^2. Note that this deck transformation does not commute with the $SO(2)$ part of the obvious transitive $SO(2) \times SO(3)$ action on $S^1 \times S^2$, so that only a residual $SO(3)$ action remains on $\mathbb{RP}^3 \uplus \mathbb{RP}^3$ whose orbits are 2-spheres except for two $\mathbb{RP}^2$s.[31] It is known to be the only example of a (closed orientable) 3-manifold that is a proper connected sum and covered by a prime.[32]

The fundamental group is the free product of two $\mathbb{Z}_2$:

$$\pi_1(\mathbb{RP}^3 \uplus \mathbb{RP}^3) \cong \mathbb{Z}_2 * \mathbb{Z}_2 = \langle a, b : a^2 = b^2 = 1 \rangle . \tag{4.6}$$

Two loops representing the generators a and b are shown in Fig. 7. Their product,

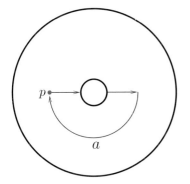 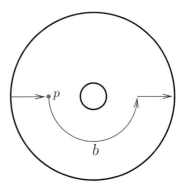

FIGURE 7. Loops a (left picture) and b (right picture) in $\mathbb{RP}^3 \uplus \mathbb{RP}^3$ that generate the fundamental group $\mathbb{Z}_2 * \mathbb{Z}_2 = \langle a, b : a^2 = b^2 = 1 \rangle$ based at point p. The product loop, $c := ab$ (first a then b), is homotopic to the loop ℓ in Fig. 6, that is, to a circle fiber.

$ab =: c$, is homotopic to a circle fiber. Replacing the generator b by ac (recall $a^2 = 1$), the presentation (4.6) can now be written in terms of a and c:

$$\pi_1(\mathbb{RP}^3 \uplus \mathbb{RP}^3) = \langle a, c : a^2 = 1, aca^{-1} = c^{-1} \rangle \cong \mathbb{Z} \rtimes \mathbb{Z}_2 . \tag{4.7}$$

where $\mathbb{Z}_2$ (generated by a) acts as the automorphism $c \mapsto c^{-1}$ on the generator c of $\mathbb{Z}$. This corresponds to the structure of $\mathbb{RP}^3 \uplus \mathbb{RP}^3$ as S^1-fiber bundle over $\mathbb{RP}^2$

---

[30] $\mathbb{RP}^3 \uplus \mathbb{RP}^3$ is an example of a Seifert fibered space without exceptional fibers, that was already mentioned explicitly in Seifert's thesis [73].

[31] $\mathbb{RP}^3$ is trivially homogeneous, being $SO(3)$. The punctured space $\mathbb{RP}^3 - \{\text{point}\}$ is also homogeneous, since it may be identified with the space of all hyperplanes (not necessarily through the origin) in $\mathbb{R}^3$, on which the group $E_3 = \mathbb{R}^3 \rtimes SO(3)$ of Euclidean motions clearly acts transitively with stabilizers isomorphic to E_2; hence $\mathbb{RP}^3 - \{\text{point}\} \cong E_3 / E_2$.

[32] Clearly, no proper connected sum ($\pi_2 \neq 0$) can be covered by an irreducible prime ($\pi_2 = 0$).

with $\pi_1(\text{base})$ acting on $\pi_1(\text{fibre})$. Algebraically, the normal subgroup $\mathbb{Z}$ generated by c is just the subgroup of words in a, b containing an even number of letters.

The generators of mapping classes are the (unique) exchange, ω, the slide μ_{12} of prime 2 through prime 1 (there is only one generator of π_1 for each prime and hence a unique generating slide through each prime), and the slide μ_{21} of prime 1 through prime 2. The relations between them are $\omega^2 = \mu_{12}^2 = \mu_{21}^2 = 1$ and $\omega\mu_{12}\omega^{-1} = \mu_{21}$. There are no other relations, as one may explicitly check using the Fouxe-Rabinovitch relations [62]. The particle group G^P is just $\mathbb{Z}_2$, generated by ω, and the slide subgroup G^S is $\mathbb{Z}_2 * \mathbb{Z}_2$, generated by μ_{12} and μ_{21}. We have $\mathcal{G}_F(\Sigma) = G^S \rtimes G^P = (\mathbb{Z}_2 * \mathbb{Z}_2) \rtimes \mathbb{Z}_2$, where $\mathbb{Z}_2$ acts on $- \mathbb{Z}_2 * \mathbb{Z}_2$ by permuting the factors.

Instead, we may get a presentation in terms of just two generators, ω and $\mu := \mu_{12}$, by dropping μ_{21} and the relation $\omega\mu_{12}\omega^{-1} = \mu_{21}$. This gives[33]

$$\mathcal{G}_F(\mathbb{RP}^3 \uplus \mathbb{RP}^3) = \langle \omega, \mu : \omega^2 = \mu^2 = 1 \rangle \cong \mathbb{Z}_2 * \mathbb{Z}_2 \cong \mathbb{Z} \rtimes \mathbb{Z}_2, \qquad (4.8)$$

where the last isomorphism follows as above (cf.(4.6,4.7)). The normal $\mathbb{Z}$ is given by the subgroup of words in ω, μ with an even number of letters. A visualization of the generators ω and μ is attempted in Fig. 8. It relies on the picture developed in Fig. 6, where the region inside the sphere S is identified with prime 1 and the region outside S with prime 2.

Following [3], the set of inequivalent unitary irreducible representations (UIR's) can be determined directly as follows. First observe that under any UIR $\rho : (\omega, \mu) \mapsto (\hat{\omega}, \hat{\mu})$ the algebra generated by $\hat{\omega}, \hat{\mu}$—we call it the 'representor algebra'—contains $\hat{\omega}\hat{\mu} + \hat{\mu}\hat{\omega}$ in its center. To verify this, just observe that left multiplication of that element by $\hat{\omega}$ equals right multiplication by $\hat{\omega}$ (use $\hat{\omega}^2 = 1$), and likewise with $\hat{\mu}$. Hence, since ρ is irreducible, this central element must be a multiple of the unit $\hat{1}$ (by Schur's Lemma). This implies that the representor algebra is spanned by $\{\hat{1}, \hat{\omega}, \hat{\mu}, \hat{\omega}\hat{\mu}\}$, i.e. it is four dimensional. Burnside's theorem (see e.g. § 10 in [82]) then implies that ρ is at most 2-dimensional. There are four obvious one-dimensional UIR's:

$$\rho_1 : \hat{\omega} = 1, \quad \hat{\mu} = 1, \qquad (4.9a)$$

$$\rho_2 : \hat{\omega} = 1, \quad \hat{\mu} = -1, \qquad (4.9b)$$

$$\rho_3 : \hat{\omega} = -1, \quad \hat{\mu} = 1, \qquad (4.9c)$$

$$\rho_4 : \hat{\omega} = -1, \quad \hat{\mu} = -1. \qquad (4.9d)$$

The first two are bosonic the last two fermionic, either of them appears with any of the possible slide symmetries. The two dimensional representations are determined as follows: expand $\hat{\omega}$ and $\hat{\mu}$ in terms of $\{1, \sigma_1, \sigma_2, \sigma_3\}$, where the σ_i are the standard Pauli matrices. That $\hat{\omega}$ and $\hat{\mu}$ each square to $\hat{1}$ means that $\hat{\omega} = \vec{x} \cdot \vec{\sigma}$ and $\hat{\mu} = \vec{y} \cdot \vec{\sigma}$ with $\vec{x} \cdot \vec{x} = \vec{y} \cdot \vec{y} = 1$. Using equivalences we may diagonalize $\hat{\omega}$ so that $\hat{\omega} = \sigma_3$.

---

[33]This shows $(\mathbb{Z}_2 * \mathbb{Z}_2) \rtimes \mathbb{Z}_2 \cong \mathbb{Z}_2 * \mathbb{Z}_2$. This is no surprise. The normal subgroup isomorphic to $\mathbb{Z}_2 * \mathbb{Z}_2$ of index 2 is given by the set of words in the letters ω, μ_{12}, and μ_{21} containing an even number of the letter ω.

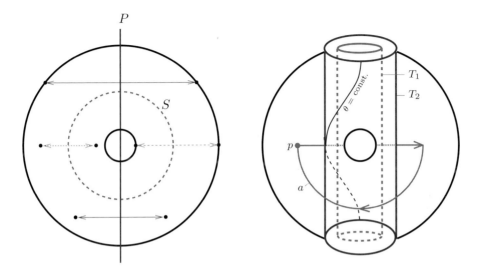

FIGURE 8. The left picture shows how the exchange generator ω can be represented by a combination of two reflections: the first reflection is at the connecting 2-sphere S (dashed), whose action is exemplified by the dashed arrowed lines. This already exchanges the primes, though it is orientation reversing. In order to restore preservation of orientation we add a reflection at a vertical plane, P, whose action is exemplified by the solid arrowed lines. The right picture shows how the slide generator μ may be represented as the transformation of the form (3.8), where the lines ($\theta = $ const) are now helical with a relative π-rotation between top and bottom, in order to be closed (due to the antipodal identification on S_2). We also draw the generator a for the fundamental group on which the slide acts by conjugating it with b. The straight part pierces both tori whereas the curved part runs in front of them as seen from the observer.

The remaining equivalences are then uniquely fixed by eliminating y_2 and ensuring $y_1 > 0$. Writing $y_1 = \sin \tau$ and $y_3 = \cos \tau$ we thus have :

$$\rho_\tau(\omega) = \hat{\omega} = \begin{pmatrix} 1 & 0 \\ 0 & -1 \end{pmatrix}, \qquad \rho_\tau(\mu) = \hat{\mu} = \begin{pmatrix} \cos \tau & \sin \tau \\ \sin \tau & -\cos \tau \end{pmatrix}, \qquad \tau \in (0, \pi).$$

$$(4.10)$$

This continuum of inequivalent 2-dimensional UIRs has interesting properties as regards the statistics types it describes. The angle τ mixes the bosonic and fermionic sector. This mixing is brought about by slides, which physically correspond to transformation where the two 'lumps of topology' (geons) truly penetrate each other. Hence these geons have fixed statistics as long they do not come too close (at low energies), but cease to do so when deep scattering is involved. For further discussion we refer to [2] and especially [79], where the case of more than two geons is considered.

In other non-linear field theories in which to each kink-solution[34] there exists an anti-kink solution, general statements can be made regarding spin-statistics correlations [77]. One would not expect such a relation to generalize to gravity without further specifications, since there is no such thing as an anti-geon. This follows immediately from the uniqueness of the prime decomposition.

It has been verified that spinorial manifolds do in fact give rise to half-integer angular momentum states, e.g. in the kinematical Hilbert space of loop quantum gravity [4] (see also [72] for fractional spin in 2+1 dimensions). Hence a natural question is whether the existence or non-existence of spin-statistics violating states throws some light on the different schemes for the construction of states in quantum gravity. This has e.g. been looked at in [12], where it was shown that the sum over histories approach does exclude fermionic quantization of lens spaces (which are non-spinorial, as we have seen). This touches upon the general question of whether topology changing amplitudes are necessary in order to avoid an embarrassing abundance of (unphysical?) sectors, like those violating standard spin-statistics correlations.

5. Further remarks on the general structure of $\mathcal{G}_F(\Sigma)$

Generically, the group $\mathcal{G}_F(\Sigma)$ is non-abelian and of infinite order; hence it will not be an easy task to understand its structure. We anticipated that the space of inequivalent UIRs label sectors in quantum gravity. However, this seems to only make sense if the group $\mathcal{G}_F(\Sigma)$ is of type I (see [56, 57]), since only then can we *uniquely* decompose an unitary representation into irreducibles. It is known that a countable discrete group is of type I iff it contains an abelian normal subgroup of finite index [80]. This was indeed the case in our example above, where $\mathbb{Z}_2 * \mathbb{Z}_2 \cong \mathbb{Z} \rtimes \mathbb{Z}_2$, hence $\mathbb{Z}$ is normal and of finite index. But, generically, being type I will be rather exceptional.

Another important point is the following: suppose we argue (as e.g. the authors of [79] do) that 'internal' state spaces should be finite dimensional, in which case we would only be interested in finite dimensional representations of $\mathcal{G}_F(\Sigma)$. Are these sufficient to 'make use' of each element of $\mathcal{G}_F(\Sigma)$? In other words: is any non-trivial element $\mathcal{G}_F(\Sigma)$ non-trivially represented in some finite dimensional representation? If not, the intersection of all kernels of finite dimensional representations would lead to a non-trivial normal subgroup and instead of $\mathcal{G}_F(\Sigma)$ we would only 'see' its corresponding quotient. This question naturally leads to the general notion of residual finiteness.

Definition 5.1. A group G is *residually finite* iff for any non-trivial g in G there is a homomorphism ρ_g into a finite group F such that $\rho_g(g)$ is non-trivial in F.

---

[34]We avoid the name soliton since we do not wish to imply that these objects are dynamically stable.

Equivalently, for each non-trivial g in G there exists a normal subgroup N_g of G (the kernel of ρ_g) of finite index such that $g \notin N_g$.

Residual finiteness is carried forward by various constructions. For example:

1.) A subgroup of a residually finite group is residually finite The proof is elementary and given in the appendix (Proposition 6.1).

2.) Let G be the free product $G = G_1 * \cdots * G_n$. Then G is residually finite iff each G_i is. For a proof see e.g. [35].

3.) Let G be finitely generated. If G contains a residually finite subgroup of finite index then G is itself residually finite. The proof is given in the appendix (Proposition 6.6).

4.) Let G be finitely generated and residually finite. Then $\mathrm{Aut}(G)$ is residually finite. Again a proof is given in the appendix (Proposition 6.7).

Note that 3.) implies that finite *upward* extensions of residually finite groups (which are the normal subgroups) are residually finite. But, unfortunately, it is *not* likewise true that finite *downward* extensions (i.e. now the finite group is the normal subgroup) of residually finite groups are always residually finite (see e.g. [48]), not even if the extending group is as simple as $\mathbb{Z}_2$.[35]

We already mentioned above that if a group is residually finite the set of finite dimensional representations, considered as functions on the group, separate the group. Many other useful properties are implied by residual finiteness. For example, proper quotients of a residually finite group G are never isomorphic to G. In other words: any surjective homomorphism of G onto G is an isomorphism (such groups are called 'Hopfian'). Most importantly, any residually finite group has a solvable word problem; see Proposition 6.8 in the appendix.

Large classes of groups share the property of residual finiteness. For example, all free groups are residually finite. Moreover, any group that has a faithful finite-dimensional representation in $GL(n, \mathbb{F})$, where $\mathbb{F}$ is a commutative field, i.e. any matrix group over a commutative field, is residually finite. On the other hand, it is also not difficult to define a group that is *not* residually finite. A famous example is the group generated by two symbols a, b and the single relation $a^{-1}b^2a = b^3$. Generally speaking, there are strong group-cohomological obstructions against residual finiteness [49]. We refer to [58] for an introductory survey and references on residual finiteness.

Now we recall that the mapping class group $\mathcal{G}_F(\Sigma)$ is a finite downward extension of $h_F\big(\mathrm{Aut}(\pi_1(\Sigma))\big)$, where $\pi_1(\Sigma) = \pi_1(P_1) * \cdots * \pi_1(P_N)$. Suppose each $\pi_1(P_i)$ $(1 \le i \le N)$ is residually finite, then so is $\pi_1(\Sigma)$ by 2.), $\mathrm{Aut}(\pi_1(\Sigma))$ by

---

[35] A simple (though not finitely generated) example is the central product of countably infinite copies of the 8 element dihedral group $D_8 := \langle a, b : a^4 = b^2 = (ab)^2 = e \rangle \cong \mathbb{Z}_4 \rtimes \mathbb{Z}_2$, which can be thought of as the symmetry group of a square, where a is the generator of the $\mathbb{Z}_4$ of rotations and b is a reflection. Its center is isomorphic to $\mathbb{Z}_2$ and generated by a^2, the π-rotation of the square. In the infinite central product, where all centers of the infinite number of copies are identified to a single $\mathbb{Z}_2$, every normal subgroup of finite index contains the center. I thank Otto Kegel for pointing out this example.

4.), and $h_F(\text{Aut}(\pi_1(\Sigma))$ by 1.). So we must ask: are the fundamental groups of prime 3-manifolds residually finite? They trivially are for spherical space forms and the handle. For the fundamental group of Haken manifolds residual finiteness has been shown in [44]. Hence, by 3.), it is also true for 3-manifolds which are virtually Haken, i.e. finitely covered by Haken manifolds. As already stated, it is conjectured that every irreducible 3-manifold with infinite fundamental group is virtually Haken. If this were the case, this would show that all prime 3-manifolds, and hence all 3-manifolds[36], have residually finite fundamental group.

Assuming the validity of the 'virtually-Haken' conjecture (or else discarding those primes which violate it) we learn that $h_F(\text{Aut}(\pi_1(\Sigma))$ is residually finite. This almost proves residually finiteness for $\mathcal{G}_F(\Sigma)$ in case our primes also satisfy the homotopy-implies-isotopy property, since then $\mathcal{G}_F(\Sigma)$ is just a (downward) central $\mathbb{Z}_2^{n_s+m}$ extension of $h_F(\text{Aut}(\pi_1(\Sigma))$, where n_s is the number of spinorial primes and m the number of handles. It can be shown that the handle twists (right picture in Fig. 2, which account for m of the $\mathbb{Z}_2$s combine with $h_F(\text{Aut}(\pi_1(\Sigma))$ into a semi-direct product (the extension splits with respect to $\mathbb{Z}_2^m$). Hence the result is residually finite by 3.). On the other hand, the remaining extension by the neck-twists (left picture in Fig. 2) certainly does not split and I do not know of a general argument that would also show residual finiteness in this case, though I would certainly expect it to hold.

6. Summary and outlook

We have seen that mapping-class groups of 3-manifolds enter naturally in the discussion of quantum general relativity and, more generally, in any diffeomorphism invariant quantum theory. Besides being a fascinating mathematical topic in its own right, there are intriguing aspects concerning the physical implications of diffeomorphism invariance in the presence of non-trivial spatial topologies. Everything we have said is valid in any canonical approach to quantum gravity, may it be formulated in metric or loop variables. These approaches use 3-manifolds of fixed topology as fundamental entities out of which spaces of states and observables are to be built in a diffeomorphism (3-dimensional) equivariant way. Neither the spatial topology nor the spatial diffeomorphisms are replaced any anything discrete or quantum. Therefore the rich structures of mapping-class groups are carried along into the quantum framework. John Friedman and Rafael Sorkin were the first to encourage us to take this structure seriously from a physical perspective. Their work remind us on the old idea of Clifford, Riemann, Tait, and others, that otherwise empty space has enough structure to define localized matter-like properties: quasiparticles out of lumps of topology, an idea that was revived in the 1950s and 60s by John Wheeler and collaborators [83].

The impact of diffeomorphism invariance is one of the central themes in all approaches to quantum gravity. The specific issue of mapping-class groups is

---

[36]Recall that we restrict to compact and orientable(the latter being inessential here) manifolds.

clearly just a tiny aspect of it. But this tiny aspect serves very well to give an idea of the range of possible physical implications, which is something that we need badly in a field that, so far, is almost completely driven by formal concepts. For example, the canonical approach differs in its wealth of sectors, deriving from 3-dimensional mapping classes, from the sum-over-histories approach. In the latter, the mapping classes of the bounding 3-manifold do not give rise to extra sectors if they are annihilated after taking the quotient with respect to the normal closure of those diffeomorphisms that extend to the bulk; see e.g. [36][31]. Other examples are the spin-statistics violating sectors, which have been shown to disappear in the sum-over-histories approach in specific cases [12]. However, whether the wealth of sectors provided by the canonical approach does indeed impose an 'embarrassment of richness' from a physical point of view remains to be seen.

Appendix: Elements of residual finiteness

For the readers convenience this appendix collects some of the easier proofs for the standard results on residual finiteness that where used in the main text. We leave out the proof for the result that a free product is residually finite iff each factor is, which is too involved to be presented here. The standard reference is [35].

In the following $H < G$ or $G > H$ indicates that H is a subgroup of G and $H \lhd G$ or $G \rhd H$ that H is normal. The order of the group G is denoted by $|G|$ and the index of H in G by $[G : H]$. The group identity will usually be written as e. The definition of residual finiteness was already given in Definition 5.1, so that we can already start with the first.

Proposition 6.1. *A subgroup of a residually finite group is again residually finite.*

Proof. Let G be residually finite and $G' < G$. Hence, for $e \neq g' \in G'$ there exists a $K \lhd G$ of finite index such that $g' \notin K$. Then $K' := K \cap G'$ is clearly a normal subgroup of G' which does not contain g'. It is also of finite index in G' since the cosets of K' in G' are given by the intersections of the cosets of K in G with G'. To see the latter, note that for $g' \in G'$ one has $g'K' = g'(K \cap G') = (g'K) \cap G'$, since $g'k \in G'$ iff $k \in G'$. $\square$

Lemma 6.2. *Let G be a group and H_i, $i = 1, \cdots, n$ a finite number of subgroups of finite indices. Then the intersection $H := \bigcap_{i=1}^{n} H_i$ is itself of finite index.*

Proof. It suffices to prove this for two subgroups H_1 and H_2. We consider the left cosets of H, H_1 and H_2 and set $|G/H_i| = n_i$ for $i = 1, 2$. Elements $g, h \in G$ are in the same H-coset iff $h^{-1}g \in H$, which is equivalent to $h^{-1}g \in H_i$ for $i = 1, 2$. Hence the H-cosets are given by the non-trivial intersections of the H_1-cosets with the H_2-cosets, of which there are at most $n_1 \cdot n_s$. $\square$

Lemma 6.3. *Let G be finitely generated group, then the number of subgroups of a given finite index, say n, is finite.*

Proof. We essentially follow Chapter III in [5]. Let $H < G$ be of index n and let $\rho(1), \cdots, \rho(n)$ a complete set of left-coset representatives, where without loss of generality we may choose $\rho(1) \in H$. The left cosets are then denoted by $\rho(i)H$ for $i = 1, \cdots, n$. We now consider the (H-dependent) homomorphism $\varphi : G \to S_n$, $g \mapsto \varphi_g$, of G into the symmetric group of degree n, defined through $g(\rho(i)H) =: \rho(\varphi_g(i))H$. Note that this just corresponds to the usual left action of G on the left cosets, which we identified with the numbers $1, \cdots, n$ via the choice of coset representatives. Since $gH = H \Leftrightarrow g \in H$ we have $\mathrm{stab}_\varphi(1) := \{g \in G : \varphi_g(1) = 1\} = H$. Now suppose $H' < G$ is also of index n. Repeating the construction above with left-coset representatives $g'(1), \cdots, g'(n)$ of H', where $g'(1) \in H'$, we obtain another homomorphism $\varphi' : H' \to S_n$ with $\mathrm{stab}_{\varphi'}(1) = H'$. Hence $H \neq H' \Rightarrow \varphi \neq \varphi'$. But since G can be generated by a finite number of elements, say m, there are at most $(n!)^m$ different homomorphisms of G into S_n, and hence at most $(n!)^m$ different subgroups of G with index n. $\qquad\square$

For the following we recall that a subgroup $H < G$ is called *characteristic*, iff it is invariant under *any* automorphism of G. Note that in case the group allows for non-trivial outer automorphisms this is a strictly stronger requirement than that of *normality* which just requires invariance under inner automorphisms. We define

$$G_n := \bigcap \{H < G : [G : H] = n\}, \tag{6.1a}$$

$$\bar{G}_n := \bigcap \{H < G : [G : H] \leq n\}, \tag{6.1b}$$

i.e., the intersections of all subgroups of index n or index $\leq n$ respectively. Lemma 6.3 ensure that there are only finitely many groups to intersect and Lemma 6.2 guarantees that the intersection is again a group of finite index. Moreover, since an automorphism maps a subgroup of index n to a subgroup of index n it also leaves invariant the sets of subgroups of index n or index $\leq n$ respectively. Hence we have

Lemma 6.4. *G_n and $\bar{G}_n$ are characteristic subgroups of finite index.*

This can be used to give a convenient alternative characterization of residual finiteness for finitely generated groups:

Proposition 6.5. *Let G be finitely generated. G is residually finite $\Leftrightarrow$*

$$\hat{G} := \bigcap \{H < G : [G : H] < \infty\} = \{e\} \tag{6.2}$$

Proof. "$\Rightarrow$": Residual finiteness implies that $g \neq e$ in G is not contained in some normal subgroup of finite index. Hence it is not contained in $\hat{G}$.
"$\Leftarrow$": For $g \neq e$ we have $g \notin \hat{G}$ and hence $g \notin G_n$ for some n. Since G_n is in particular normal and of finite index, G is residually finite. $\qquad\square$

The following proposition is a conditional converse to Proposition 6.1:

Proposition 6.6. *Let G be finitely generated. If G contains a residually finite subgroup G' of finite index then G is itself residually finite.*

Proof. Let $[G : G'] = k$, then $\bar{G}_n \subseteq G'$ for all $n \geq k$. Hence

$$\hat{G} = \bigcap \{H < G : [G : H] < \infty\} = \bigcap \{H < G' : [G' : H] < \infty\} = \{e\}, \quad (6.3)$$

where in the last step we applied Proposition 6.5 to G'. This is allowed if G' is finitely generated, which is indeed the case since G is finitely generated and G' is of finite index. $\square$

Note that this proposition implies that finite upward extensions (cf. footnote 16) of residually finite groups are again residually finite. This is not true for finite downward extensions (see below).

There are no analogs of Propositions 6.1 and 6.6 for quotient groups. First, quotient groups of residually finite groups need not be residually finite. An example is provided by the free group F_2 on two generators, which is residually finite (as is any free group F_n, cf. [35]), but not its quotient group $\langle a, b : ab^2a^{-1} = b^3 \rangle$ (see [58], p. 307,308). Second, finite downward extensions of residually finite groups also need not be residually finite [48].

We now turn to other important instances where residually finiteness is inherited.

Proposition 6.7. *Let G be a finitely generated and residually finite group. Then $\mathrm{Aut}(G)$ is residually finite.*

Proof. We follow Section 6.5 of [59]. Assuming $\mathrm{Aut}(G)$ is non-trivial, let α be a non-trivial automorphism. Hence there exists a $g_* \in G$ such that $h := g_*^{-1}\alpha(g_*) \neq e$. Residual finiteness of G implies the existence of a $K \lhd G$ of finite index, say n, not containing h so that $h \notin G_n$ (cf.(6.1a)). On the other hand, since G_n is characteristic, we have a natural homomorphism $\sigma : \mathrm{Aut}(G) \to \mathrm{Aut}(G/G_n)$, simply given by $\sigma(\varphi)(gG_n) := \varphi(g)G_n$, with kernel

$$\mathrm{kernel}(\sigma) = \{\varphi \in \mathrm{Aut}(G) : g^{-1}\varphi(g) \in G_n \, \forall g \in G\} . \quad (6.4)$$

By Lemma 6.4 G/G_n is finite, and so $\mathrm{Aut}(G/G_n)$ and $\mathrm{image}(\sigma) = \mathrm{Aut}(G)/\mathrm{kernel}(\sigma)$ are finite, too. $h \notin G_n$ now implies $\alpha \notin \mathrm{kernel}(\sigma)$. Hence $\mathrm{Aut}(G/G_n))$ is the sought for finite homomorphic image of $\mathrm{Aut}(G)$ into which α maps non-trivially via σ. $\square$

Finally we mention one of the most important consequences of residual finiteness:

Proposition 6.8. *Let G be finitely presented. If G is residually finite, it has a soluble word problem.*

Proof. The idea is to construct two Turing machines, T_1 and T_2, which work as follows: Given a word w on the finite set of generators, T_1 simply checks all consequences of the finite number of relations and stops if w is transformed into e. So if w indeed defines the neutral element in G, T_1 will eventually stop. In contrast, T_2 is now so constructed that it stops it w is *not* the neutral element. Using residual finiteness, this is possible as follows: T_2 writes down the image of w under all homomorphisms of G into all finite groups and stops if this image is not trivial. To do this it lists all finite groups and all homomorphisms into them in a 'diagonal' (Cantor) fashion. To list all homomorphisms it lists all mappings from the finite set of generators of G into that of the finite group, checking each time whether the relations are satisfied (i.e. whether the mapping defines a homomorphism). If w does not define the neutral element, we know by residual finiteness that it will have a non-trivial image in some finite group and hence T_2 will stop after a finite number of steps. Now we run T_1 and T_2 simultaneously. After a finite number of steps either T_1 or T_2 will stops and we will know whether w defines the neutral element in G or not.   □

References

[1] Luis Alvarez-Gaumé and Paul Ginsparg. The structure of gauge and gravitational anomalies. *Annals of Physics*, 161:423–490, 1985.

[2] Charilaos Aneziris et al. Aspects of spin and statistics in general covariant theories. *International Journal of Modern Physics A*, 14(20):5459–5510, 1989.

[3] Charilaos Aneziris et al. Statistics and general relativity. *Modern Physics Letters A*, 4(4):331–338, 1989.

[4] Matthias Arnsdorf and Raquel S. Garcia. Existence of spinorial states in pure loop quantum gravity. *Classical and Quantum Gravity*, 16:3405–3418, 1999.

[5] Gilbert Baumslag. *Topics in Combinatorial Group Theory*. Lectures in Mathematics - ETH Zürich. Birkhäuser, Basel, 1993.

[6] Robert Beig and Niall Ó Murchadha. The Poincaré group as symmetry group of canonocal general relativity. *Annals of Physics (NY)*, 174:463–498, 1987.

[7] Theodor Bröcker and Klaus Jänich. *Einführung in die Differentialtopologie*. Springer Verlag, Berlin, 1973.

[8] Alejandro Cabo and Domingo Louis-Martinez. On Dirac's conjecture for Hamiltonian systems with first-class constraints. *Physical Review D*, 42(8):2726–2736, 1990.

[9] John H. Conway et al. *ATLAS of Finite Groups*. Oxford University Press, Oxford, 1985.

[10] Bryce S. DeWitt. Quantum theory of gravity. I. The canonical theory. *Physical Review*, 160(5):1113–1148, 1967.

[11] Paul Dirac. *Lectures on Quantum Mechanics*. Belfer Graduate School of Science, Monographs Series Number Two. Yeshiva University, New York, 1964.

[12] Fay Dowker and Rafael Sorkin. A spin-statistics theorem for certain topological geons. *Classical and Quantum Gravity*, 15:1153–1167, 1998.

[13] Arthur E. Fischer. The theory of superspace. In M. Carmeli, S.I. Fickler, and L. Witten, editors, *Relativity*, Proceedings of the Relativity Conference in the Midwest, Cincinnati, Ohio, June 2-6 1969, pages 303–357. Plenum Press, New York, 1970.

[14] Arthur E. Fischer. Resolving the singularities in the space of Riemannian geometries. *Journal of Mathematical Physics*, 27:718–738, 1986.

[15] D.I. Fouxe-Rabinovitch. Über die Automorphismengruppen der freien Produkte I. *Matematicheskii Sbornik*, 8:265–276, 1940.

[16] D.I. Fouxe-Rabinovitch. Über die Automorphismengruppen der freien Produkte II. *Matematicheskii Sbornik*, 9:297–318, 1941.

[17] Charles Freifeld. One-parameter subgroups do not fill a neighborhood of the identity in an infinite-dimensional lie (pseudo-) group. In Cecile M. DeWitt and John A. Wheeler, editors, *Battelle Rencontres*, 1967 Lectures in Mathematics and Physics, pages 538–543. W.A. Benjamin, New York, 1968.

[18] John Friedman and Rafael Sorkin. Spin 1/2 from gravity. *Physical Review Letters*, 44:1100–1103, 1980.

[19] John Friedman and Donald Witt. Homotopy is not isotopy for homeomorphisms of 3-manifolds. *Topology*, 25(1):35–44, 1986.

[20] Gary W. Gibbons. The elliptic interpretation of black holes and quantum mechanics. *Nuclear Physics*, B 98:497–508, 1986.

[21] Nick D. Gilbert. Presentations of the automorphims group of a free product. *Proceedings of the London Mathematical Society*, 54:115–140, 1987.

[22] Domenico Giulini. On the possibility of spinorial quantization in the skyrme model. *Modern Physics Letters A*, 8(20):1917–1924, 1993.

[23] Domenico Giulini. 3-manifolds for relativists. *International Journal of Theoretical Physics*, 33:913–930, 1994.

[24] Domenico Giulini. Diffeomorphism invariant states in Witten's 2+1 quantum gravity on $R \times T^2$. *Classical and Quantum Gravity*, 12:2735–2745, 1995.

[25] Domenico Giulini. On the configuration-space topology in general relativity. *Helvetica Physica Acta*, 68:86–111, 1995.

[26] Domenico Giulini. Quantum mechanics on spaces with finite fundamental group. *Helvetica Physica Acta*, 68:439–469, 1995.

[27] Domenico Giulini. What is the geometry of superspace? *Physical Review D*, 51(10):5630–5635, 1995.

[28] Domenico Giulini. The group of large diffeomorphisms in general relativity. *Banach Center Publications*, 39:303–315, 1997.

[29] Domenico Giulini. On the construction of time-symmetric black hole initial data. In F.W. Hehl, C. Kiefer, and R. Metzler, editors, *Black Holes: Theory and Observation*, volume 514 of *Lecture Notes in Physics*, pages 224–243. Springer Verlag, Berlin, 1998.

[30] Domenico Giulini. That strange procedure called quantisation. In D. Giulini, C. Kiefer, and C. Lämmerzahl, editors, *Quantum Gravity: From Theory to Experimental Search*, volume 631 of *Lecture Notes in Physics*, pages 17–40. Springer Verlag, Berlin, 2003.

[31] Domenico Giulini and Jorma Louko. Theta-sectors in spatially flat quantum cosmology. *Physical Review D*, 46:4355–4364, 1992.

[32] Alfred S. Goldhaber. Connection of spin and statistics for charge-monopole composites. *Physical Review Letters*, 36(19):1122–1125, 1976.

[33] Mark J. Gotay. On the validity of Dirac's conjecture regarding first-class secondary constraints. *Journal of Physics A: Mathematical and General*, 16:L141–L145, 1983.

[34] Mark J. Gotay and James M. Nester. Apartheid in the Dirac theory of constraints. *Journal of Physics A: Mathematical and General*, 17:3063–3066, 1984.

[35] Karl W. Grünberg. Residual properties of infinite soluble groups. *Proceedings of the London Mathematical Society*, 7:29–62, 1957.

[36] James Hartle and Donald Witt. Gravitational θ-states and the wave function of the universe. *Physical Review D*, 37(10):2833–2837, 1988.

[37] Peter Hasenfratz and Gerard 't Hooft. Fermion-boson puzzle in a gauge theory. *Physical Review Letters*, 36(19):1119–1122, 1976.

[38] Allan E. Hatcher. Linearization in 3-dimensional topology. In Olli Lehto, editor, *Proceedings of the International Congress of Mathematicians*, volume 1, pages 463–468, Helsinki, 1978. American Mathematical Society (1980).

[39] Allen E. Hatcher. Notes on basic 3-manifold topology. Online available at www.math.cornell.edu/~hatcher/3M/3Mdownloads.html.

[40] Allen E. Hatcher. Homeomorphisms of sufficiently large P^2–irreducible 3-manifolds. *Topology*, 15:343–347, 1976.

[41] Allen E. Hatcher. A proof of the Smale conjecture, $\mathrm{Diff}(S^3) \simeq O(4)$. *Annals of Mathematics*, 117:553–607, 1983.

[42] Stephen W. Hawking and George F.R. Ellis. *The Large Scale Structure of Spacetime*. Cambridge University Press, Cambridge, 1973.

[43] John Hempel. *3-Manifolds*. Princeton University Press, Princeton, New Jersey, 1976.

[44] John Hempel. Residual finiteness for 3-manifolds. In S.M. Gersten and J.R. Stallings, editors, *Combinatorial Group Theory and Topology*, volume 111 of *Annals of Mathematics Studies*, pages 379–396. Princeton University Press, Princeton, New Jersey, 1987.

[45] Harrie Hendriks. Application de la théorie d'obstruction en dimension 3. *Mémoires de la Société Mathématique de France*, 53:81–196, 1977. Online available at www.numdam.org.

[46] Harrie Hendriks and Darryl McCullough. On the diffeomorphism group of a reducible manifold. *Topology and its Applications*, 26:25–31, 1987.

[47] Marc Henneaux and Claudio Teitelboim. *Quantization of Gauge Systems*. Princeton University Press, Princeton, New Jersey, 1992.

[48] Paul R. Hewitt. Extensions of residually finite groups. *Journal of Algebra*, 163(1):757–771, 1994.

[49] Paul R. Hewitt. An army of cohomology against residual finiteness. In C.M. Campbell, T.C. Hurley, E.F. Robertson, S.J. Tobin, and J.J. Ward, editors, *Groups '93*, volume 212 of *London Mathematical Society, Lecture Notes*, pages 305–313. Cambridge University Press, Cambridge, 1995.

[50] Christopher J. Isham. *Theta*–states induced by the diffeomorphism group in canonically quantized gravity. In J.J. Duff and C.J. Isham, editors, *Quantum Structure of Space and Time*, Proceedings of the Nuffield Workshop, August 3-21 1981, Imperial College London, pages 37–52. Cambridge University Press, London, 1982.

[51] John Kalliongis and Darryl McCullough. Isotopies of 3-manifolds. *Topology and its Applications*, 71(3):227–263, 1996.

[52] Claus Kiefer. *Quantum Gravity*, volume 124 of *International Series of Monographs on Physics*. Clarendon Press, Oxford, 2004.

[53] Hellmuth Kneser. Geschlossene Flächen in dreidimensionalen Mannigfaltigkeiten. *Jahresberichte der deutschen Mathematiker Vereinigung*, 38:248–260, 1929.

[54] Martin D. Kruskal. Maximal extension of schwarzschild metric. *Physical Review*, 119(5):1743–1745, 1960.

[55] Jorma Louko. Single-exterior black holes. In J. Kowalski-Glikman, editor, *Towards Quantum Gravity*, volume 541 of *Lecture Notes in Physics*, pages 188–202. Springer Verlag, Berlin, 2000.

[56] George W. Mackey. Infinite-dimensional group representations. *Bulletin of the American Mathematical Society*, 69:628–686, 1963.

[57] George W. Mackey. *Unitary Group Representations in Physics, Probability, and Number Theory*. Advanced Book Classics. Addison-Wesley Publishing Company, Redwood City, California, 1989. Originally published in 1978 as part of the Mathematics Lecture Notes Series by the Benjamin/Cummings Publishing Company.

[58] Wilhelm Magnus. Residually finite groups. *Bulletin of the American Mathematical Society*, 75(2):305–316, 1969.

[59] Wilhelm Magnus, Abraham Karrass, and Donald Solitar. *Combinatorial Group Theory: Presentations of Groups in Terms of Generators and Relations*, volume XIII of *Pure and Applied Mathematics*. Interscience Publishers, John Wiley and Sons Inc., New York, 1966. There is a Dover reprint edition of this book.

[60] G.S. McCarty. Homeotopy groups. *Transactions of the American Mathematical Society*, 106:293–303, 1963.

[61] Darryl McCullough. Topological and algebraic automorphisms of 3-manifolds. In Renzo Piccinini, editor, *Groups of Homotopy Equivalences and Related Topics*, volume 1425 of *Springer Lecture Notes in Mathematics*, pages 102–113. Springer Verlag, Berlin, 1990.

[62] Darryl McCullough and Andy Miller. Homeomorphisms of 3-manifolds with compressible boundary. *Memoirs of the American Mathematical Society*, 61(344), 1986.

[63] Dusa McDuff. The lattice of normal subgroups of the group of diffeomorphisms or homeomorphisms of an open manifold. *Journal of the London Mathematical Society*, 18 (2. series):353–364, 1978.

[64] John W. Milnor. Groups which act on S^n without fixed points. *American Journal of Mathematics*, 79(3):623–630, 1957.

[65] John W. Milnor. A unique decomposition theorem for 3-manifolds. *American Journal of Mathematics*, 84(1):1–7, 1962.

[66] John W. Milnor. *Introduction to Algebraic K-Theory*, volume 72 of *Annals of Mathematics Studies*. Princeton University Press, Princeton, New Jersey, 1971.

[67] Charles Misner and John A. Wheeler. Classical physics as geometry: Gravitation, electromagnetism, unquantized charge, and mass as properties of curved empty space. *Annals of Physics (NY)*, 2:525–660, 1957.

[68] Philip Nelson and Luis Alvarez-Gaumé. Hamiltonian interpretation of anomalies. *Communications in Mathematical Physics*, 99(1):103–114, 1985.

[69] Steven Plotnick. Equivariant intersection forms, knots in S^4 and rotations in 2-spheres. *Transactions of the American Mathematical Society*, 296(2):543–575, 1986. Online available at www.jstor.org.

[70] Kurt Reidemeister. Homotopieringe und Linsenräume. *Abhandlungen des Mathematischen Seminars der Universität Hamburg*, 11:102–109, 1935.

[71] Wolfgang Rindler. Elliptic Kruskal-Schwarzschild space. *Physical Review Letters*, 15(26):1001–1002, 1965.

[72] Joseph Samuel. Fractional spin from gravity. *Physical Review Letters*, 71(2):215–218, 1993.

[73] Herbert Seifert. Topologie dreidimensionaler gefaserter Räume. *Acta Mathematica*, 60:147–288, 1933.

[74] Tony Hilton Royle Skyrme. Kinks and the Dirac equation. *Journal of Mathematical Physics*, 12(8):1735–1743, 1971.

[75] Stephen Smale. Diffeomorphisms of the 2-sphere. *Proceedings of the American Mathematical Society*, 10(4):621–626, 1959.

[76] Rafael Sorkin. Introduction to topological geons. In P.G. Bergmann and V. De Sabbata, editors, *Topological Properties and Global Structure of Space-Time*, volume B138 of *NATO Advanced Study Institutes Series*, page 249. D. Reidel Publishing Company, Dordrecht-Holland, 1986.

[77] Rafael Sorkin. A general relation between kink-exchange and kink-rotation. *Communications in Mathematical Physics*, 115:421–434, 1988.

[78] Rafael Sorkin. Classical topology and quantum phases: Quantum geons. In S. De Filippo, M. Marinaro, G. Marmo, and G. Vilasi, editors, *Geometrical and Algebraic Aspects of Nonlinear Field Theory*, pages 201–218. Elsevier Science Publishers B.V., Amsterdam, 1989.

[79] Rafael Sorkin and Sumati Surya. An analysis of the representations of the mapping class group of a multi-geon three-manifold. *International Journal of Modern Physics A*, 13(21):3749–3790, 1998.

[80] Elmar Thoma. Über unitäre Darstellungen abzählbarer, diskreter Gruppen. *Mathematische Annalen*, 153(2):111–138, 1964. Online available via www.digizeitschriften.de.

[81] Charles B. Thomas. *Elliptic Structures on 3-Manifolds*, volume 104 of *London Mathematical Society Lecture Notes Series*. Cambridge University Press, Cambridge, 1986.

[82] Hermann Weyl. *Gruppentheorie und Quantenmechanik*. Wissenschaftliche Buchgesellschaft, Darmstadt, 1981.

[83] John A. Wheeler. *Geometrodynamics*. Academic Press, New York, 1962.

[84] John Henry Constantine Whitehead. On incidence matrices, nuclei and homotopy types. *Annals of Mathematics*, 42(5):1197–1239, 1941.

[85] Donald Witt. Symmetry groups of state vectors in canonical quantum gravity. *Journal of Mathematical Physics*, 27(2):573–592, 1986.

[86] Donald Witt. Vacuum space-times that admit no maximal slices. *Physical Review Letters*, 57(12):1386–1389, 1986.

[87] Joseph A. Wolf. *Spaces of Constant Curvature*. MacGraw-Hill Book Company, New York, 1967.

Domenico Giulini
Physikalisches Institut
Universität Freiburg
Hermann-Herder-Straße 3
D-79104 Freiburg
Germany
e-mail: `giulini@physik.uni-freiburg.de`

Quantum Gravity
B. Fauser, J. Tolksdorf and E. Zeidler, Eds., 203–219

Kinematical Uniqueness of Loop Quantum Gravity

Christian Fleischhack

Abstract. We review uniqueness results for the kinematical part of loop quantum gravity. After sketching the general loop formalism, the holonomy-flux and the Weyl algebras are introduced. In both cases, then, diffeomorphism invariant representations are described.

Mathematics Subject Classification (2000). Primary 83C45; Secondary 22E67.

Keywords. Loop quantum gravity, kinematical uniqueness, Ashtekar variables, loop variables, holonomy-flux ∗-algebras, diffeomorphism invariance.

1. Introduction

This Blaubeuren workshop has been dedicated to one of the greatest mysteries of modern physics – the unification of gravity and quantum theory. As can be seen from this edited volume, there are several different attempts to disclose at least glimpses of the merged theory. In general, there are two distinct strategies: either some radically new ideas are presented to formulate quantum gravity from scratch, or one focuses on fundamental results inside existing approaches. In this review, we will concentrate on the second issue. We will less discuss the general problem of quantum gravity itself, but study how far mathematical consistency may lead us to deeper insights into the conceptual foundations of one of the major possible routes to quantum gravity – loop quantum gravity.

The main achievement of loop quantization is to quantize gravity as it is – geometry. No additional structures are involved. In some sense, it is a minimalistic quantization. On the other hand, it does not include other interactions in nature. It may, of course, be questioned whether a quantum theory of gravity, or better a quantum theory of general relativity has to contain all existing forces. Indeed, classical gravity itself can be seen as some sort of derived interaction as it is determined intrinsically and purely geometrically by the principle of general relativity. There are approaches, first and foremost string theory, that imitate some sort of such an emergence strategy. However, certain features of gravity – in particular, its geometric origin – are usually quite hidden there. Loop quantum gravity tries to keep track of them. First of all, it admits a rigorous and nonperturbative treatment of diffeomorphism invariance being one of the most important and most peculiar implications of general relativity.

The overall conceptual framework, which loop quantum gravity provides, is reviewed in the articles by Ashtekar, Lewandowski [6] and Ashtekar [5]. In the present article, we are going to focus on a special conceptual point: the current status of diffeomorphism invariance. We will see that this symmetry to a large extent uniquely fixes the kinematical framework of loop quantization. Note, however, that the theorems below are not to be seen as ultimate general statements on some kinematical structure of quantum gravity. They are applicable under the assumption only, that instead of metric and extrinsic curvature, parallel transports of Ashtekar connections and densitized dreibein fields are the canonical variables of gravity to be quantized. It is not the purpose of the present paper to decide whether these are indeed the appropriate classical variables to be quantized. Instead, we will discuss how unique the quantization of general relativity is if formulated in these terms.

The paper is organized as follows: First, Ashtekar's formulation of gravity and the loop variables used there are recalled. Then we study the configuration space with its compactification and calculate the basic Poisson brackets. In the main part of this article, we review the fundamental uniqueness results for diffeomorphism invariant representations of the holonomy-flux $*$-algebra and the Weyl algebra. We conclude with a discussion.

2. Ashtekar variables

Using Ashtekar variables [1], canonical pure gravity can be written in terms of an $SU(2)$ connection A_a^i and a densitized dreibein field E_j^b fulfilling three types of constraints: Gauß, diffeomorphism and Hamilton constraint. The standard ADM formulation can be re-obtained by canonical transformations and symplectic reduction of the system w.r.t. the constraint surface given by the Gauß constraint. As for other gauge theories, this constraint generates the gauge transformations, here in some $SU(2)$ bundle over a Cauchy surface. The remaining constraints encode the invariance of general relativity w.r.t. diffeomorphisms of space-time. The so-called diffeomorphism[1] constraint generates diffeomorphisms within the Cauchy surface, while the Hamilton constraint governs the "dynamics" of the theory. Whereas the latter one is only very poorly understood to date, the Gauß and diffeomorphism constraints are well implemented as will be reviewed in this paper.

---

[1] More precisely, one should speak about the spatial diffeomorphism constraint.

3. Loop variables

The very first step to canonically quantize a classical system is to choose some appropriate set of classical variables to be quantized. This is a highly nontrivial task, since already this selection very much restricts possible quantizations. If too many classical variables get quantized, van Howe arguments may show that there is no nontrivial quantization at all. If, on the other hand, too little variables are in the game, information about certain physical degrees of freedom may be lost.

3.1. Parallel transports

If aiming at a functional integral quantization, one has to construct measures on the configuration space. Since we are dealing with Ashtekar's canonical gravity in the present case, we have to study gauge field theories. There, the configuration space is the space $\mathcal{A}/\mathcal{G}$ of (smooth) connections modulo gauge transformations. This space, however, is mathematically quite delicate: it is non-affine, non-compact, not a manifold and by far not finite-dimensional, whence measure theory is very complicated. As we will see, this problem can be solved using nonlocal parallel transport variables instead of connections themselves. The parallel-transport description is not only equivalent to that using connections, parallel transports even have a much nicer transformation behaviour w.r.t. bundle automorphisms, in particular, gauge transformations. In fact, instead of (locally)

$$A \longmapsto g^{-1}Ag + g^{-1}dg$$

we have

$$h_A(\gamma) \longmapsto g_{\gamma(0)}^{-1} h_A(\gamma) g_{\gamma(1)} \tag{1}$$

for all connections A, all sufficiently smooth paths γ in the Cauchy slice and all gauge transformations g. Here, $h_A(\gamma) \equiv h_\gamma(A)$ denotes the parallel transport w.r.t. A along γ. These variables now comprise the "position" variables to be quantized.[2]

3.2. Fluxes

The generalized momenta in Ashtekar gravity are densitized dreibein fields. In order to have non-singular classical Poisson brackets with parallel transports being smeared along one-dimensional objects, one has to smear the dreibein fields along one-codimensional objects, i.e. along hypersurfaces. More precisely, one first uses the antisymmetric symbol to turn the E into a $\mathfrak{g}^*$-valued $(d-1)$-form E' on some hypersurface S and then smears it. The resulting flux variables $E_{S,f} := \int_S E_i' f^i$ with always S being some hypersurface and $f : S \longrightarrow \mathfrak{g}$ some function, are the "momentum" variables selected for quantization.

---

[2] In the gauge invariant case, only closed paths γ are to be considered. These loops gave loop quantum gravity its name.

4. Configuration space

Any classical system is determined by the Poisson brackets between its basic variables. To quantize the system, first classical variables turn into abstract operators and Poisson brackets into commutators (or anti-commutators). Different quantizations then correspond to different representations of the algebras formed by (appropriate functions of) the abstract operators subject to the commutation relations. In our case, these algebras are the holonomy-flux ∗-algebra $\mathfrak{a}$ [17] and the Weyl algebra $\mathfrak{A}$ [14], to be described below. In the former case, the parallel transports and the fluxes themselves generate the algebra; in the latter one, the fluxes are exponentiated. We will see that diffeomorphism invariance selects (to a large extent) unique representations of these algebras. Hence, quantization in the loop framework is unique at the kinematical level.

Before getting to the representation theory part, we have to study the Poisson relations and at least some parts of the algebras. They naturally lead, in particular, to a completion of the classical configuration space. Since, in the following, the results do not require the Ashtekar connection to be $\mathfrak{su}(2)$-valued and the Cauchy slice to be three-dimensional, we will deal with the general case of a pure gauge field theory over a principal fibre bundle $P(M, \mathbf{G})$ with M being some manifold and $\mathbf{G}$ being some structure Lie group. We will only assume that M is at least two-dimensional and that $\mathbf{G}$ is connected and compact. Now the configuration space consists of all smooth connections (modulo gauge transforms[3]) in P.

4.1. Semianalytic structures

As we will see below, the basic Poisson bracket $\{(h_\gamma)_n^m, E_{S,f}\}$ with m and n being matrix indices, is completely determined by the intersection behaviour between the geometric ingredients γ and S. This is no surprise, since diffeomorphism invariance reduces the theory essentially to topology. The intersection between graphs and hypersurfaces, however, may be quite wild in general. Already in the smooth category, there may exist accumulation points of intersections dividing paths into infinitely many parts. The only natural way out of this seems to always use real-analytic objects for paths, hypersurfaces and diffeomorphisms. However, there are two important bars. First, the intersection of analytic hypersurfaces, as arising for Poisson brackets between fluxes, is far from being analytic again. And second –and most importantly–, the use of analytic objects only, is very unphysical. In fact, gravity is a local theory. Already Einstein's hole argument uses the fact that gravity is invariant w.r.t. diffeomorphisms that are nontrivial only on some small compact neighbourhood in M. Analyticity, on the other hand, is very nonlocal: If you modify an analytic object, you modify it globally. In other words, we have to choose some other smoothness category, reconciling locality and analyticity. Fortunately, this is possible: The semianalytic category [16, 18, 8] provides us with

---

[3]For simplicity, by "configuration space" we will mean the space of connections without factorization.

the framework we need. Sloppily, there we are working with piecewise analytic objects. Analytic submanifolds are replaced by semianalytic sets, which themselves may have some kinks or creases, but at the same time consist of locally finitely many analytic submanifolds. For example, just imagine a sheet of a newspaper: It consist of three "analytic" parts, i.e., the two two-dimensional sheets with the columns and one one-dimensional fold. Unless the newspaper's fold is straightened, the full sheet is only semianalytic instead of analytic.

In what follows, we will always assume that we are working with semianalytic objects.

4.2. Cylindrical functions

Cylindrical functions, in general, are functions on the configuration space depending only on a finite number of values of the basic variables. Using here parallel transports rather than connections themselves, the algebra Cyl of cylindrical functions is now generated by all matrix elements $(T_{\phi,\gamma})_n^m := \phi(h_\gamma)_n^m$ of parallel transports h_γ, where γ runs over all paths in M, ϕ runs over all (equivalence classes of) irreducible representations of $\mathbf{G}$, and m and n over all the corresponding matrix indices. Finite products of $(T_{\phi,\gamma})_n^m$ are called spin network functions [7][4] provided the underlying paths are distinct and form a graph. They already span Cyl as a vector space.

4.3. Generalized connections

As we will see, the cylindrical functions form a subalgebra of both the holonomy-flux *-algebra $\mathfrak{a}$ and the Weyl algebra $\mathfrak{A}$. Therefore, representations of these algebras are automatically representations of Cyl. Since, on the other hand, Cyl is unital abelian, they are completely characterized by the Gelfand-Naimark theorem. In fact, denote the spectrum of the completion of Cyl by $\overline{\mathcal{A}}$. Then each representation of Cyl is the direct sum of representations of Cyl by multiplication operators on $L_2(\overline{\mathcal{A}}, \mu)$ for certain measures[5] μ on $\overline{\mathcal{A}}$. Of course, $\overline{\mathcal{A}}$ is compact and Hausdorff, and $\overline{\text{Cyl}}$ is isomorphic to the continuous functions on $\overline{\mathcal{A}}$. Moreover, by separation properties of parallel transports, $\mathcal{A}$ is even densely embedded into $\overline{\mathcal{A}}$. The elements of $\overline{\mathcal{A}}$ are, therefore, called generalized (or distributional) connections. W.l.o.g., we may consider $\overline{\mathcal{A}}$ as configuration space instead of $\mathcal{A}$ [2].

4.4. Projective limit

Equivalently, $\overline{\mathcal{A}}$ may be described using projective limits [3]. For this, observe

$$\overline{\mathcal{A}} \cong \text{Hom}(\mathcal{P}, \mathbf{G})$$

where $\mathcal{P}$ denotes the groupoid[6] of paths in M. Indeed, each $h \in \text{Hom}(\mathcal{P}, \mathbf{G})$ defines a multiplicative functional $h_{\overline{A}}$ on $\overline{\text{Cyl}}$ via $h_{\overline{A}}((T_{\phi,\gamma})_n^m) := \phi(h(\gamma))_n^m$, which implies

---

[4]Note that, in the quoted reference, spin network functions have been defined a slightly different way in order to implement gauge invariance.

[5]From now on, all measures are assumed to be normalized, regular, and Borel.

[6]The groupoid structure is induced by the standard concatenation of paths modulo reparametrization and deletion/insertion of immediate retracings.

$h_{\overline{A}} \in \overline{A}$. Now, each finite graph γ in M defines a continuous projection

$$\pi_\gamma : \overline{A} \longrightarrow \mathrm{Hom}(\mathcal{P}_\gamma, \mathbf{G}) \cong \mathbf{G}^n$$

with $n := \#\gamma$ via

$$\pi_\gamma(h_{\overline{A}}) \quad := \quad h_{\overline{A}}|_{\mathcal{P}_\gamma} \ \widehat{=} \ h_{\overline{A}}(\gamma) \ \equiv \ \left(h_{\overline{A}}(\gamma_1), \ldots, h_{\overline{A}}(\gamma_n) \right),$$

where $\mathcal{P}_\gamma$ denotes the paths in γ. Note that the edges $\gamma_1, \ldots, \gamma_n$ of γ generate $\mathcal{P}_\gamma$ freely. Using the natural subgraph relation and defining

$$\pi_\gamma^\delta : \mathrm{Hom}(\mathcal{P}_\delta, \mathbf{G}) \longrightarrow \mathrm{Hom}(\mathcal{P}_\gamma, \mathbf{G})$$

for $\gamma \leq \delta$ again by restriction, we get a projective system over the set of all finite graphs in M, whose projective limit $\varprojlim_\gamma \mathrm{Hom}(\mathcal{P}_\gamma, \mathbf{G})$ is $\overline{A}$ again. We remark finally that for every $\psi \in \mathrm{Cyl}$ there is a graph γ in M and some continuous ψ_γ with $\psi = \psi_\gamma \circ \pi_\gamma$.

4.5. Ashtekar-Lewandowski measure

The main mathematical advantage of $\overline{A}$ is its compactness. This opens the door to many structures from measure theory on, in particular, projective limits. In fact, if any sequence of measures μ_γ on $\mathrm{Hom}(\mathcal{P}_\gamma, \mathbf{G}) \cong \mathbf{G}^{\#\gamma}$ is given with γ running over all graphs, such that the compatibility conditions

$$\mu_\gamma \quad = \quad (\pi_\gamma^\delta)_* \, \mu_\delta$$

are fulfilled for all $\gamma \leq \delta$, then there is a unique measure μ on $\overline{A}$ with $(\pi_\gamma)_* \mu = \mu_\gamma$ for all γ. This way, it is rather easy to define such measures. The most obvious choice provides us with the Ashtekar-Lewandowski measure μ_0 [4]. One simply demands that each measure μ_γ equals the Haar measure. Due to the Peter-Weyl theorem, μ_0 is the only measure on $\overline{A}$, where all nontrivial spin network functions have zero integral. The relevance of μ_0 will become clear below.

4.6. Gauge transforms and diffeomorphisms

Since $\overline{A}$ equals $\mathrm{Hom}(\mathcal{P}, \mathbf{G})$, we can naturally extend the action of gauge transforms from $\mathcal{A}$ to $\overline{A}$. Simply generalize (1). [3] Diffeomorphisms can be implemented equally easily. The action of diffeomorphisms on M is naturally lifted to the set of paths and graphs, whence to $\overline{A}$ and $C(\overline{A})$ as well. For instance, for any cylindrical function $\psi_\gamma \circ \pi_\gamma$ over some graph γ, we have $\alpha_\varphi(\psi_\gamma \circ \pi_\gamma) = \psi_\gamma \circ \pi_{\varphi(\gamma)}$. Here, α_φ denotes the action of the diffeomorphism φ on Cyl. Note that both gauge transformations and diffeomorphisms leave the Ashtekar-Lewandowski measure on $\overline{A}$ invariant.

5. Poisson brackets

By means of the Poisson bracket, the fluxes may be regarded as a derivation [17] on the algebra Cyl of cylindrical functions:

$$X_{S,f} \, \psi \quad := \quad \{\psi, E_{S,f}\}. \tag{2}$$

In particular, the result is always a cylindrical function and depends only on the intersection behaviour between the hypersurface S and the graph γ underlying the cylindrical function ψ. More precisely, if the transversal intersections of γ and S are always vertices in γ, then $X_{S,f}(\psi_\gamma \circ \pi_\gamma)$ is again a cylindrical function over γ, essentially given by left and right Lie derivatives on ψ_γ. The explicit formula is given most easily using the Weyl operators being the exponentiated derivations.

5.1. Weyl operators

Let S be some oriented semianalytic subset in M, and let $d : S \longrightarrow \mathbf{G}$ be some function. The intersection properties of S with paths γ are encoded in some function $\sigma_S(\gamma)$. For its definition, first observe that, due to semianalyticity, each path can be decomposed into a finite number of paths whose interior is either fully contained in S (internal path) or disjoint to S (external path). If γ is external path starting non-tangent at S, then $\sigma_S(\gamma)$ is $+1$ (or, resp., -1), if γ starts to above (or, resp., below) S. Of course, "above" and "below" refer to the orientation of S. In any other case, we have $\sigma_S(\gamma) = 0$. Now, one checks very easily [11] that there is a unique map $\Theta_{S,d} : \overline{\mathcal{A}} \longrightarrow \overline{\mathcal{A}}$, with

$$h_{\Theta_{S,d}(\overline{A})}(\gamma) \;\; = \;\; d(\gamma(0))^{\sigma_S(\gamma)} \, h_{\overline{A}}(\gamma) \, d(\gamma(1))^{-\sigma_S(\gamma^{-1})} \tag{3}$$

for all external or internal $\gamma \in \mathcal{P}$ and all $\overline{A} \in \overline{\mathcal{A}}$.

Since $\Theta_{S,d}$ is a homeomorphism, its pull-back $w_{S,d}$ is an isometry on $C(\overline{\mathcal{A}})$. By the translation invariance of Haar measures, $\Theta_{S,d}$ even preserves the Ashtekar-Lewandowski measure μ_0, turning $w_{S,d}$ into a unitary operator on $L_2(\overline{\mathcal{A}}, \mu_0)$. The operators $w_{S,d}$ are called Weyl operators [14].

5.2. Flux derivations

Now, it is straightforward to write down the Poisson bracket (2) more explicitly:

$$X_{S,f}\, \psi \;\; = \;\; \frac{\mathrm{d}}{\mathrm{d}t}\Big|_{t=0} \psi \circ \Theta_{S,e^{tf}} \, .$$

Note that $X_{S,f}$ maps Cyl to Cyl, although for its description again $\overline{\mathcal{A}}$ instead of $\mathcal{A}$ has been used and $\Theta_{S,d}$ generally fails to preserve $\mathcal{A}$. Note, moreover, that $X_{S,f}$ is linear in f [17].

5.3. Higher codimensions

Note that we did not restrict ourselves to the case of genuine hypersurfaces S, i.e., subsets of codimension 1. Although the Poisson bracket is originally given just for this case, the extension to higher codimensions can be justified easily. In fact, observe that the Weyl operator of the disjoint union of hypersurfaces equals the product of the (mutually commuting) Weyl operators of the single hypersurfaces. This way, e.g., the Weyl operator for an equator can either be defined directly as above or obtained from the Weyl operator of the full sphere times the inverses of the Weyl operators corresponding to the upper and lower hemisphere. Both options give the same result. [14]

6. Holonomy-flux $*$-algebra

As we have seen above, the fluxes can be regarded as a derivation on the algebra
Cyl of cylindrical functions. The cylindrical functions, on the other hand, are in a
natural way multiplication operators on Cyl. Therefore, both the position and the
momentum variables may be seen as operators on Cyl.

6.1. Definition

Consider the vector space [17]

$$\mathfrak{a}_{\text{class}} \quad := \quad \text{Cyl} \times \Gamma_1^0(\overline{\mathcal{A}})$$

where the vector space $\Gamma_1^0(\overline{\mathcal{A}})$ of complexified generalized vector fields is generated
by all the $X_{S,f} : \text{Cyl} \longrightarrow \text{Cyl}$ and given a Cyl-module structure. Moreover, $\mathfrak{a}_{\text{class}}$
is equipped with a Lie algebra structure by

$$\left\{(\psi_1, Y_1), (\psi_2, Y_2)\right\} \quad = \quad -\left(Y_1\psi_2 - Y_2\psi_1, [Y_1, Y_2]\right).$$

The quantum **holonomy-flux $*$-algebra** $\mathfrak{a}$ [17] is the $*$-algebra of all words in $\mathfrak{a}_{\text{class}}$
with concatenation as multiplication $\cdot$ and factorized by the canonical commuta-
tion relations

$$a \cdot b - b \cdot a \quad = \quad \text{i} \left\{a, b\right\}, \tag{4}$$

induced by the Poisson brackets, and by the Cyl-module relations

$$\psi \cdot c + c \cdot \psi \quad = \quad 2\psi\, c.$$

Here, $a, b, c \in \mathfrak{a}_{\text{class}}$ and $\psi \in \text{Cyl}$. We indicate the corresponding equivalence
classes by a hat. It turns out that the relations above do not impose additional
relations on Cyl, whence Cyl is embedded into $\mathfrak{a}$. Therefore, we may drop the hats
there. Moreover, note that $\mathfrak{a}$ is generated by all (equivalence classes corresponding
to the) products of a cylindrical function with any, possibly vanishing number
of flux derivations. The flux derivations are invariant w.r.t. the involution $*$; for
cylindrical functions, we have $\psi^* = \overline{\psi}$. Finally, in a natural way, gauge transforms
and diffeomorphisms act covariantly and by automorphisms on $\mathfrak{a}$.

6.2. Symmetric state

The representation theory of $\mathfrak{a}$, responsible for the superselection theory of loop
quantum gravity, can be reduced via GNS to the study of states ω on $\mathfrak{a}$. Of particular interest are states that are symmetric w.r.t. certain algebra automorphisms
α, i.e., we have $\omega = \omega \circ \alpha$ for all such α.

There is a state ω_0 on $\mathfrak{a}$ which is invariant w.r.t. all bundle automorphisms
on P, i.e., it is invariant w.r.t. all gauge transformations and all diffeomorphisms.
It is given [17] by

$$\omega_0(a \cdot \widehat{Y}) \quad = \quad 0$$

and

$$\omega_0(\psi) \;=\; \int_{\overline{\mathcal{A}}} \psi \, \mathrm{d}\mu_0$$

for all $a \in \mathfrak{a}$, $Y \in \Gamma_1^0(\overline{\mathcal{A}})$, and $\psi \in \mathrm{Cyl}$. The invariance heavily relies on the gauge and diffeomorphism invariance of the Ashtekar-Lewandowski measure.

As proven by Lewandowski, Okołów, Sahlmann and Thiemann, for some reasonable technical assumptions, ω_0 is even *the only one* state on $\mathfrak{a}$, that is invariant w.r.t. all bundle automorphisms on P. In other words, there is only one possible quantization implementing gauge and diffeomorphism invariance. The proof of this theorem is outlined in the next subsection; the first two steps are similar to [17].

6.3. Uniqueness proof

Let ω be some state on $\mathfrak{a}$ with the desired invariance properties. Assume further that all smearing functions f are semianalytic and have compact support.

First, we prove $\omega(\widehat{X}^*\widehat{X}) = 0$ for all flux operators $X = X_{S,f}$. Using trivializations, partitions of unity, and linearity in f, we may restrict ourselves to the case that S is given by the intersection of some cube around the origin in $\mathbb{R}^n$ with some linear hyperspace and that $f = f_\| \, \tau$ with supp $f_\| \subseteq S$ and $\tau \in \mathfrak{g}$. Consider now the sesquilinear form

$$(f_1, f_2) \;:=\; \omega\big(\widehat{X}^*_{S,f_1\tau} \, \widehat{X}_{S,f_2\tau}\big)$$

for f_1, f_2 having supports as above. Choose some function $f_\perp$ on the perpendicular to S through the origin with $f_\perp(0) = 1$, such that $\chi := f_\| \otimes f_\perp$ has support in the cube. Fix some vector $\vec{e}$ in S and define

$$\varphi_\lambda \;:=\; \mathrm{id} + \lambda\chi\vec{e}.$$

If $\lambda \in \mathbb{R}$ is sufficiently small, then φ_λ is a diffeomorphism on M preserving S and being the identity outside the cube above. Now one immediately checks, that every function F on M with $F(\vec{x}) = \vec{e} \cdot \vec{x}$ on supp χ fulfills

$$\varphi_\lambda^* F \;=\; F + \lambda f_\|$$

on S, whence we get

$$(F, F) \;=\; (\varphi_\lambda^* F, \varphi_\lambda^* F) \;=\; (F, F) + 2\lambda \, \mathrm{Re} \, (F, f_\|) + \lambda^2 \, (f_\|, f_\|)$$

for all small λ by diffeomorphism invariance. This implies, as desired, $(f_\|, f_\|) = 0$.

Second, the GNS construction for ω yields a Hilbert space $\mathfrak{H} := \overline{\mathfrak{a}/\mathfrak{i}}$ with scalar product $\langle [a], [b] \rangle = \omega(a^*b)$ and a representation π_ω of $\mathfrak{a}$ on $\mathfrak{H}$ with $\pi_\omega(a)[b] = [ab]$ for $a, b \in \mathfrak{a}$. Here, $\mathfrak{i}$ is the left ideal given by the elements $a \in \mathfrak{a}$ with $\omega(a^*a) = 0$. As just seen, all flux derivations are contained in $\mathfrak{i}$. Hence, ω vanishes on all products of cylindrical functions with one or more flux derivations, as ω_0 does.

Finally, we show that ω equals ω_0 for the remaining generators of $\mathfrak{a}$—all spin network functions. By the $*$-invariance of flux derivations and by relation (4), we have for all cylindrical functions ψ and all flux derivations X

$$\omega(X\psi) \;=\; -\mathrm{i}\,\omega(\widehat{X} \cdot \psi) \;=\; -\mathrm{i}\,\overline{\omega(\overline{\psi} \cdot \widehat{X})} \;=\; 0.$$

Let us denote the homeomorphism on $\overline{\mathcal{A}}$ associated to X by Θ_t. By inspection, we see that $\psi \circ \Theta_t$ is always a finite linear combination of cylindrical functions, whereas the t-dependence (and differentiability information) is completely contained in the coefficients. Therefore, we may exchange ω and differentiation to get for all t_0

$$\frac{\mathrm{d}}{\mathrm{d}t}\Big|_{t=t_0} \omega(\psi \circ \Theta_t) = \omega\big(X(\psi \circ \Theta_{t_0})\big) = 0.$$

Hence, $\omega(\psi \circ \Theta_t) = \omega(\psi)$ for all t. Let now ψ be a nontrivial spin network function. We may decompose it into $(T_{\phi,\gamma})_n^m T$ with some edge γ and some nontrivial ϕ, where T is a (possibly trivial) spin network function. Moreover, choose some hypersurface S dividing γ transversally at some point x into γ_1 and γ_2, without intersecting any other edge in the graph underlying ψ. Finally, for each $g \in \mathbf{G}$ choose some smearing function f, such that $\mathrm{e}^{2f(x)} = g$. Directly from definition (3), we get (for the appropriate orientation of S)

$$\psi \circ \Theta_{S,\mathrm{e}f} = \frac{\phi(g)_s^r}{\dim \phi} (T_{\phi,\gamma_1})_r^m (T_{\phi,\gamma_2})_n^s T.$$

Now, the nontriviality of ϕ gives the desired equation:

$$\omega(\psi) = \int_{\mathbf{G}} \omega(\psi) \, \mathrm{d}\mu_{\mathrm{Haar}}(g) = \int_{\mathbf{G}} \omega(\psi \circ \Theta_{S,\mathrm{e}f}) \, \mathrm{d}\mu_{\mathrm{Haar}}(g)$$

$$= \omega\big((T_{\phi,\gamma_1})_r^m (T_{\phi,\gamma_2})_n^s T\big) \int_{\mathbf{G}} \frac{\phi(g)_s^r}{\dim \phi} \, \mathrm{d}\mu_{\mathrm{Haar}}(g) = 0 = \omega_0(\psi).$$

The proof concludes with $\omega(\mathbf{1}) = 1 = \omega_0(\mathbf{1})$.

7. Weyl algebra

Roughly speaking, the holonomy-flux $*$-algebra contains exponentiated positions (parallel transports), but non-exponentiated momenta (fluxes). Working with unitary Weyl operators instead of self-adjoint flux derivations being their generators, allows us to study both positions and momenta in their exponentiated versions. Together they form the Weyl algebra of loop quantum gravity. This is similar to the Weyl algebra [10, 19], studied by Stone and von Neumann in quantum mechanics.

7.1. Definition

Recall that the cylindrical functions are bounded multiplication operators and the Weyl operators unitary operators on $\mathfrak{H}_0 := L_2(\overline{\mathcal{A}}, \mu_0)$. There, the diffeomorphisms act by unitaries as well.

Now, the C^*-subalgebra $\mathfrak{A}$ of $\mathcal{B}(\mathfrak{H}_0)$, generated by cylindrical functions and Weyl operators for constant smearing functions, is called **Weyl algebra** [14]. Its natural representation on $\mathfrak{H}_0$ will be denoted by π_0. Sometimes, we will consider the C^*-subalgebra $\mathfrak{A}_{\mathrm{Diff}}$ of $\mathcal{B}(\mathfrak{H}_0)$, generated by the Weyl algebra $\mathfrak{A}$ and the diffeomorphism group $\mathcal{D}$. One immediately sees that $\mathcal{D}$ acts covariantly on $\mathfrak{A}$.

7.2. Irreducibility

It is easy to see that $\mathfrak{A}$ is irreducible [14, 12, 13]. Let $f \in \mathfrak{A}'$. First, by $C(\overline{\mathcal{A}}) \subseteq \mathfrak{A}$, we have $\mathfrak{A}' \subseteq C(\overline{\mathcal{A}})' = L_\infty(\overline{\mathcal{A}}, \mu_0)$. Second, by unitarity of Weyl operators w, we have $f = w^* \circ f \circ w = w^*(f)$. Therefore, $\langle T, f \rangle = \langle T, w^*(f) \rangle = \langle w(T), f \rangle$ for every spin network function T. Each nontrivial T may be decomposed into $T = (T_{\phi,\gamma})_n^m \, T'$ with some edge γ and nontrivial ϕ, where T' is a (possibly trivial) spin network function. There are two cases: Either ϕ is abelian or nonabelian.

If ϕ is abelian, choose some hypersurface S intersecting γ, but no edge underlying T'. Then[7] $w_{S,g}(T) = \phi(g^2) \, T$ for all $g \in \mathbf{G}$, whence

$$\langle T, f \rangle \;=\; \langle w_{S,g}(T), f \rangle \;=\; \overline{\phi(g^2)} \, \langle T, f \rangle.$$

Since ϕ is nontrivial, there is some $g \in \mathbf{G}$ with $\phi(g^2) \neq 1$. Hence, $\langle T, f \rangle = 0$.

If ϕ is nonabelian, then $\mathrm{tr}\,\phi$ has a zero [15]. Since square roots exist in any compact connected Lie group, there is a $g \in \mathbf{G}$ with $\mathrm{tr}\,\phi(g^2) = 0$. Choose now infinitely many mutually disjoint surfaces S_i intersecting γ, but no edge used for T'. A straightforward calculation yields for $i \neq j$

$$\langle w_{S_i,g}(T), w_{S_j,g}(T) \rangle \;=\; \left| \frac{\mathrm{tr}\,\phi(g^2)}{\dim \phi} \right|^2 \;=\; 0.$$

Now, $\langle w_{S_i,g}(T), f \rangle = \langle T, f \rangle = \langle w_{S_j,g}(T), f \rangle$ implies $\langle T, f \rangle = 0$ again.

Altogether, $\langle T, f \rangle = 0$ for all nontrivial spin network functions T, whence f is constant. Consequently, $\mathfrak{A}'$ consists of scalars only.

7.3. Diffeomorphism invariant representation

Beyond irreducibility, the natural representation π_0 of $\mathfrak{A}$ has some special properties. First, it is diffeomorphism invariant, i.e., there is a diffeomorphism invariant vector in $\mathfrak{H}_0$ (the constant function) and the diffeomorphisms act covariantly on $\mathfrak{A}$. Second, this vector is even cyclic. Third, π_0 is regular, i.e., it is weakly continuous w.r.t. the Weyl operator smearings g. Now, as in the case of the holonomy-flux $*$-algebra, these properties already (to a large extent) distinguish π_0 among the C^*-algebra representations of $\mathfrak{A}$. [14]

7.4. Uniqueness proof

Let $\pi : \mathfrak{A} \longrightarrow \mathcal{B}(\mathfrak{H})$ be some regular representation of $\mathfrak{A}$ on some Hilbert space $\mathfrak{H}$. Assume π diffeomorphism invariant, i.e., π is the restriction of some representation π_{Diff} of $\mathfrak{A}_{\mathrm{Diff}}$ on $\mathfrak{H}$ having some diffeomorphism invariant vector. Moreover, let this vector be cyclic for π. Technically, let us assume that the dimension of M is at least three and that all the hypersurfaces used for the definition of Weyl operators are "reasonably" triangulizable. Finally, let the diffeomorphisms act naturally (as to be explained below). We are going to sketch the proof [14, 13] that π equals π_0.

First, observe that $C(\overline{\mathcal{A}})$, by the denseness of cylindrical functions, is contained in $\mathfrak{A}$. Now, by the general theory of C^*-algebras, the restriction of π to $C(\overline{\mathcal{A}})$ is the direct sum of canonical representations π_ν of $C(\overline{\mathcal{A}})$ by multiplication

---

[7] We shortly write $w_{S,g}$ instead of $w_{S,d}$ if $d(x)$ equals $g \in \mathbf{G}$ everywhere on S.

operators on $L_2(\overline{\mathcal{A}}, \mu_\nu)$ for some measures μ_ν. The constants $\mathbf{1}_\nu$ are cyclic for π_ν. We may choose some $\mathbf{1}_c$ to be diffeomorphism invariant and cyclic for π.

Second, for simplicity, let us assume $\mathbf{G}$ abelian. Fix $\varepsilon > 0$. Let ψ be some nontrivial spin network function, i.e., $\psi = (h_\gamma)^n T$ for some spin network function T and some $n \neq 0$. Assume $\langle \mathbf{1}_c, \pi(\psi)\mathbf{1}_c \rangle_{\mathfrak{H}} \neq 0$. Now, define for some "cubic" hypersurface S

$$w_t := w^S_{e^{it}/2} \quad \text{and} \quad v_t := \frac{1}{2^m} \sum_{k=1}^{2^m} \alpha_{\varphi_k}(w_t).$$

Here, each φ_k is a diffeomorphism winding γ, such that it has exactly m punctures with S. Each k corresponds to a sequence of m signs $+$ or $-$ denoting the relative orientations of S and $\varphi_k(\gamma)$ at the m punctures. (The windings are the reason for M to be at least three-dimensional.) Then

$$v_t(\psi) = \left(\frac{e^{int} + e^{-int}}{2}\right)^m \cdot \psi$$

and

$$\left\|\left(v_t - e^{-\frac{1}{2}m(nt)^2}\right)\psi\right\|_\infty \leq O(m(nt)^4)\,\|\psi\|_\infty.$$

Hence, for any small t, there is some $m = m(nt, \varepsilon)$ with

$$\begin{aligned}
\varepsilon \;&<\; O(m(nt)^2)\,|\langle \mathbf{1}_c, \pi(\psi)\mathbf{1}_c\rangle_{\mathfrak{H}}| - O(m(nt)^4)\,\|\psi\|_\infty \\
&\leq\; |(1 - e^{-\frac{1}{2}m(nt)^2})\langle \mathbf{1}_c, \pi(\psi)\mathbf{1}_c\rangle_{\mathfrak{H}}| - |\langle \mathbf{1}_c, \pi[(v_t - e^{-\frac{1}{2}m(nt)^2})\psi]\mathbf{1}_c\rangle_{\mathfrak{H}}| \\
&\leq\; |\langle \mathbf{1}_c, \pi[(v_t - 1)\psi]\mathbf{1}_c\rangle_{\mathfrak{H}}|.
\end{aligned}$$

Now, for each small t, there is a diffeomorphism φ with

$$\begin{aligned}
\varepsilon \;&\leq\; |\langle \mathbf{1}_c, \pi[(w_t - 1)(\alpha_\varphi(\psi))]\mathbf{1}_c\rangle_{\mathfrak{H}}| \\
&\leq\; 2\,\|\mathbf{1}_c\|_{\mathfrak{H}}\,\|(\pi(w_t) - 1)\mathbf{1}_c\|_{\mathfrak{H}}\,\|\pi(\alpha_\varphi(\psi))\|_{\mathcal{B}(\mathfrak{H})} \\
&=\; 2\,\|\mathbf{1}_c\|_{\mathfrak{H}}\,\|(\pi(w_t) - 1)\mathbf{1}_c\|_{\mathfrak{H}}\,\|\psi\|_\infty,
\end{aligned}$$

by diffeomorphism invariance. The final term, however, does not depend on φ, whence by regularity it goes to zero for $t \to 0$ giving a contradiction. Consequently, $\langle\psi\rangle_{\mu_c} \equiv \langle \mathbf{1}_c, \pi(\psi)\mathbf{1}_c\rangle_{\mathfrak{H}} = 0$ implying $\mu_c = \mu_0$ or, for simplicity, $c = 0$.

Finally, let w be a Weyl operator assigned to some ball or simplex S, possibly having higher codimension. Observe that, for w commuting with α_φ, we have

$$\langle\pi(\alpha_\varphi(\psi))\mathbf{1}_0, \pi(w)\mathbf{1}_0\rangle_{\mathfrak{H}} = \langle\pi(\psi)\mathbf{1}_0, \pi(w)\mathbf{1}_0\rangle_{\mathfrak{H}}.$$

for all spin-network functions ψ. Next, for nontrivial ψ, we choose infinitely many diffeomorphisms φ_i that leave S invariant, but move the respective graph underlying ψ to mutually distinct ones. Then each φ_i commutes with w and we have

$$\delta_{ij} = \langle\alpha_{\varphi_i}(\psi), \alpha_{\varphi_j}(\psi)\rangle_{\mathfrak{H}_0} \equiv \langle\pi(\alpha_{\varphi_i}(\psi))\mathbf{1}_0, \pi(\alpha_{\varphi_j}(\psi))\mathbf{1}_0\rangle_{\mathfrak{H}}.$$

This is possible, unless the ball or simplex has dimension 1 or 2. In fact, there the graph underlying ψ may coincide with S or its boundary. In these cases, the argumentation is technically more involved; here, we only refer to [14]. In the other cases, we now have, with P_0 being the canonical projection from $\mathfrak{H}$ to $\mathfrak{H}_c \cong \mathfrak{H}_0$,

$$0 \;=\; \langle \pi(\psi)\mathbf{1}_0, \pi(w)\mathbf{1}_0 \rangle_{\mathfrak{H}} \;=\; \langle \psi, P_0\pi(w)\mathbf{1}_0 \rangle_{\mathfrak{H}_0}.$$

Thus, $P_0\pi(w)\mathbf{1}_0 = c(w)\,\mathbf{1}_0$ with $c(w) \in \mathbb{C}$, whence also $(\mathbf{1} - P_0)\pi(w)\mathbf{1}_0$ generates $L_2(\overline{\mathcal{A}}, \mu_0)$. The naturality[8] of π w.r.t. the action of diffeomorphisms implies that $\pi(w)\mathbf{1}_0$ is diffeoinvariant itself. Since S is assumed to be a ball or simplex (with lower dimension than M), there is a semianalytic diffeomorphism mapping S to itself, but inverting its orientation. Now, we have

$$\pi(w)^2 \mathbf{1}_0 \;=\; \pi(w)\pi(\alpha_\varphi)\pi(w)^*\pi(\alpha_\varphi)^*\mathbf{1}_0 \;=\; \mathbf{1}_0$$

implying $\pi(w)\mathbf{1}_0 = \mathbf{1}_0$ by taking the square root of the smearing. The proof is completed using triangulizability (disjoint unions of semianalytic subsets correspond to products of Weyl operators) and cyclicity.

8. Conclusions

Let us compare the two main results reviewed above.

8.1. Theorem – self-adjoint case

Let $\mathfrak{a}$ be given as in Subsection 6.1, whereas only those flux derivations are used that correspond to oriented, one-codimensional, semianalytic C^k hypersurfaces S and to compactly supported, semianalytic C^k smearing functions f on S, with some $k > 0$. Moreover, let M be at least two-dimensional and let $\mathbf{G}$ be connected, compact and nontrivial. Then, ω_0 is the only state on $\mathfrak{a}$ that is invariant w.r.t. all semianalytic bundle transformations, acting covariantly on $\mathfrak{a}$.

8.2. Theorem – unitary case

Let $\mathfrak{A}$ be given as in Subsection 7.1, whereas only those Weyl operators are used that correspond to oriented, widely[9] triangulizable, at least one-codimensional, semianalytic C^0 subsets S and to constant smearing functions d on S. Moreover, let M be at least three-dimensional and let $\mathbf{G}$ be connected, compact and nontrivial. Then π_0 is the only regular representation of $\mathfrak{A}$ having a cyclic and diffeomorphism invariant vector, whereas the semianalytic C^0 diffeomorphisms act naturally and covariantly on $\mathfrak{A}$.

---

[8]A representation π is called natural w.r.t. the action of diffeomorphisms iff, for each decomposition of $\pi|_{C(\overline{\mathcal{A}})}$ into cyclic components π_ν, the diffeomorphism invariance of the $C(\overline{\mathcal{A}})$-cyclic vector for π_{ν_1} implies that for π_{ν_2}, provided the measures μ_1 and μ_2 underlying these representations coincide.

[9]A triangulation (K, χ) is called wide iff for every $\sigma \in K$ there is some open chart in M containing the closure of $\chi(\sigma)$ and mapping it to a simplex in that chart.

8.3. Comparison

Mathematically, both theorems look quite related. In fact, for instance, in the case of Lie theory in finite dimensions, the representations of a compact Lie group are always determined by the representations of the corresponding Lie algebra. At most, it may happen that some of the algebra representations do not extend to group representations by some global "discrete" restrictions. Here, the situation seems similar. In the Lie case, the self-adjoint generators of the Lie group unitaries form the Lie algebra; here, the self-adjoint flux derivations in the holonomy-flux *-algebra $\mathfrak{a}$ are the generators of Weyl operators in the Weyl algebra $\mathfrak{A}$. Hence, the uniqueness result for $\mathfrak{a}$ should imply a uniqueness results for $\mathfrak{A}$. However, appearances are deceiving. The main point is the hugeness of the algebras in the game, whence on the self-adjoint level domain issues are to be taken seriously. In the Lie case, it can be proven that they are waived; here, however, we have to deal with infinite-dimensional objects, even nonseparable Hilbert spaces, whence general results are quite scarce. Indeed, by now, there is no direct mathematical relation between these two uniqueness results known.

Moreover, there are quite some other differences between the two settings. For instance, in the holonomy-flux case, the smearings are always compactly supported, whereas in the Weyl case they are constant. The former idea allowed us to use the linearity of the flux derivations w.r.t. the smearing functions; the latter one opened the road to use triangulations into geometrically simpler objects to get rid of the non-linearities of the smearings. It is not known how far either theorems remain valid in the other cases.

A disadvantage of the Weyl case is to need at least three dimensions. The main advantage of the Weyl case, of course, is to circumvent all domain problems ubiquitous in the holonomy-flux case. The only remnant is the regularity of the Weyl operators themselves. However, this only requires that every Weyl operator has a self-adjoint generator with some dense domain; in the holonomy-flux case, these domains all have to coincide. On the other hand, the naturality of the action of diffeomorphisms is only required in the Weyl case, whereas the scope of this assumption is not known yet. Whether this might be cancelled by assuming not only diffeomorphism invariance, but also gauge invariance as in the holonomy-flux case is not known.

8.4. Discussion

Physically, nevertheless, both results are far-reaching. Up to the technical issues mentioned in the previous paragraphs, they show that there is essentially only one quantization of Ashtekar gravity within the loop formalism. Therefore, for this framework, they approach the relevance of the Stone-von Neumann theorem for quantum mechanics. This celebrated result, proven some 75 years ago, is responsible for the (to a large extent) uniqueness of quantization of classical mechanics.

In fact, let us consider one-dimensional classical mechanics. For quantization, it is assumed that the position and momentum variables x and p turn into self-adjoint operators that fulfill $[\hat{x}, \hat{p}] = i$ as induced by the Poisson brackets. Now, these operators generate weakly continuous one-parameter subgroups of unitaries: $U(\sigma) := e^{i\sigma\hat{x}}$ and $V(\lambda) := e^{i\lambda\hat{p}}$. The commutation relation above turns into

$$U(\sigma)V(\lambda) \;=\; e^{i\sigma\lambda}\,V(\lambda)U(\sigma). \tag{5}$$

The Stone-von Neumann theorem now tells us the following [19]: Each pair (U, V) of unitary representations of $\mathbb{R}$ on some Hilbert space that satisfies the commutation relations (5) for all $\sigma, \lambda \in \mathbb{R}$, is equivalent to multiples of the Schrödinger representation

$$U(\sigma) \;=\; e^{i\sigma x}. \quad \text{and} \quad V(\lambda) \;=\; L^*_\lambda$$

by multiplication and pulled-back translation operators on $\mathbb{R}$. The desired uniqueness now follows from irreducibility. In other words, assuming continuity and irreducibility, all "pictures" of quantum mechanics are equivalent. They are physically indistinguishable.

Although the relevance of the loop quantum gravity theorems may indeed be related to the Stone-von Neumann theorem, mathematically there is a major difference between them. In the latter case, both the position and the momentum operators are assumed to be regular, i.e., weakly continuous. In the gravity case, however, the position operators are no longer subject to this requirement. Even more, they are *not* weakly continuous. In fact, only the parallel transports turn into well-defined quantum operators; the connections, in some sense their original generators, are ill defined at the quantum level. The continuity is lost, when the cylindrical functions have been used to form basic variables. Of course, since the continuity is lost already at the level of the algebra and not only at that of representations, this does not weaken the results reviewed in the present article.

Incidently, when the regularity assumption is dropped in the case of quantum mechanics, other representations appear that are non-equivalent to the Schrödinger representation. One of them is given by almost-periodic functions, which lead to the Bohr compactification of the real line. The Hilbert space basis is given by $\{|x\rangle \mid x \in \mathbb{R}\}$, and the operators U and V act by

$$U(\sigma)|x\rangle \;=\; e^{i\sigma x}|x\rangle \quad \text{and} \quad V(\lambda)|x\rangle \;=\; |x + \lambda\rangle.$$

Obviously, V is not continuous. Hence, $\hat{p}$ is not defined, but the operator $V(\lambda)$ corresponding to $e^{i\lambda p}$ only. This type of representation fits much more the pattern we described above. It is very remarkable that just this Bohr-type representation reappears in loop quantum cosmology and leads to a resolution of the big bang singularity [9].

Acknowledgements

The author is very grateful to Bertfried Fauser, Jürgen Tolksdorf and Eberhard Zeidler for the kind invitation to the Blaubeuren workshop on quantum gravity.

References

[1] Abhay Ashtekar: New Variables for classical and quantum gravity. *Phys. Rev. Lett.* **57** (1986) 2244–2247.

[2] Abhay Ashtekar and Chris J. Isham: Representations of the holonomy algebras of gravity and nonabelian gauge theories. *Class. Quant. Grav.* **9** (1992) 1433–1468. e-print: hep-th/9202053.

[3] Abhay Ashtekar and Jerzy Lewandowski: Projective techniques and functional integration for gauge theories. *J. Math. Phys.* **36** (1995) 2170–2191. e-print: gr-qc/9411046.

[4] Abhay Ashtekar and Jerzy Lewandowski: Representation theory of analytic holonomy C^* algebras. In: *Knots and Quantum Gravity* (Riverside, CA, 1993), edited by John C. Baez, pp. 21–61, Oxford Lecture Series in Mathematics and its Applications 1 (Oxford University Press, Oxford, 1994). e-print: gr-qc/9311010.

[5] Abhay Ashtekar: Gravity and the Quantum. *New J. Phys.* **7** (2005) 189. e-print: gr-qc/0410054.

[6] Abhay Ashtekar and Jerzy Levandowski: Background Independent Quantum Gravity: A Status Report. *Class. Quant. Grav.* **21** (2004) R53. e-print:gr-qc/0404018.

[7] John C. Baez: Spin network states in gauge theory. *Adv. Math.* **117** (1996) 253–272. e-print: gr-qc/9411007.

[8] Edward Bierstone and Pierre D. Milman: Semianalytic and subanalytic sets. *Publ. Math. IHES* **67** (1988) 5–42.

[9] Martin Bojowald: Absence of Singularity in Loop Quantum Cosmology. *Phys. Rev. Lett.* **86** (2001) 5227–5230. e-print: gr-qc/0102069.

[10] Ola Bratteli and Derek W. Robinson: *Operator Algebras and Quantum Statistical Mechanics, vol. 2 (Equilibrium States, Models in Quantum Statistical Mechanics).* Springer-Verlag, New York, 1996.

[11] Christian Fleischhack: Construction of Generalized Connections. MPI-MIS preprint 105/2005. e-print: math-ph/0601005.

[12] Christian Fleischhack: Irreducibility of the Weyl Algebra in Loop Quantum Gravity. *Preprint.*

[13] Christian Fleischhack: Quantization Restrictions for Diffeomorphism Invariant Gauge Theories. In: *Complex Analysis, Operator Theory and Applications to Mathematical Physics* (Wien, 2005), edited by Friedrich Haslinger, Emil Straube, and Harald Upmeier. *(to appear)*

[14] Christian Fleischhack: Representations of the Weyl Algebra in Quantum Geometry. e-print: math-ph/0407006.

[15] Patrick X. Gallagher: Zeros of group characters. *Math. Zeitschr.* **87** (1965) 363–364.

[16] R. M. Hardt: Stratification of Real Analytic Mappings and Images. *Invent. Math.* **28** (1975) 193–208.

[17] Jerzy Lewandowski, Andrzej Okołów, Hanno Sahlmann, and Thomas Thiemann: Uniqueness of diffeomorphism invariant states on holonomy-flux algebras. e-print: gr-qc/0504147.

[18] Stanisław Łojasiewicz: Triangulation of semi-analytic sets. *Ann. Scuola Norm. Sup. Pisa* **18** (1964) 449–474.

[19] Jonathan Rosenberg: A selective history of the Stone-von Neumann theorem. *Contemp. Math.* **365** (2004) 331–353.

Christian Fleischhack
Max-Planck-Institut für Mathematik in den Naturwissenschaften
Inselstraße 22–26
04103 Leipzig
Germany
e-mail: `chfl@mis.mpg.de`

Quantum Gravity
B. Fauser, J. Tolksdorf and E. Zeidler, Eds., 221–235

Topological Quantum Field Theory as Topological Quantum Gravity

Kishore Marathe

Abstract. We discuss a special case of the gauge theory to string theory correspondence in the Euclidean version of the theories, where exact results are available. We show how the Witten-Reshetikhin-Turaev invariant in $SU(n)$ Chern-Simons theory on S^3 is related via conifold transition to the all-genus generating function of the topological string amplitudes on a Calabi-Yau manifold. This result can be thought of as an interpretation of TQFT (Topological Quantum Field Theory) as TQG (Topological Quantum Gravity).

Mathematics Subject Classification (2000). Primary 81T45; Secondary 58D29.

Keywords. Chern-Simons theory, string amplitudes, conifold transition, Witten-Reshetikhin-Turaev invariants, knot invariants, topological field theory, topological quantum gravity.

1. Introduction

This talk[1] and in fact, our conference premise ignores the strong admonition by Galileo against[2] "disputar lungamente delle massime questioni senza conseguir verità nissuna". However, I do follow his most important advice: "To read the book of Universe which is wide open in front of our eyes, you must know the language in which it is written. This is the language of Geometry." I will discuss the geometric setting which has had great success in the study of fundamental particles and forces. I will also follow Gallileo's preference and discuss "a small truth" which shows a beautiful and quite unexpected relationship between topological quantum field theory and string theory amplitudes, both calculated in the Euclidean version of the theories. The result is thus primarily of interest in geometric topology. Quantum group computations initiated by Reshetikhin and Turaev and Kohno's

---

[1]Given at the workshop on Mathematical and Physical Aspects of Quantum Gravity
[2]lengthy discussions about the greatest questions that fail to lead to any truth whatever.

special functions corresponding to representations of mapping class groups in the space of conformal blocks also lead to the same result.

Topological quantum field theory was ushered in by Witten in his 1989 paper [4] "QFT and the Jones' polynomial". WRT (Witten–Reshetikhin–Turaev) invariants arose as a byproduct of the quantization of Chern-Simons theory used to characterize the Jones' polynomial. At this time, it is the only known geometric characterization of the Jones' polynomial, although the Feynman integrals used by Witten do not yet have a mathematically acceptable definition. Space-time manifolds in such theories are compact Riemannian manifolds. They are referred to as Euclidean theories in the physics literature. Their role in physically interesting theories is not clear at this time and they should be regarded as toy models.

The unification of electric and magnetic fields by Maxwell is one of the most important chapters in mathematical physics. It is the only field theory which has had great predictive success in classical physics and an extension to the quantum domain. Predictions of Quantum Electrodynamics are in agreement with experimental observations to a very high degree of accuracy. Yang-Mills theory predicted massless particles and was unused for over two decades until the mechanism of symmetry breaking led to the electro-weak theory and the standard model. String theory and certain other supersymmetric theories seem to be the most promising candidates to lead to the so called grand unification of all four fundamental forces. Unifying different string theories into a single theory (such as M-theory) would seem to be the natural first step. Here even the physical foundations are not yet clear. From a mathematical point of view we would be lucky if in a few years we know what are the right questions to ask.

This year we are celebrating a number of special years. The Gauss' year and the 100th anniversary of Einstein's "Annus Mirabilis" (the miraculous year) are the most important among these. Indeed, Gauss' "Disquisitiones generale circa superficies curvas" was the basis and inspiration for Riemann's work which ushered in a new era in geometry. It is an extension of this geometry that is the cornerstone of relativity theory. More recently, we have witnessed the marriage between Gauge Theory and the Geometry of Fiber Bundles from the sometime warring tribes of Physics and Mathematics. Marriage brokers were none other than Chern and Simons. The 1975 paper by Wu and Yang [5] can be regarded as the announcement of this union. It has led to many wonderful offspring. The theories of Donaldson, Chern-Simons, Floer-Fukaya, Seiberg-Witten, and TQFT are just some of the more famous members of their extended family. Quantum Groups, CFT, Supersymmetry, String Theory, Gromov-Witten theory and Gravity also have close ties with this family. In this talk we will discuss one particular relationship between gauge theory and string theory, that has recently come to light. The qualitative aspects of Chern-Simons theory as string theory were investigated by Witten [17] almost ten years ago. Before recounting the main idea of this work we review the Feynman path integral method of quantization which is particularly suited for studying topological quantum field theories. For general background on gauge theory and geometric topology see, for example, [12, 13, 14].

In section 2 we introduce the Feynman path integral approach to QFT. The Euclidean version of this theory is applied in section 3 to the Chern-Simons Lagrangian to obtain the skein relations for the Jones-Witten polynomial of a link in S^3. A by product of this is the family of WRT invariants of 3-manifolds. They are discussed in section 4. Sections 5 to 7 are devoted to studying the relation between WRT invariants of S^3 with gauge group $SU(n)$ and the open and closed string amplitudes in generalized Calabi-Yau manifolds. This result is a special case of the general program introduced by Witten in [17]. A realization of this program even within the Euclidean field theory promises to be a rich and rewarding area of research. We have given some indication of this at the end of section 7. Various formulations of Einstein's equations for gravitational field are discussed in section 8. They also make a surprising appearance in Perelman's proof of Thurston's Geometrization conjecture. No physical application of Euclidean gravitational instantons is known at this time. So their appearance in supersymmetric string theory does not give the usual field equations of gravitation. However, topological amplitudes calculated in this theory can be thought of as Euclidean TQG. We give some further arguments in the concluding section 9 in support of supersymmetric string theory as a candidate for the unification of all four fundamental forces.

2. Quantum Observables

A **quantum field theory** may be considered as an assignment of the **quantum expectation** $< \Phi >_\mu$ to each gauge invariant function $\Phi : \mathcal{A}(M) \to \mathbb{C}$, where $\mathcal{A}(M)$ is the space of gauge potentials for a given gauge group G and the base manifold (space-time) M. Φ is called a **quantum observable** or simply an **observable** in quantum field theory. Note that the invariance of Φ under the group of gauge transformations $\mathcal{G}$ implies that Φ descends to a function on the moduli space $\mathcal{B} = \mathcal{A}/\mathcal{G}$ of gauge equivalence classes of gauge potentials. In the Feynman path integral approach to quantization the quantum or vacuum expectation $< \Phi >_\mu$ of an observable is given by the following expression.

$$< \Phi >_\mu = \frac{\int_{\mathcal{B}(M)} e^{-S_\mu(\omega)} \Phi(\omega) \mathcal{DB}}{\int_{\mathcal{B}(M)} e^{-S_\mu(\omega)} \mathcal{DB}}, \qquad (1)$$

where $e^{-S_\mu} \mathcal{DB}$ is a suitably defined measure on $\mathcal{B}(M)$. It is customary to express the quantum expectation $< \Phi >_\mu$ in terms of the **partition function** Z_μ defined by

$$Z_\mu(\Phi) := \int_{\mathcal{B}(M)} e^{-S_\mu(\omega)} \Phi(\omega) \mathcal{DB}. \qquad (2)$$

Thus we can write

$$< \Phi >_\mu = \frac{Z_\mu(\Phi)}{Z_\mu(1)}. \qquad (3)$$

In the above equations we have written the quantum expectation as $< \Phi >_\mu$ to indicate explicitly that, in fact, we have a one-parameter family of quantum expectations indexed by the coupling constant μ in the action. There are several examples of gauge invariant functions. For example, primary characteristic classes evaluated on suitable homology cycles give an important family of gauge invariant functions. The instanton number and the Yang-Mills action are also gauge invariant functions. Another important example is the Wilson loop functional well known in the physics literature.

Wilson loop functional: Let ρ denote a representation of G on V. Let $\alpha \in \Omega(M, x_0)$ denote a loop at $x_0 \in M$. Let $\pi : P(M, G) \to M$ be the canonical projection and let $p \in \pi^{-1}(x_0)$. If ω is a connection on the principal bundle $P(M, G)$, then the parallel translation along α maps the fiber $\pi^{-1}(x_0)$ into itself. Let $\hat{\alpha}_\omega : \pi^{-1}(x_0) \to \pi^{-1}(x_0)$ denote this map. Since G acts transitively on the fibers, $\exists g_\omega \in G$ such that $\hat{\alpha}_\omega(p) = p g_\omega$. Now define

$$\mathcal{W}_{\rho,\alpha}(\omega) := Tr[\rho(g_\omega)] \ \forall \omega \in \mathcal{A}. \tag{4}$$

We note that g_ω and hence $\rho(g_\omega)$, change by conjugation if, instead of p, we choose another point in the fiber $\pi^{-1}(x_0)$, but the trace remains unchanged. We call these $\mathcal{W}_{\rho,\alpha}$ the Wilson loop functionals associated to the representation ρ and the loop α. In the particular case when $\rho = Ad$ the adjoint representation of G on $\mathbf{g}$, our constructions reduce to those considered in physics. If $L = (\kappa_1, \ldots, \kappa_n)$ is an oriented link with component knots κ_i, $1 \leq i \leq n$ and if ρ_i is a representation of the gauge group associated to κ_i, then we can define the quantum observable $\mathcal{W}_{\rho,L}$ associated to the pair (L, ρ), where $\rho = (\rho_1, \ldots, \rho_n)$ by

$$\mathcal{W}_{\rho,L} = \prod_{i=1}^{n} \mathcal{W}_{\rho_i, \kappa_i} \ .$$

3. Link Invariants

In the 1980s, Jones discovered his polynomial invariant $V_\kappa(q)$, called the **Jones polynomial**, while studying Von Neumann algebras and gave its interpretation in terms of statistical mechanics. These new polynomial invariants have led to the proofs of most of the Tait conjectures. As with most of the earlier invariants, Jones' definition of his polynomial invariants is algebraic and combinatorial in nature and was based on representations of the braid groups and related Hecke algebras. The Jones polynomial $V_\kappa(t)$ of κ is a Laurent polynomial in t (polynomial in t and t^{-1}) which is uniquely determined by a simple set of properties similar to the well known axioms for the Alexander-Conway polynomial. More generally, the Jones polynomial can be defined for any oriented link L as a Laurent polynomial in $t^{1/2}$.

A geometrical interpretation of the Jones' polynomial invariant of links was provided by Witten by applying ideas from QFT to the Chern-Simons Lagrangian constructed from the Chern-Simons action

$$\mathcal{A}_{CS} = \frac{k}{4\pi} \int_M tr(A \wedge dA + \frac{2}{3} A \wedge A \wedge A),$$

where A is the gauge potential of the $SU(n)$ connection ω. The Chern-Simons action is not gauge invariant. Under a gauge transformation g the action transforms as follows:

$$\mathcal{A}_{CS}(A^g) = \mathcal{A}_{CS}(A) + 2\pi k \mathcal{A}_{WZ}, \tag{5}$$

where $\mathcal{A}_{WZ}$ is the **Wess-Zumino action functional**. It can be shown that the Wess-Zumino functional is integer valued and hence, if the Chern-Simons coupling constant k is taken to be an integer, then the partition function Z defined by

$$Z(\Phi) := \int_{\mathcal{B}(M)} e^{-i\mathcal{A}_{CS}(\omega)} \Phi(\omega) \mathcal{DB}$$

is gauge invariant.

We denote the Jones polynomial of L simply by V. Recall that there are 3 standard ways to change a link diagram at a crossing point. The Jones polynomials of the corresponding links are denoted by V_+, V_- and V_0 respectively. To verify the defining relations for the Jones' polynomial of a link L in S^3, Witten [4] starts by considering the Wilson loop functionals for the associated links L_+, L_-, L_0. Witten obtains the following skein relation for the polynomial invariant V of the link

$$t^{n/2}V_+ - t^{-n/2}V_- = (t^{1/2} - t^{-1/2})V_0 \tag{6}$$

where we have put

$$<\Phi> = V(t), \text{ and } t = e^{2\pi i/(k+n)}.$$

We note that the result makes essential use of 3-manifolds with boundary and the Verlinde fusion rules in $2d$ conformal field theory.

For $SU(2)$ Chern-Simons theory, equation (6) is the skein relation that defines a variant of the original Jones' polynomial. This variant also occurs in the work of Kirby and Melvin [9] where the invariants are studied by using representation theory of certain Hopf algebras and the topology of framed links. It is not equivalent to the Jones polynomial. In an earlier work [14] I had observed that under the transformation $\sqrt{t} \to -1/\sqrt{t}$, it goes over into the equation which is the skein relation characterizing the Jones polynomial. The Jones polynomial belongs to a different family that corresponds to the negative values of the level. Note that the coefficients in the skein relation (6) are defined for positive values of the level k. To extend them to negative values of the level we must also note that the shift in k by the dual Coxeter number would now change the level $-k$ to $-k - n$. If in equation (6) we now allow negative values of n and take t to be a formal variable, then the extended family includes both positive and negative levels.

Let $V^{(n)}$ denote the Jones-Witten polynomial corresponding to the skein relation (6), (with $n \in \mathbb{Z}$) then the family of polynomials $\{V^{(n)}\}$ can be shown to be equivalent to the two variable HOMFLY polynomial $P(\alpha, z)$ which satisfies the following skein relation

$$\alpha P_+ - \alpha^{-1}P_- = zP_0. \tag{7}$$

If we put $\alpha = t^{-1}$ and $z = (t^{1/2} - t^{-1/2})$ in equation (7) we get the skein relation for the original Jones polynomial V. If we put $\alpha = 1$ we get the skein relation for the Alexander-Conway polynomial.

To compare our results with those of Kirby and Melvin we note that they use q to denote our t and t to denote its fourth root. They construct a modular Hopf algebra U_t as a quotient of the Hopf algebra $U_q(sl(2, \mathbb{C}))$ which is the well known q-deformation of the universal enveloping algebra of the Lie algebra $sl(2, \mathbb{C})$. Jones polynomial and its extensions are obtained by studying the representations of the algebras U_t and U_q.

4. WRT invariants

If $Z_k(1)$ exists, it provides a numerical invariant of M. For example, for $M = S^3$ and $G = SU(2)$, using the Chern-Simons action Witten obtains the following expression for this partition function as a function of the level k

$$Z_k(1) = \sqrt{\frac{2}{k+2}} \sin\left(\frac{\pi}{k+2}\right). \tag{8}$$

This partition function provides a new family of invariants for $M = S^3$, indexed by the level k. Such a partition function can be defined for a more general class of 3-manifolds and gauge groups. More precisely, let G be a compact, simply connected, simple Lie group and let $k \in \mathbb{Z}$. Let M be a 2-framed closed, oriented 3-manifold. We define the **Witten invariant** $\mathcal{T}_{G,k}(M)$ of the triple (M, G, k) by

$$\mathcal{T}_{G,k}(M) := Z(1) := \int_{\mathcal{B}(M)} e^{-i\mathcal{A}_{CS}} \mathcal{DB}, \tag{9}$$

where $e^{-i\mathcal{A}_{CS}} \mathcal{DB}$, is a suitable measure on $\mathcal{B}(M)$. We note that no precise definition of such a measure is available at this time and the definition is to be regarded as a formal expression. Indeed, one of the aims of TQFT is to make sense of such formal expressions. We define the **normalized Witten invariant** $\mathcal{W}_{G,k}(M)$ of a 2-framed, closed, oriented 3-manifold M by

$$\mathcal{W}_{G,k}(M) := \frac{\mathcal{T}_{G,k}(M)}{\mathcal{T}_{G,k}(S^3)}. \tag{10}$$

If G is a compact, simply connected, simple Lie group and M, N be two 2-framed, closed, oriented 3-manifolds. Then we have the following results:

$$\mathcal{T}_{G,k}(S^2 \times S^1) = 1 \tag{11}$$

$$\mathcal{T}_{SU(2),k}(S^3) = \sqrt{\frac{2}{k+2}} \sin\left(\frac{\pi}{k+2}\right) \tag{12}$$

$$\mathcal{W}_{G,k}(M \# N) = \mathcal{W}_{G,k}(M)\mathcal{W}_{G,k}(N) \tag{13}$$

In his work Kohno [10] defined a family of invariants $\Phi_k(M)$ of a 3-manifold M by its Heegaard decomposition along a Riemann surface Σ_g and representations of its mapping class group in the space of conformal blocks. Similar results were also obtained, independently, by Crane [6]. The agreement of these results (up to normalization) with those of Witten may be regarded as strong evidence for the usefulness of the ideas from TQFT and CFT in low dimensional geometric topology. We remark that a mathematically precise definition of the Witten invariants via solutions of the Yang-Baxter equations and representations of the corresponding quantum groups was given by Reshetikhin and Turaev. For this reason, we now refer to them as Witten-Reshetikhin-Turaev or WRT invariants. The invariant is well defined only at roots of unity. But in special cases it can be defined near roots of unity by a perturbative expansion in Chern-Simons theory. A similar situation occurs in the study of classical modular functions and Ramanujan's mock theta functions. Ramanujan had introduced his mock theta functions in a letter to Hardy in 1920 (the famous last letter) to describe some power series in variable $q = e^{2\pi i z}, z \in \mathbb{C}$. He also wrote down (without proof, as was usual in his work) a number of identities involving these series which were completely verified only in 1988 . Recently, Lawrence and Zagier [11] have obtained several different formulas for the Witten invariant $\mathcal{W}_{SU(2),k}(M)$ of the Poincaré homology sphere $M = \Sigma(2,3,5)$. They show how the Witten invariant can be extended from integral k to rational k and give its relation to the mock theta function. This extension is obtained by a mathematical procedure, Its physical meaning is not yet understood. For integral k they obtain the following fantastic formula, a la Ramanujan, for the Witten invariant of the Poincaré homology sphere

$$\mathcal{W} = 1 + \sum_{n=1}^{\infty} x^{-n^2}(1+x)(1+x^2)\ldots(1+x^{n-1})$$

where $x = e^{\pi i/(k+2)}$. We note that the series on the right hand side of this formula terminates after $k + 2$ terms[3].

5. Chern-Simons and String Theory

The general question "what is the relationship between gauge theory and string theory?" is not meaningful at this time. However, interesting special cases where such relationship can be established are emerging. For example, Witten [17] has argued that Chern-Simons gauge theory on a 3-manifold M can be viewed as a string theory constructed by using a topological sigma model with target space T^*M. The perturbation theory of this string will coincide with Chern-Simons perturbation theory, in the form discussed by Axelrod and Singer [1]. The coefficient of k^{-r} in the perturbative expansion of $SU(n)$ theory in powers of $1/k$ comes from Feynman diagrams with r loops. Witten shows how each diagram can be replaced by a Riemann surface Σ of genus g with h holes (boundary components) with $g = (r - h + 1)/2$. Gauge theory would then give an invariant $\Gamma_{g,h}(M)$ for every topological type of Σ. Witten shows that this invariant would equal the corresponding string partition function $Z_{g,h}(M)$. We now give an example of gauge theory

---

[3]I would like to thank Don Zagier for bringing this work to my attention

to string theory correspondence relating the non-perturbative WRT invariants in Chern-Simons theory with gauge group $SU(n)$ and topological string amplitudes which generalize the GW (Gromov-Witten) invariants of Calabi-Yau 3-folds. The passage from real 3 dimensional Chern-Simons theory to the 10 dimensional string theory and further onto the 11 dimensional M-theory can be schematically represented by the following:

$$
\begin{aligned}
3 + 3 &= 6 \text{ (real symplectic 6-manifold)} \\
&= 6 \text{ (conifold in } \mathbb{C}^4 \text{)} \\
&= 6 \text{ (Calabi-Yau manifold)} \\
&= 10 - 4 \text{ (string compactification)} \\
&= (11 - 1) - 4 \text{ (M-theory)}
\end{aligned}
$$

We now discuss the significance of the various terms of the above equation array. Recall that string amplitudes are computed on a 6-dimensional manifold which in the usual setting is a complex 3-dimensional Calabi-Yau manifold obtained by string compactification. This is the most extensively studied model of passing from the 10-dimensional space of supersymmetric string theory to the usual 4-dimensional space-time manifold. However, in our work we do allow these so called extra dimensions to form an open or a symplectic Calabi-Yau manifold. We call these the generalized Calabi-Yau manifolds. The first line suggests that we consider open topological strings on such a generalized Calabi-Yau manifold, namely, the cotangent bundle T^*S^3, with Dirichlet boundary conditions on the zero section S^3. We can compute the open topological string amplitudes from the $SU(n)$ Chern-Simons theory. Conifold transition [16] has the effect of closing up the holes in open strings to give closed strings on the Calabi-Yau manifold obtained by the usual string compactification from 10 dimensions. Thus we recover a topological gravity result starting from gauge theory. In fact, as we discussed earlier, Witten had anticipated such a gauge theory string theory correspondence almost ten years ago. Significance of the last line is based on the conjectured equivalence of M-theory compactified on S^1 to type IIA strings compactified on a Calabi-Yau threefold. We do not consider this aspect here. The crucial step that allows us to go from a real, non-compact, symplectic 6-manifold to a compact Calabi-Yau manifold is the conifold or geometric transition. Such a change of geometry and topology is expected to play an important role in other applications of string theory as well.

6. Conifold Transition

To understand the relation of the WRT invariant of S^3 for $SU(n)$ Chern-Simons theory with open and closed topological string amplitudes on "Calabi-Yau" manifolds we need to discuss the concept of conifold transition. From the geometrical point of view this corresponds to symplectic surgery in six dimensions. It replaces a vanishing Lagrangian 3-sphere by a symplectic S^2. The starting point of the construction is the observation that T^*S^3 minus its zero section is symplectomorphic to the cone $z_1^2 + z_2^2 + z_3^2 + z_4^2 = 0$ minus the origin in $\mathbb{C}^4$, where each manifold is taken with its standard symplectic structure. The complex singularity at the origin can be smoothed out by the manifold M_τ defined by $z_1^2 + z_2^2 + z_3^2 + z_4^2 = \tau$ producing a Lagrangian S^3 vanishing cycle. There are also two so called small resolutions $M^\pm$ of the singularity with exceptional set $\mathbb{C}P^1$.

They are defined by

$$M^\pm := \left\{ z \in \mathbb{C}^4 \mid \frac{z_1 + iz_2}{z_3 \pm iz_4} = \frac{-z_3 \pm iz_4}{z_1 - iz_2} \right\} .$$

Note that $M_0 \setminus \{0\}$ is symplectomorphic to each of $M^\pm \setminus \mathbb{C}P^1$. Blowing up the exceptional set $\mathbb{C}P^1 \subset M^\pm$ gives a resolution of the singularity which can be expressed as a fiber bundle F over $\mathbb{C}P^1$. Going from the fiber bundle T^*S^3 over S^3 to the fiber bundle F over $\mathbb{C}P^1$ is referred to in the physics literature as the conifold transition. We note that the holomorphic automorphism of $\mathbb{C}^4$ given by $z_4 \mapsto -z_4$ switches the two small resolutions $M^\pm$ and changes the orientation of S^3. Conifold transition can also be viewed as an application of mirror symmetry to Calabi-Yau manifolds with singularities. Such an interpretation requires the notion of symplectic Calabi-Yau manifolds and the corresponding enumerative geometry. The geometric structures arising from the resolution of singularities in the conifold transition can also be interpreted in terms of the symplectic quotient construction of Marsden and Weinstein.

7. WRT invariants and topological string amplitudes

To find the relation between the large n limit of $SU(n)$ Chern-Simons theory on S^3 to a special topological string amplitude on a Calabi-Yau manifold we begin by recalling the formula for the partition function (vacuum amplitude) of the theory $\mathcal{T}_{SU(n),k}(S^3)$ or simply $\mathcal{T}$. Upto a phase, it is given by

$$\mathcal{T} = \frac{1}{\sqrt{n(k+n)^{(n-1)}}} \prod_{j=1}^{n-1} \left[2\sin\left(\frac{j\pi}{k+n}\right) \right]^{n-j} . \tag{14}$$

Let us denote by $F_{(g,h)}$ the amplitude of an open topological string theory on T^*S^3 of a Riemann surface of genus g with h holes. Then the generating function for the free energy can be expressed as

$$-\sum_{g=0}^{\infty} \sum_{h=1}^{\infty} \lambda^{2g-2+h} n^h F_{(g,h)} \tag{15}$$

This can be compared directly with the result from Chern-Simons theory by expanding the $\log \mathcal{T}$ as a double power series in λ and n.

Instead of that we use the conifold transition to get the topological amplitude for a closed string on a Calabi-Yau manifold. We want to obtain the large n

expansion of this amplitude in terms of parameters λ and τ which are defined in terms of the Chern-Simons parameters by

$$\lambda = \frac{2\pi}{k+n}, \tau = n\lambda = \frac{2\pi n}{k+n}. \tag{16}$$

The parameter λ is the string coupling constant and τ is the 't Hooft coupling $n\lambda$ of the Chern-Simons theory. The parameter τ entering in the string amplitude expansion has the geometric interpretation as the Kähler modulus of a blown up S^2 in the resolved $M^{\pm}$. If $F_g(\tau)$ denotes the amplitude for a closed string at genus g then we have

$$F_g(\tau) = \sum_{h=1}^{\infty} \tau^h F_{(g,h)} \tag{17}$$

So summing over the holes amounts to filling them up to give the closed string amplitude.

The large n expansion of $\mathcal{T}$ in terms of parameters λ and τ is given by

$$\mathcal{T} = \exp\left[-\sum_{g=0}^{\infty} \lambda^{2g-2} F_g(\tau)\right], \tag{18}$$

where F_g defined in (17) can be interpreted on the string side as the contribution of closed genus g Riemann surfaces. For $g > 1$ the F_g can be expressed in terms of the Euler characteristic χ_g and the Chern class c_{g-1} of the Hodge bundle of the moduli space $\mathcal{M}_g$ of Riemann surfaces of genus g as follows

$$F_g = \int_{\mathcal{M}_g} c_{g-1}^3 - \frac{\chi_g}{(2g-3)!} \sum_{n=1}^{\infty} n^{2g-3} e^{-n(\tau)}. \tag{19}$$

The integral appearing in the formula for F_g can be evaluated explicitly to give

$$\int_{\mathcal{M}_g} c_{g-1}^3 = \frac{(-1)^{(g-1)}}{(2\pi)^{(2g-2)}} 2\zeta(2g-2)\chi_g. \tag{20}$$

The Euler characteristic is given by the Harer-Zagier [7] formula

$$\chi_g = \frac{(-1)^{(g-1)}}{(2g)(2g-2)} B_{2g}, \tag{21}$$

where B_{2g} is the $(2g)$-th Bernoulli number. We omit the special formulas for the genus 0 and genus 1 cases. The formulas for F_g for $g \geq 0$ coincide with those of the g-loop topological string amplitude on a suitable Calabi-Yau manifold. The change in geometry that leads to this calculation can be thought of as the result of coupling to gravity. Such a situation occurs in the quantization of Chern-Simons theory. Here the classical Lagrangian does not depend on the metric, however, coupling to the gravitational Chern-Simons term is necessary to make it TQFT.

We have mentioned the following four approaches that lead to the WRT invariants.

1. Witten's QFT calculation of the Chern-Simons partition function

2. Quantum group (or Hopf algebraic) computations initiated by Reshetikhin and Turaev

3. Kohno's special functions corresponding to representations of mapping class groups in the space of conformal blocks and a similar approach by Crane

4. open or closed string amplitudes in suitable Calabi-Yau manifolds

These methods can also be applied to obtain invariants of links, such as the Jones polynomial. Indeed, this was the objective of Witten's original work. WRT invariants were a byproduct of this work. Their relation to topological strings came later.

The WRT to string theory correspondence has been extended by Gopakumar, Ooguri, Vafa by using string theoretic arguments to show that the expectation value of the quantum observables defined by the Wilson loops in the Chern-Simons theory also has a similar interpretation in terms of a topological string amplitude. This leads them to conjecture a correspondence between certain knot invariants (such as the Jones polynomial) and Gromov-Witten type invariants of generalized Calabi-Yau manifolds. A knot should correspond to a Lagrangian D-brane on the string side and the knot invariant would then give a suitably defined count of compact holomorphic curves with boundary on the D-brane. Special cases of the conjecture have been verified. To understand a proposed proof, recall first that a categorification of an invariant I is the construction of a suitable homology such that its Euler characteristic equals I. A well known example of this is Floer's categorification of the Casson invariant.

Recently Khovanov [8] has obtained a categorification of the Jones polynomial $V_\kappa(q)$ by constructing a bi-graded $sl(2)$-homology $H_{i,j}$ determined by the knot κ. Its quantum or graded Euler characteristic equals the Jones polynomial. i.e.

$$V_\kappa(q) = \sum_{i,j}(-1)^j q^i \dim H_{i,j} .$$

Now let L_κ be the Lagrangian submanifold corresponding to the knot κ of a fixed Calabi-Yau space X. Let r be a fixed relative integral homology class of the pair (X, L_κ). Let $\mathcal{M}_{g,r}$ denote the moduli space of pairs (Σ_g, A), where Σ_g is a compact Riemann surface in the class r with boundary S^1 and A is a flat $U(1)$ connection on Σ_g. This data together with the cohomology groups $H^k(\mathcal{M}_{g,r})$ determines a tri-graded homology. It generalizes the Khovanov homology. Its Euler characteristic is a generating function for the BPS states' invariants in string theory and these can be used to obtain the Gromov-Witten invariants. Taubes has given a construction of the Lagrangians in the Gopakumar-Vafa conjecture. We note that counting holomorphic curves with boundary on a Lagrangian manifold was introduced by Floer in his work on the Arnold conjecture.

The tri-graded homology is expected to unify knot homologies of the Khovanov type as well as knot Floer homology constructed by Ozswáth and Szabó [15] which provides a categorification of the Alexander polynomial. Knot Floer homology is defined by counting pseudo-holomorphic curves and has no known combinatorial description. An explicit construction of a tri-graded homology for certain torus knots has been recently given by Dunfield, Gukov and Rasmussen [math.GT/0505662].

8. Strings and gravity

Recall that in string theory, an elementary particle is identified with a vibrational mode of a string. Different particles correspond to different harmonics of vibration. The Feynman diagrams of the usual QFT are replaced by fat graphs or Riemann surfaces that are generated by moving strings splitting or joining together. The particle interactions described by these Feynman diagrams are built into the basic structure of string theory. The appearance of Riemann surfaces explains the relation to conformal field theory. We have already discussed Witten's argument relating gauge and string theories. It now forms a small part of the program of relating quantum group invariants and topological string amplitudes. In general, the string states are identified with fields. The ground state of the closed string turns out to be a massless spin two field which may be interpreted as a graviton. In the large distance limit, (at least at the lower loop levels) string theory includes the vacuum equations of Einstein's general relativity theory. String theory avoids the ultraviolet divergences that appear in conventional attempts at quantizing gravity. In physically interesting string models one expects the string space to be a non-trivial bundle over a Lorentzian space-time M with compact or non-compact fibers.

Relating the usual Einstein's equations with cosmological constant with the Yang-Mills equations requires the ten dimensional manifold $\Lambda^2(M)$ of differential forms of degree two. There are several differences between the Riemannian functionals used in theories of gravitation and the Yang-Mills functional used to study gauge field theories. The most important difference is that the Riemannian functionals are dependent on the bundle of frames of M or its reductions, while the Yang-Mills functional can be defined on any principal bundle over M. However, we have the following interesting theorem [2].

Theorem: Let (M, g) be a compact, 4-dimensional, Riemannian manifold. Let $\Lambda_+^2(M)$ denote the bundle of self-dual 2-forms on M with induced metric G_+. Then the Levi-Civita connection λ_g on M satisfies the Euclidean gravitational instanton equations if and only if the Levi-Civita connection λ_{G_+} on $\Lambda_+^2(M)$ satisfies the Yang-Mills instanton equations.

We note that the usual Einstein's equations with cosmological constant can also be formulated as operator equations on the ten dimensional manifold $\Lambda^2(M)$ with linear fibers over the space-time manifold M as follows. We define the gravitational tensor W_g, by

$$W_g := R + g \times_c T, \tag{22}$$

where R is the Riemann-Christoffel curvature tensor, T is the energy-momentum tensor and $\times_c$ is the Petrov product. We also denote by W_g the linear transformation of $\Lambda_x^2(M)$ induced by W_g. Then g satisfies the generalized field equations of gravitation if W_g commutes with the Hodge operator J_x or equivalently W_g is a complex linear transformation of the complex vector space $\Lambda_x^2(M)$ with almost complex structure J_x. Proof of this as well as other geometric formulations may be found in [3].

We would like to add that the full Einstein equations with dilaton field as source play a fundamental role in Perelman's work on the Poincaré conjecture. The Ricci flow is perturbed by a scalar field which corresponds in string theory to the dilaton. It is supposed to determine the overall strength of all interactions. The value of the dilaton field can be thought of as the size of an extra dimension of space. This would give the space 11 dimensions as required in the M-theory. The low energy effective action of the dilaton field is given by the functional $\mathcal{F}(g, f) = \int_M (R + |\nabla f|^2)e^{-f}$. The corresponding variational equations are

$$R_{ij} - \frac{1}{2}Rg_{ij} = -(\nabla_i\nabla_j f - \frac{1}{2}(\Delta f)g_{ij}).$$

These are the usual Einstein equations with the energy-momentum tensor of the dilaton field as source. They lead to the decoupled evolution equations

$$(g_{ij})_t = -2(R_{ij} + \nabla_i\nabla_j f), f_t = -R - \Delta f.$$

After applying a suitable diffeomorphism these equations lead to the gradient flow equations. This modified Ricci flow can be pushed through the singularities by surgery and rescaling. A detailed case by case analysis is then used to prove Thurston's geometrization conjecture. This includes as a special case the classical Poincaré conjecture.

9. Conclusion

We have seen that QFT calculations have their counterparts in string theory. One can speculate that this is a topological quantum gravity (TQG) interpretation of a result in TQFT, in the Euclidean version of the theories. If modes of vibration of a string are identified with fundamental particles, then their interactions are already built into the theory. Consistency with known physical theories requires string theory to include supersymmetry. While supersymmetry has had great success in mathematical applications, its physical verification is not yet available. However, there are indications that it may be the theory that unifies fundamental forces in the standard model at energies close to those at currently existing and planned accelerators. Perturbative supersymmetric string theory (at least up to lower loop levels) avoids the ultraviolet divergences that appear in conventional attempts at quantizing gravity. Recent work relating the Hartle-Hawking wave function to string partition function can be used to obtain a wave function for the metric fluctuations on S^3 embedded in a Calabi-Yau manifold. This may be a first step in a realistic quantum cosmology relating the entropy of certain black holes with the topological string wave function. While a string theory model unifying all fundamental forces is not yet available, a number of small results (some of which we have discussed in this chapter) are emerging to suggest that supersymmetric string theory could play a fundamental role in constructing such a model. Developing a theory and phenomenology of 4-dimensional string vacua and relating them to experimental physics and cosmological data would be a major step in this direction.

Acknowledgements

This work was supported in part by a research fellowship of the Max Planck Gesellschaft at the Max Planck Institute for Mathematics in the Sciences, Leipzig. I would like to thank the organizers, Prof. Dr. Eberhard Zeidler, Dr. Fauser and Dr. Tolksdorf for inviting me to speak at the Blaubcuren 2005 workshop and for their continued interest in my work. I benefited from discussions with Dr. Fleischhack and Dr. Paschke. Dr. Paschke's careful reading of the paper helped me avoid a number of errors of omission and commission.

References

[1] S. Axelrod and I. Singer. Chern-Simons Perturbation Theory. *J. Diff. Geom.*, 39:787–902, 1994.

[2] A. Besse. *Einstein Manifolds*. Springer-Verlag, Berlin, 1986.

[3] K. B. Marathe. The Mean Curvature of Gravitational Fields. *Physica*, 114A:143–145,1982.

[4] E. Witten. Quantum field theory and the Jones polynomial. *Prog. Math.*, 133:637–678, 1995.

[5] T. T. Wu and C. N. Yang. Concept of Nonintegrable Phase Factors and Global Formulation of Gauge Fields. *Phys. Rev.*, 12D:3845–3857, 1975.

[6] L. Crane. 2-d physics and 3-d topology. *Comm. Math. Phys.*, 135:615–640, 1991.

[7] J. Harer and D. Zagier. The Euler characteristic of the moduli space of curves. *Inven. Math.*, 85:457–485, 1986.

[8] M. Khovanov. A categorification of the Jones Polynomial. *Duke Math. J.*, 101:359–426, 2000.

[9] R. Kirby and P. Melvin. The 3-manifold invariants of Witten and Reshetikhin-Turaev for $sl(2, \mathbb{C})$. *Inven. Math.*, 105:473–545, 1991.

[10] T. Kohno. Topological invariants for three manifolds using representations of the mapping class groups I. *Topology*, 31:203–230, 1992.

[11] R. Lawrence and D. Zagier. Modular forms and quantum invariants of 3-manifolds. *Asian J. Math.*, 3:93–108, 1999.

[12] K. B. Marathe and G. Martucci. *The Mathematical Foundations of Gauge Theories*. Studies in Mathematical Physics, vol. 5. North-Holland, Amsterdam, 1992.

[13] K. B. Marathe, G. Martucci, and M. Francaviglia. Gauge Theory, Geometry and Topology. *Seminario di Matematica dell'Università di Bari*, 262:1–90, 1995.

[14] Kishore B. Marathe. *A Chapter in Physical Mathematics: Theory of Knots in the Sciences*, pages 873–888. Springer-Verlag, Berlin, 2001.

[15] P. Ozsváth and Z. Szabó. On knot Floer homology and the four-ball genus. *Geom. Topol.*, 7:225–254, 2003.

[16] I. Smith, R¿ P¿ Thomas, and S.-T. Yau. Symplectic conifold transitions. *J. Diff. Geom.*, 62:209–242, 2002.

[17] E. Witten. Chern-Simons gauge theory as string theory. *Prog. Math.*, 133:637–678, 1995.

Kishore Marathe
Department of Mathematics
Brooklyn College, CUNY
Brooklyn, NY 11210
USA
e-mail: `kmarathe@brooklyn.cuny.edu`

Quantum Gravity
B. Fauser, J. Tolksdorf and E. Zeidler, Eds., 237–262

Strings, Higher Curvature Corrections, and Black Holes

Thomas Mohaupt

Abstract. We review old and recent results on subleading contributions to black hole entropy in string theory.

Mathematics Subject Classification (2000). Primary 81T30; Secondary 83C57; 83E30.

Keywords. String theory, black holes, entropy, microstates, higher curvature corrections, area law, Hartle-Hawking wave function, variational principles.

1. Introduction

The explanation of black hole entropy in terms of microscopic states is widely regarded as one of the benchmarks for theories of quantum gravity. The analogy between the laws of black hole mechanics and the laws of thermodynamics, combined with the Hawking effect, suggests to assign to a black hole of area A the 'macroscopic' (or 'thermodynamic') entropy[1]

$$\mathcal{S}_{\mathrm{macro}} = \frac{A}{4} \, . \tag{1}$$

$\mathcal{S}_{\mathrm{macro}}$ depends on a small number of parameters which can be measured far away from the black hole and determine its 'macroscopic' state: the mass M, the angular momentum J and its charges Q with respect to long range gauge forces. A theory of quantum gravity should be able to specify and count the microstates of the black hole which give rise to the same macrostate. If there are N states corresponding to a black hole with parameters M, J, Q, then the associated 'microscopic' or 'statistical' entropy is

$$\mathcal{S}_{\mathrm{micro}} = \log N \, . \tag{2}$$

---

[1] We work in Planckian units, where $c = \hbar = G_N = 1$. We also set $k_B = 1$.

By the analogy to the relation between thermodynamics and statistical mechanics, it is expected that the macroscopic and microscopic entropies agree.[2]

The characteristic feature of string theory is the existence of an infinite tower of excitations with ever-increasing mass. Therefore it is natural to take the fundamental strings themselves as candidates for the black hole microstates [1, 2]. In the realm of perturbation theory, which describes strings moving in a flat background space-time at asymptotically small string coupling one has access to the number of states with a given mass. The asymptotic number of states at high mass is given by the famous formula of Hardy-Ramanujan. Taking the open bosonic string for definiteness, the mass formula is $\alpha' M^2 = N - 1$, where α' is the Regge slope (the only independent dimensionful constant of string theory) and $N \in \mathbb{N}$ is the excitation level. For large N the number of states grows like $\exp(\sqrt{N})$, so that the statistical entropy grows like

$$\mathcal{S}_{\text{micro}} \approx \sqrt{N} \, . \tag{3}$$

It is clear that when increasing the mass (at finite coupling), or, alternatively, when increasing the coupling while keeping the mass fixed, the backreaction of the string onto its ambient space-time should lead to the formation of a black hole. Roughly, this happens when the Schwarzschild radius r_S of the string equals the string length $l_S = \sqrt{\alpha'}$. Using the relation $l_P = l_S g_S$ between the string length, the Planck length l_P and the dimensionless string coupling g_S, together with the fact that a black hole of mass $\alpha' M^2 \approx N$ has Schwarzschild radius $r_S \approx G_N M \approx \sqrt{\alpha'} g_S^2 \sqrt{N}$ and entropy $\mathcal{S}_{\text{macro}} \approx \frac{A}{G_N} \approx g_S^2 N$ one finds that [1, 2][3]

$$r_S \approx l_S \Leftrightarrow g_S^2 \sqrt{N} \approx 1 \, . \tag{4}$$

It is precisely in this regime that the entropy of string states $\mathcal{S}_{\text{micro}} \approx \sqrt{N}$ equals the entropy $\mathcal{S}_{\text{macro}} \approx g_S^2 N$ of a black hole with the same mass, up to factors of order unity. The resulting scenario of a string – black hole correspondence, where strings convert into black holes and vice versa at a threshold in mass/coupling space is quite appealing. In particular, it applies to Schwarzschild-type black holes and makes a proposal for the final state of black hole evaporation, namely the conversion into a highly excited string state of the same mass and entropy. However, this picture is very qualitative and one would like to have examples where one can make a quantitative comparison or even a precision test of the relation between macroscopic and microscopic entropy.

Such examples are available, if one restricts oneself to supersymmetric states, also called BPS states. We will consider four-dimensional black holes, where the

---

[2] As we will see later, there are examples where they agree in leading order of a semiclassical expansion, but disagree at the subleading level. This is a success rather than a problem because the discrepancies can be explained: the entropies that one compares correspond to different statistical ensembles.

[3] In this paragraph we have reconstructed the dimensionful quantities G_N and α' for obvious reasons. All approximate identities given here hold up to multiplicative constants of order unity and up to subleading additive corrections.

setup is as follows: one considers a string compactification which preserves a four-dimensional supersymmetry algebra with central charges Z. Then there exist supermultiplets on which part of the superalgebra is realized trivially. These multiplets are smaller than generic massive supermultiplets (hence also called 'short multiplets'), they saturate a mass bound of the form $M \geq |Z|$, and many of their properties are severely restricted by the supersymmetry algebra. By counting all states of given mass and charges, one obtains the statistical entropy $\mathcal{S}_{\text{micro}}$. This can now be compared to the entropy $\mathcal{S}_{\text{macro}}$ of a black hole which has the same mass, carries the same charges and is invariant under the same supertransformations. As above, the underlying idea is that by increasing the string coupling we can move from the regime of perturbation theory in flat space to a regime where the backreaction onto space-time has led to the formation of a black hole. This regime can be analyzed by using the low-energy effective field theory of the massless string modes, which encodes all long range interactions. The corresponding effective action can be constructed using string perturbation theory and is valid for small (but finite) coupling $g_S \leq 1$ and for space-time curvature which is small in units of the string length. One then constructs supersymmetric black hole solutions with the appropriate mass and charge. A black hole solution is called supersymmetric if it has Killing spinors, which are the 'fermionic analogues' of Killing vectors. More precisely, if we denote the supersymmetry transformation parameter by $\epsilon(x)$, the fields collectively by $\Phi(x)$, and the particular field configuration under consideration by $\Phi_0(x)$, then $\epsilon(x)$ is a Killing spinor and $\Phi_0(x)$ is a supersymmetric (or BPS) configuration, if the supersymmetry transformation with parameter $\epsilon(x)$ vanishes in the background $\Phi_0(x)$:[4]

$$\left.\left(\delta_{\epsilon(x)}\Phi\right)\right|_{\Phi_0(x)} = 0 \ . \tag{5}$$

We consider black holes which are asymptotically flat. Therefore it makes sense to say that a black hole is invariant under 'the same' supertransformations as the corresponding string states. In practise, the effective action is only known up to a certain order in g_S and α'. Thus we need to require that the string coupling and the curvature at the event horizon are small. This can be achieved by taking the charges and, hence, the mass, to be large.

Having constructed the black hole solution, we can extract the area of the event horizon and the entropy $\mathcal{S}_{\text{macro}}$ and then compare to the result of state counting, which yields $\mathcal{S}_{\text{micro}}$. Both quantities are measured in different regimes, and therefore it is not clear a priori that the number of states is preserved when interpolating between them. For BPS states, there exist only two mechanism which can eliminate or create them when changing parameters: (i) at lines of marginal stability a BPS multiplet can decay into two or more other BPS multiplets, (ii) BPS multiplets can combine into non-BPS multiplets. It is not yet clear whether these processes really play a role in the context of black hole state counting, but in principle one needs to deal with them. One proposal is that the quantity which

---

[4]This is completely analogous to the concept of a Killing vector $\xi(x)$, which generates an isometry of a given metric $g(x)$, i.e., $(L_{\xi(x)}g)(x) = 0$.

should be compared to the entropy is not the state degeneracy itself, but a suitable weighted sum, a supersymmetric index [3, 4, 5]. We will ignore these subtleties here and take an 'experimental' attitude, by just computing $\mathcal{S}_{\mathrm{macro}}$ and $\mathcal{S}_{\mathrm{micro}}$ in the appropriate regimes and comparing the results. In fact, we will see that the agreement is spectacular, and extends beyond the leading order.[5] In particular we will see that higher derivative terms in the effective action become important and that one can discriminate between the naive area law and Wald's generalized definition of black hole entropy for generally covariant theories of gravity with higher derivative terms. Thus, at least for supersymmetric black holes, one can make precision tests which confirm that the number of microstates agrees with the black hole entropy.

Besides fundamental strings, string theory contains other extended objects, which are also important in accounting for black hole microstates. One particular subclass are the D-branes. In string perturbation theory they appear as submanifolds of space-time, on which open strings can end. In the effective field theory they correspond to black p-brane solutions, which carry a particular kind of charge, called Ramond-Ramond charge, which is not carried by fundamental strings. In string compactifications one can put the spatial directions of p-branes along the compact space and thereby obtain black holes in the lower-dimensional space-time. D-branes gave rise to the first successful quantitative matching between state counting and black hole entropy [6]. Here 'quantitative' means that the leading contribution to $\mathcal{S}_{\mathrm{micro}}$ is precisely $\frac{A}{4}$, i.e, the prefactor comes out exactly.

When D-branes and other extended objects enter the game, the state counting becomes more complicated, but the basic ideas remain as explained above. Also note that instead of a flat background space-time one can consider other consistent string backgrounds. In particular one can count the microstates of four-dimensional black holes which arise in string compactifications on tori, orbifolds and Calabi-Yau manifolds.

2. The black hole attractor mechanism

In this section we discuss BPS black hole solutions of four-dimensional $N = 2$ supergravity. This is the most general setup which allows supersymmetric black hole solutions and arises in various string compactifications, including compactifications of the heterotic string on $K3 \times T^2$ and of the type-II superstring on Calabi-Yau threefolds. The $N = 2$ supergravity multiplet is a supersymmetric version of Einstein-Maxwell theory: it contains the graviton, a gauge field called the graviphoton, and a doublet of Majorana gravitini. The extreme Reissner-Nordström black

---

[5]Of course, at some level the question whether the black hole entropy corresponds to the true state degeneracy or to an index becomes relevant. However, none of the examples analyzed in [4, 5] appears to be conclusive.

hole is a solution of this theory and provides the simplest example of a supersymmetric black hole [7, 8].[6]

In string compactifications the gravity multiplet is always accompanied by matter multiplets. The only type of matter which is relevant for our discussion is the vector multiplet, which contains a gauge field, a doublet of Majorana spinors, and a complex scalar. We will consider an arbitrary number n of vector multiplets. The resulting Lagrangian is quite complicated, but all the couplings are encoded in a single holomorphic function, called the prepotential [9, 10]. In order to understand the structure of the entropy formula for black holes, we need to review some more details.

First, let us note that the fields which are excited in black hole solutions are only the bosonic ones. Besides the metric there are n scalar fields z^A, $A = 1, \ldots, n$ and $n + 1$ gauge fields $F^I_{\mu\nu}$. The field equations are invariant under $Sp(2n + 2, \mathbb{R})$ transformations, which generalize the electric-magnetic duality rotations of the Maxwell theory. These act linearly on the gauge fields and rotate the $F^I_{\mu\nu}$ among themselves and into their duals. Electric charges q_I and magnetic charges p^I are obtained from flux integrals of the dual field strength and of the field strength, respectively. They form a symplectic vector (p^I, q_J). While the metric is inert, the action on the scalars is more complicated. However, it is possible to find a parametrization of the scalar sector that exhibits a simple and covariant behaviour under symplectic transformations. The scalar part of the Lagrangian is a non-linear sigma-model, and the scalar fields can be viewed as coordinates on a complex n-dimensional manifold M. The geometry of M is restricted by supersymmetry, and the resulting geometry is known as 'special geometry' [10]. In the context of the superconformal calculus, the coupling of n vector multiplets to Poincaré supergravity is constructed by starting with $n+1$ superconformal vector multiplets, and imposing suitable gauge conditions which fix the additional symmetries. As already mentioned, the vector multiplet Lagrangian is encoded in a single function, the prepotential, which depends holomorphically on the lowest components of the superconformal vector multiplets. A consistent coupling to supergravity further requires the prepotential to be homogeneous of degree 2:

$$F(\lambda Y^I) = \lambda^2 F(Y^I) \,. \tag{6}$$

Here, the complex fields Y^I, $I = 0, 1, \ldots, n$ are the lowest components of the superconformal vector multiplets.[7] The physical scalar fields z^A are given by the independent ratios, $z^A = \frac{Y^A}{Y^0}$. Geometrically, the Y^I are coordinates of a complex cone $C(M)$ over the scalar manifold M. The existence of a prepotential is equivalent to the existence of a holomorphic Lagrangian immersion of $C(M)$ into the complex symplectic vector space $\mathbb{C}^{2n+2}$ [11]. If (Y^I, F_J) are coordinates on

---

[6]See for example [25] for a pedagogical treatment.

[7]In the context of black hole solutions, it is convenient to work with rescaled variables. The fields Y^I used in this paper are related to the lowest components X^I of the superconformal vector multiplets by $Y^I = \overline{Z} X^I$, where $Z = p^I F_I - q_I X^I$ is the central charge. Using that F_I is homogeneous of degree one has $F_I(Y) = \overline{Z} F_I(X)$. See [15, 13] for more details.

$\mathbb{C}^{2n+2}$, then, along the immersed $C(M)$, the second half of the coordinates can be expressed in terms of the first half as $F_I = \frac{\partial F}{\partial Y^I}$, where F is the generating function of the Lagrangian immersion, i.e., the prepotential.[8] Under symplectic transformations (Y^I, F_J) transforms as a vector. Therefore it is convenient to use it instead of z^A to parametrize the scalar fields.

Let us now turn to static, spherically symmetric, supersymmetric black hole solutions. In $N = 2$ supergravity, which has eight independent real supercharges, one can have 8 or 4 or 0 Killing spinors. Solutions with 8 Killing spinors preserve as many supersymmetries as flat space-time and are regarded as supersymmetric vacua. Besides $\mathbb{R}^4$ the only supersymmetric vacua are $AdS^2 \times S^2$ and planar waves [12].[9] Supersymmetric black holes are solutions with 4 Killing spinors. Since they preserve half as many supersymmetries as the vacuum, they are called $\frac{1}{2}$ BPS solutions. Since these solutions are asymptotically flat, the number of Killing spinors doubles if one goes to infinity.

One difference between supersymmetric black holes in theories with vector multiplets and the extreme Reissner-Nordström black hole of pure $N = 2$ supergravity is that there are several gauge fields, and therefore several species of electric and magnetic charges. The other difference is that we now have scalar fields which can have a non-trivial dependence on the radial coordinate. A black hole solution is parametrized by the magnetic and electric charges (p^I, q_J), which are discrete quantities (by Dirac quantization) and by the asymptotic values of the scalar fields in the asymptotically flat region, $z^A(\infty)$, which can be changed continuously. In particular, the mass of a black hole can be changed continuously by tuning the values of the scalar fields at infinity. The area of the horizon and hence $\mathcal{S}_{\text{macro}}$ depends on the charges and on the values of the scalar fields at the horizon. If the latter could be changed continuously, this would be at odds with the intended interpretation in terms of state counting.

What comes to the rescue is the so-called black hole attractor mechanism [14]: if one imposes that the solution is supersymmetric and regular at the horizon, then the values of the scalars at the horizon, and also the metric, are determined in terms of the charges. Thus the scalars flow from arbitrary initial values at infinity to fixed point values at the horizon. The reason behind this behaviour is that if the horizon is to be finite then the number of Killing spinors must double on the horizon. This fixes the geometry of the horizon to be of the form $AdS^2 \times S^2$, with fixed point values for the scalars. In the notation introduced above, the values of the scalar fields can be found from the following black hole attractor equations

---

[8]It is assumed here that the immersion is generic, so that that the Y^I are coordinates on $C(M)$. This can always be arranged by applying a symplectic rotation.

[9]Presumably this is still true in the presence of neutral matter and including higher curvature corrections. In [13] the most general *stationary* vacuum solution for this case was shown to be $AdS^2 \times S^2$.

[14]: [10]

$$
\begin{aligned}
(Y^I - \overline{Y}^I)_{\text{Horizon}} &= ip^I \, , \\
(F_I - \overline{F}_I)_{\text{Horizon}} &= iq_I \, .
\end{aligned}
\tag{7}
$$

For a generic prepotential F it is not possible to solve this set of equations for the scalar fields in closed form. However, explicit solutions have been obtained for many physically relevant examples, where either the prepotential is sufficiently simple, or for non-generic configurations of the charges (i.e., when switching off some of the charges) [16, 17].[11]

The entropy of the corresponding solution is

$$
\mathcal{S}_{\text{macro}} = \frac{A}{4} = \pi |Z|^2_{\text{Horizon}} = \pi(p^I F_I - q_I Y^I)_{\text{Horizon}} \, .
\tag{8}
$$

Here Z is a particular, symplectically invariant contraction of the fields with the charges, which gives the central charge carried by the solution when evaluated at infinity. At the horizon, this quantity sets the scale of the $AdS^2 \times S^2$ space and therefore gives the area A.

Let us take a specific example. We consider the prepotential of a type-II Calabi-Yau compactification at leading order in both the string coupling g_S and the string scale α'. If we set half of the charges to zero, $q_A = p^0 = 0$, then the attractor equations can be solved explicitly. To ensure weak coupling and small curvature at the horizon, the non-vanishing charges must satisfy $|q_0| \gg p^A \gg 1$.[12] The resulting entropy is [16]:

$$
\mathcal{S}_{\text{macro}} = 2\pi \sqrt{\tfrac{1}{6}|q_0|C_{ABC}p^A p^B p^C} \, .
\tag{9}
$$

Here C_{ABC} are geometrical parameters (triple intersection numbers) which depend on the specific Calabi-Yau threefold used for compactification.

For this example the state counting has been performed using the corresponding brane configuration. The result is [19, 20]:

$$
\mathcal{S}_{\text{micro}} = 2\pi \sqrt{\tfrac{1}{6}|q_0|(C_{ABC}p^A p^B p^C + c_{2A}p^A)} \, ,
\tag{10}
$$

where c_{2A} is another set of geometrical parameters of the underlying Calabi-Yau threefolds (the components of the second Chern class with respect to a homology basis). Since this formula contains a subleading term, which is not covered by the macroscopic entropy (9), this raises the question how one can improve the treatment of the black hole solutions. Since we interpret supergravity actions as effective actions coming from string theory, the logical next step is to investigate the effects of higher derivative terms in the effective action, which are induced by quantum and stringy corrections.

---

[10] We use the notation of [15], where the scalar fields X^I used in [14] have been rescaled in the way explained above.

[11] Also note that if one can solve the attractor equations, then one can also find the solution away from the horizon [18].

[12] In our conventions $q_0 < 0$ under the conditions stated in the text.

3. Beyond the area law

There is a particular class of higher derivative terms for which the $N = 2$ supergravity action can be constructed explicitly [21] (see also [22] for a review). These terms are encoded in the so-called Weyl multiplet and can be taken into account by giving the prepotential a dependence on an additional complex variable Υ, which is proportional to the lowest component of the Weyl multiplet. The equations of motion relate Υ to the (antiselfdual part of the) graviphoton field strength. The generalized prepotential is holomorphic and homogeneous of degree 2:

$$F(\lambda Y^I, \lambda^2 \Upsilon) = \lambda^2 F(Y^I, \Upsilon) . \tag{11}$$

Expanding in Υ as

$$F(Y^I, \Upsilon) = \sum_{g=0}^{\infty} F^{(g)}(Y^I) \Upsilon^g , \tag{12}$$

one gets an infinite sequence of coupling functions $F^{(g)}(Y^I)$. While $F^{(0)}(Y^I)$ is the prepotential, the $F^{(g)}(Y^I)$, $g \geq 1$, are coefficients of higher derivative terms. Among these are terms of the form

$$F^{(g)}(Y^I)(C^-_{\mu\nu\rho\sigma})^2 (F^-_{\tau\lambda})^{2g-2} + \text{c.c.} , \tag{13}$$

where $C^-_{\mu\nu\rho\sigma}$ and $F^-_{\tau\lambda}$ are the antiselfdual projections of the Weyl tensor and of the graviphoton field strength, respectively. In the context of type-II Calabi-Yau compactifications the functions $F^{(g)}(Y^I)$ can be computed using a topologically twisted version of the theory [23, 24].

Starting from the generalized prepotential one can work out the Lagrangian and construct static, spherically symmetric BPS black hole solutions [15, 13].[13] It can be shown that the near horizon solution is still determined by the black hole attractor equations,[14] which now involve the generalized prepotential [15]:

$$
\begin{aligned}
(Y^I - \overline{Y}^I)_{\text{horizon}} &= ip^I , \\
(F_I(Y, \Upsilon) - \overline{F}_I(\overline{Y}, \overline{Y}))_{\text{horizon}} &= iq_I .
\end{aligned}
\tag{14}
$$

The additional variable Υ takes the value $\Upsilon = -64$ at the horizon. Since the generalized prepotential enters into the attractor equations, the area of the horizon is modified by the higher derivative terms. Moreover, there is a second modification, which concerns the very definition of the black hole entropy.

The central argument for interpreting the area of the horizon as an entropy comes from the first law of black hole mechanics, which relates the change of the

---

[13]In fact, one can construct stationary BPS solutions which generalize the IWP solutions of pure supergravity [13].

[14]More precisely, the attractor equations are necessary and sufficient for having a fully supersymmetric solution with 8 Killing spinors at the horizon. The geometry is still $AdS^2 \times S^2$, but with a modified scale.

mass of a stationary black hole to changes of the area and of other quantities (angular momentum, charges):

$$\delta M = \frac{\kappa_S}{8\pi}\delta A + \cdots , \tag{15}$$

where κ_S is the surface gravity of the black hole.[15] Comparing to the first law of thermodynamics,

$$\delta U = T\delta S + \cdots , \tag{16}$$

and taking into account that the Hawking temperature of a black hole is $T = \frac{\kappa_S}{2\pi}$, one is led to the identification $\mathcal{S}_{\mathrm{macro}} = \frac{A}{4}$. This is at least the situation in Einstein gravity. The first law can be generalized to more general gravitational Lagrangians, which contain higher derivative terms, in particular arbitrary powers of the Riemann tensor and of its derivatives [26, 27]. The basic assumptions entering the derivation are that the Lagrangian is generally covariant, and that it admits stationary black hole solutions whose horizons are Killing horizons. Then there still is a first law of the form

$$\delta M = \frac{\kappa_S}{2\pi}\delta\mathcal{S}_{\mathrm{macro}} + \cdots , \tag{17}$$

but $\mathcal{S}_{\mathrm{macro}} \neq \frac{A}{4}$ in general. Rather, $\mathcal{S}_{\mathrm{macro}}$ is given by the surface charge associated with the horizontal Killing vector field ξ:

$$\mathcal{S}_{\mathrm{macro}} = 2\pi \oint_{\mathrm{horizon}} \mathbf{Q}[\xi] , \tag{18}$$

which can be expressed in terms of variational derivatives of the Lagrangian with respect to the Riemann tensor [27]:

$$\mathcal{S}_{\mathrm{macro}} = -2\pi \oint_{\mathrm{horizon}} \frac{\delta L}{\delta R_{\mu\nu\rho\sigma}} \varepsilon_{\mu\nu}\varepsilon_{\rho\sigma} \sqrt{h}\, d^2\theta . \tag{19}$$

Here $\varepsilon_{\mu\nu}$ is the normal bivector of the horizon, normalized as $\varepsilon_{\mu\nu}\varepsilon^{\mu\nu} = -2$, and h is the pullback of the metric onto the horizon.[16] From (19) it is clear that corrections to the area law will be additive:

$$\mathcal{S}_{\mathrm{macro}} = \frac{A}{4} + \cdots . \tag{20}$$

Here the leading term comes from the variation of the Einstein-Hilbert action.

The general formula (19) can be evaluated for the special case of $N = 2$ supergravity with vector multiplets and higher derivative terms encoded in the generalized prepotential. The result is [15]:

$$\mathcal{S}_{\mathrm{macro}}(q,p) = \pi\left((p^I F_I - q_I Y^I) + 4\,\mathrm{Im}(\Upsilon F_\Upsilon)\right)_{\mathrm{horizon}} , \tag{21}$$

---

[15]See for example [22] for a review of the relevant properties of black hole horizons.
[16]In carrying out the variational derivatives one treats the Riemann tensor formally as if it was independent of the metric. At first glance this rule looks ambiguous, because one can perform partial integrations. But the underlying formalism guarantees that the integrated quantity $\mathcal{S}_{\mathrm{macro}}$ is well defined [26, 27] (see also [35] for an alternative proof).

where $F_\Upsilon = \frac{\partial F}{\partial \Upsilon}$. Thus F_Υ, which depends on the higher derivative couplings $F^{(g)}$, $g \geq 1$, encodes the corrections to the area law.

If the prepotential is sufficiently simple, one can find explicit solutions of the attractor equations [15, 35, 47].[17] In particular, we can now compare $\mathcal{S}_{\text{macro}}$ to the $\mathcal{S}_{\text{micro}}$ computed from state counting (10) [15]:

$$
\begin{aligned}
&\mathcal{S}_{\text{macro}}(q,p) \\
&= \frac{A}{4} + \text{Correction term} \\
&= 2\pi \frac{\frac{1}{6}|q_0|(C_{ABC}p^A p^B p^C + \frac{1}{2}c_{2A}p^A)}{\sqrt{\frac{1}{6}|q_0|(C_{ABC}p^A p^B p^C + c_{2A}p^A)}} + 2\pi \frac{\frac{1}{12}|q_0|c_{2A}p^A}{\frac{1}{6}|q_0|(C_{ABC}p^A p^B p^C + c_{2A}p^A)} \\
&= 2\pi \sqrt{\frac{1}{6}|q_0|(C_{ABC}p^A p^B p^C + c_{2A}p^A)} \\
&= \mathcal{S}_{\text{micro}} \,.
\end{aligned}
\tag{22}
$$

In the second line we can see explicitly how the higher derivative terms modify the area. But, when sticking to the naive area law, one finds that $\frac{A}{4}$ differs from $\mathcal{S}_{\text{micro}}$ already in the first subleading term in an expansion in large charges. In contrast, when taking into account the modification of the area law, $\mathcal{S}_{\text{macro}}$ and $\mathcal{S}_{\text{micro}}$ agree completely. In other words 'string theory state counting knows about the modification of the area law.' This provides strong evidence that string theory captures the microscopic degrees of freedom of black holes, at least of supersymmetric ones.

At this point one might wonder about the role of other types of higher derivatives terms. So far, we have only included a very particular class, namely those which can be described using the Weyl multiplet. The full string effective action also contains other higher derivative terms, including terms which are higher powers in the curvature. Naively, one would expect that these also contribute to the black hole entropy. However, as we will see in the next sections, one can obtain an even more impressive agreement between microscopic and macroscopic entropy by just using the terms encoded in the Weyl multiplet. One reason might be the close relationship between the terms described by the Weyl multiplet and the topological string, which we are going to review in the next section. There are two other observation which indicate that the Weyl multiplet encodes all contributions relevant for the entropy.[18] The first observation is that when one just adds a Gauss-Bonnet term to the Einstein Hilbert action, one obtains the same entropy formula (10) as when using the full Weyl multiplet [38, 37]. The second is that (10) can also be derived using gravitational anomalies [41, 40]. Both suggest that the black hole entropy is a robust object, in the sense that it does not seem to depend sensitively on details of the Lagrangian.

---

[17]One can also construct the solution away from the horizon, at least iteratively [13, 64].

[18]For toroidal compactifications of type-II string theory there are no R^2-corrections, but the entropy of string states is non-vanishing. This case seems to require the presence of higher derivative terms which are not captured by the Weyl multiplet. See [4] for further discussion.

One might also wonder, to which extent the matching of microscopic and macroscopic entropy depends on supersymmetry. Here it is encouraging that the derivations of (10) in [36, 37] and [40] do not invoke supersymmetry directly. Rather, [36, 37] analyses black holes with near horizon geometry $AdS^2 \times S^2$ in the context of general higher derivative covariant actions, without assuming any other specifics of the interactions. This leads to a formalism based on an entropy function, which is very similar to the one found for supersymmetric black holes some time ago [16], and which we will review in a later section. The work of [40] relates Wald's entropy formula to the AdS/CFT correspondence.

Finally, it is worth remarking that according to [36, 40] similar results should hold in space-time dimensions other than four. A particularly interesting dimension seems to be five, because there is a very close relationship between four-dimensional supersymmetric black holes and five-dimensional supersymmetric rotating black holes and black rings [42, 38], which holds in the presence of higher curvature terms.

Coming back to (22), we remark that it is intriguing that two complicated terms, the area and the correction term, combine into a much simpler expression. This suggests that, although (21) is a sum of two terms, it should be possible to express the entropy in terms of one single function. Though it is not quite obvious how to do this, it is in fact true.

4. From black holes to topological strings

The black hole entropy (21) can be written as the Legendre transform of another function $\mathcal{F}_{\mathrm{BH}}$, which is interpreted as the black hole free energy. This is seen as follows [3]. The 'magnetic' attractor equation $Y^I - \overline{Y}^I = ip^I$ can be 'solved' by setting:[19]

$$Y^I = \frac{\phi^I}{2\pi} + i\frac{p^I}{2} , \qquad (23)$$

where $\phi^I \propto \mathrm{Re} Y^I$ is determined by the remaining 'electric' attractor equation. From the gauge field equations of motion in a stationary space-time one sees that ϕ^I is proportional to the electrostatic potential (see for example [13]). Now define the free energy

$$\mathcal{F}_{\mathrm{BH}}(\phi, p) := 4\pi \mathrm{Im} F(Y, \Upsilon)_{\mathrm{horizon}} . \qquad (24)$$

Observe that the electric attractor equations $F_I - \overline{F}_I = ip^I$ are equivalent to

$$\frac{\partial \mathcal{F}_{\mathrm{BH}}}{\partial \phi^I} = q_I . \qquad (25)$$

Next note that the homogeneity property (11) of the generalized prepotential implies the Euler-type relation

$$2F = Y^I F_I + 2\Upsilon F_\Upsilon . \qquad (26)$$

---

[19]We use the notation of [15] which is slightly different from the one of [3].

Using this one easily verifies that

$$\mathcal{F}_{\mathrm{BH}}(\phi,p) - \phi^I \frac{\partial \mathcal{F}_{\mathrm{BH}}}{\partial \phi^I} = \mathcal{S}_{\mathrm{macro}}(q,p) . \tag{27}$$

Thus the black hole entropy is obtained from the black hole free energy by a (partial) Legendre transform which replaces the electric charges q_I by the electrostatic potentials ϕ^I.

This observation opens up various routes of investigation. Let us first explore the consequences for the relation between $\mathcal{S}_{\mathrm{macro}}$ and $\mathcal{S}_{\mathrm{micro}}$. The black hole partition function associated with $\mathcal{F}_{\mathrm{BH}}$ is

$$Z_{\mathrm{BH}}(\phi,p) = e^{\mathcal{F}_{\mathrm{BH}}(\phi,p)} . \tag{28}$$

Since it depends on ϕ^I rather than on q_I, it is clear that this is not a microcanonical partition function. Rather it refers to a mixed ensemble, where the magnetic charges have been fixed while the electric charges fluctuate. The electrostatic potential ϕ^I is the corresponding thermodynamic potential. However, the actual state degeneracy $d(q,p)$ should be computed in the microcanonical ensemble, where both electric and magnetic charges are fixed. Using a standard thermodynamical relation, we see that Z_{BH} and $d(q,p)$ are formally related by a (discrete) Laplace transform:

$$Z_{\mathrm{BH}}(\phi,p) = \sum_q d(q,p) e^{\phi^I q_I} . \tag{29}$$

We can solve this formally for the state degeneracy by an (inverse discrete) Laplace transform,

$$d(q,p) = \int d\phi \; e^{\mathcal{F}_{\mathrm{BH}}(\phi,p) - \phi^I q_I} \tag{30}$$

and express the microscopic entropy

$$\mathcal{S}_{\mathrm{micro}}(q,p) = \log d(q,p) \tag{31}$$

in terms of the black hole free energy. Comparing (27) to (30) it is clear that $\mathcal{S}_{\mathrm{macro}}$ and $\mathcal{S}_{\mathrm{micro}}$ will not be equal in general. Both can be expressed in terms of the free energy, but one is given through a Laplace transform and the other through a Legendre transform [3]. From statistical mechanics we are used to the fact that quantities might differ when computed using different ensembles, but we expect them to agree in the thermodynamic limit. In our context the thermodynamic limit corresponds to the limit of large charges, in which it makes sense to evaluate the inverse Laplace transform (30) in a saddle point approximation:

$$e^{\mathcal{S}_{\mathrm{micro}}(q,p)} = \int d\phi e^{\mathcal{F}_{\mathrm{BH}}(\phi,p) - \phi^I q_I} \approx e^{\mathcal{F}_{\mathrm{BH}}(\phi,p) - \phi^I \frac{\partial \mathcal{F}_{\mathrm{BH}}}{\partial \phi^I}} = e^{\mathcal{S}_{\mathrm{macro}}(q,p)} . \tag{32}$$

Since the saddle point value of the inverse Laplace transform is given by the Legendre transform, we see that both entropies agree in the limit of large charges. Note that already the first subleading correction, which comes from quadratic fluctuations around the saddle point, will in general lead to deviations. We will illustrate the relation between $\mathcal{S}_{\mathrm{macro}}$ and $\mathcal{S}_{\mathrm{micro}}$ using specific examples later on.

We now turn to another important consequence (27). As already mentioned the couplings $F^{(g)}(Y)$ of the effective $N = 2$ supergravity Lagrangian can be computed within the topologically twisted version of type-II string theory with the relevant Calabi-Yau threefold as target space. The effect of the topological twist is roughly to remove all the non-BPS states, thus reducing each charge sector to its ground state. The coupling functions can be encoded in a generating function, called the topological free energy $F_{\text{top}}(Y^I, \Upsilon)$, which equals the generalized prepotential $F(Y^I, \Upsilon)$ of supergravity up to a conventional overall constant. The associated topological partition function

$$Z_{\text{top}} = e^{F_{\text{top}}} \tag{33}$$

can be viewed as a partition function for the BPS states of the full string theory. Taking into account the conventional normalization factor between F_{top} and $F(Y^I, \Upsilon)$ one observes [3]:

$$Z_{\text{BH}} = e^{\mathcal{F}_{\text{BH}}} = e^{4\pi \text{Im} F} = e^{F_{\text{top}} + \overline{F_{\text{top}}}} = |e^{F_{\text{top}}}|^2 = |Z_{\text{top}}|^2 . \tag{34}$$

Thus there is a direct relation between the black hole entropy and the topological partition function, which suggests that the matching between macroscopic and microscopic entropy extends far beyond the leading contributions. Moreover, the relation $Z_{\text{BH}} = |Z_{\text{top}}|^2$ suggests to interpret Z_{top} as a quantum mechanical wave function and Z_{BH} as the associated probability [3]. This can be made precise as follows: Z_{top} is a function on the vector multiplet scalar manifold, which in type-IIA (type-IIB) Calabi-Yau compactifications coincides with the moduli space of complexified Kähler structures (complex structures). This manifold is in particular symplectic, and can be interpreted as a classical phase space. Applying geometric quantization one sees that Z_{top} is indeed a wave function on the resulting Hilbert space [28]. This reminds one of the minisuperspace approximations used in canonical quantum gravity. In our case the truncation of degrees of freedom is due to the topological twist, which leaves the moduli of the internal manifold as the remaining degrees of freedom. In other words the full string theory is reduced to quantum mechanics on the moduli space. One is not restricted to only discussing black holes in this framework, but, by a change of perspective and some modifications, one can approach the dynamics of flux compactifications and quantum cosmology [29].

The link between black holes and flux compactifications is provided by the observation that from the higher-dimensional point of view the near-horizon geometry of a supersymmetric black hole is $AdS^2 \times S^2 \times X^*$, where X^* denotes the Calabi-Yau threefold at the attractor point in moduli space corresponding to the charges (q^I, p_I). This can be viewed as a flux compactification to two dimensions. The flux is given by the electric and magnetic fields along $AdS^2 \times S^2$, which are covariantly constant, and compensate for the fact that the geometry is not Ricci-flat. From the two-dimensional perspective the attractor mechanism reflects that the reduction on S^2 gives rise to a gauged supergravity theory with a nontrivial scalar potential which fixes the moduli. When taking the spatial direction of AdS^2 to be compact, so that space takes the form $S^1 \times S^2 \times X^*$, then vacua with different moduli are separated by barriers of finite energy. As a consequence, the moduli, which otherwise label superselection sectors, can fluctuate. In this context Z_{top} has been interpreted as a Hartle-Hawking type wave function for flux compactifications [29], while [30] argued that string compactifications with asymptotically free gauge groups are preferred.

It should be stressed that there are many open questions concerning these proposals, both conceptually and technically. Some of these will be discussed in the next section from the point of view of supergravity and black holes. Nevertheless these ideas are very interesting because they provide a new way to approach the vacuum selection problem of string theory. Moreover there seems to be a lot in common with the canonical approach to quantum gravity and quantum cosmology. This might help to develop new ideas how to overcome the shortcomings of present day string theory concerning time-dependent backgrounds. By phrasing string theory in the language used in canonical quantum gravity, one would have a better basis for debating the merits of different approaches to quantum gravity.

5. Variational principles for black holes

We will now discuss open problems concerning the formulae (29), (30) and (34) which relate the black hole entropy to the counting of microstates. The following sections are based on [31].[20] See also [32, 33, 34] for further discussion.

Consider for definiteness (30):

$$d(q,p) = \int d\phi \ e^{\mathcal{F}_{\text{BH}}(\phi,p) - \phi^I q_I} \ , \tag{35}$$

which relates the black hole free energy to the microscopic state degeneracy. This is formally an inverse discrete Laplace transformation, but without specifying the integration contour it is not clear that the integral converges. We will not address this issue here, but treat the integral as a formal expression which we evaluate asymptotically by saddle point methods. The next issue is the precise form of the integrand. As we stressed above various quantities of the effective $N = 2$ supergravity, in particular the charges (p^I, q_J), are subject to symplectic rotations. The microscopic state degeneracy $d(q,p)$ is an observable and therefore should transform covariantly, i.e., it should be a symplectic function. Moreover, string theory has discrete symmetries, in particular T-duality and S-duality, which are realized as specific symplectic transformations in $N \geq 2$ compactifications. Since these are symmetries, $d(q,p)$ should be invariant under them. The transformation properties with respect to symplectic transformations are simple and transparent as long as one works with symplectic vectors, such as (Y^I, F_J) and its real and

---

[20]Preliminary results have already been presented at conferences, including the 'Workshop on gravitational aspects of string theory' at the Fields Institute (Toronto, May 2005) and the 'Strings 2005' (Toronto, July 2005). See http://online.kitp.ucsb.edu/online/joint98/kaeppeli/ and http://www.fields.utoronto.ca/audio/05-06/strings/wit/index.html.

imaginary parts. By the Legendre transform we now take ϕ^I and p^I as our independent variables, and these do not form a symplectic vector. Thus manifest symplectic covariance, as it is present in the entropy formula (21), has been lost. Moreover, it is clear that if $d\phi$ is the standard Euclidean measure $\prod_I d\phi^I$, then the integral cannot be expected to be symplectically invariant. From the point of view of symplectic covariance one should expect that the integration measure is symplectically invariant, while the integrand is a symplectic function. We will now outline a systematic procedure which provides a modified version of (30) which has this property.

The starting point is the observation that the entropy of supersymmetric black holes can be obtained from a variational principle [16, 31]. Define the symplectic function

$$\Sigma(q, p, Y, \overline{Y}) := -K - W - \overline{W} + 128iF_\Upsilon - 128i\overline{F}_{\overline{\Upsilon}} , \tag{36}$$

where

$$K := i(\overline{Y}^I F_I - \overline{F}_I Y^I) \quad \text{and} \quad W := q_I Y^I - p^I F_I . \tag{37}$$

One then finds that the conditions for critical points of Σ,

$$\frac{\partial \Sigma}{\partial Y^I} = 0 = \frac{\partial \Sigma}{\partial \overline{Y}^I} \tag{38}$$

are precisely the attractor equations (14). Moreover, at the attractor we find that

$$\pi\Sigma_{\text{attractor}}(q, p) = \mathcal{S}_{\text{macro}}(q, p) . \tag{39}$$

We also note that one can split the extremization procedure consistently into two steps. If one first extremizes Σ with respect to the imaginary part of Y^I, one obtains the magnetic attractor equations. Plugging these back we find

$$\pi\Sigma(\phi, q, p)_{\text{magnetic attractor}} = \mathcal{F}_{\text{BH}}(\phi, p) - \phi^I q_I \tag{40}$$

and recover the free energy of [3] at an intermediate level. Subsequent extremization with respect to $\phi^I \propto \text{Re}Y^I$ gives the electric attractor equations, and by plugging them back we find the entropy. Moreover, while the free energy $\mathcal{F}_{BH}(\phi, p)$ is related to the black hole entropy $\mathcal{S}_{\text{macro}}(p, q)$ by a partial Legendre transform, the charge-independent part of Σ, namely $-K + 128iF_\Upsilon - 128i\overline{F}_{\overline{\Upsilon}}$ is its full Legendre transform.

Since $\pi\Sigma(q, p, Y, \overline{Y})$ is a symplectic function, which equals $\mathcal{S}_{\text{macro}}$ at its critical point, it is natural to take $\exp(\pi\Sigma)$ to define a modified version of (30). This means that we should not only to integrate over $\phi^I \propto \text{Re}Y^I$, but also over the other scalar fields $\text{Im}Y^I$. What about the measure? Since it should be symplectically invariant, the natural choice is[21]

$$d\mu(Y, \overline{Y}) = \prod_{IJ} dY^I d\overline{Y}^J \det(-2i\text{Im}(F_{KL})) , \tag{41}$$

---

[21] This follows from inspection of the symplectic transformation rules [21]. Alternatively, one might note that this measure is proportional to the top exterior power of the natural symplectic form of $C(M)$, the cone over the moduli space.

where F_{KL} denotes the second derivatives of the generalized prepotential with respect to the scalar fields. Putting everything together the proposal of [31] for a modified version of (30) is:

$$d(q,p) = \mathcal{N} \int \prod_{IJ} dY^I d\overline{Y}^J \det(-2i\mathrm{Im}(F_{KL})) \exp(\pi\Sigma) , \qquad (42)$$

where $\mathcal{N}$ is a normalization factor. In order to compare to (30) it is useful to note that one can perform the saddle point evaluation in two steps. In the first step one takes a saddle point with respect to the imaginary parts of Y^I, which imposes the magnetic attractor equations. Performing the saddle point integration one obtains

$$d(q,p) = \mathcal{N}' \int \prod_I d\phi^I \sqrt{\det(-2i(\mathrm{Im}F_{KL})_{\text{magn. attractor}})} \exp(\mathcal{F}_{\text{BH}} - \phi^I q_I) , \quad (43)$$

which is similar to the original (30) but contains a non-trivial measure factor stemming from the requirement of symplectic covariance. Subsequent saddle point evaluation with respect to $\phi^I \propto \mathrm{Re}Y^I$ gives

$$d(q,p) = \exp(\pi\Sigma_{\text{attractor}}) = \exp\mathcal{S}_{\text{macro}} . \qquad (44)$$

Let us next comment on another issue concerning (30). So far we have been working with a holomorphic prepotential $F(Y^I, \Upsilon)$, which upon differentiation yields effective gauge couplings that are holomorphic functions of the moduli. However, it is well known that the physical couplings extracted from string scattering amplitudes are not holomorphic. This can be understood purely in terms of field theory (see for example [45] for a review): if a theory contains massless particles, then the (quantum) effective action (the generating functional of 1PI Greens function) will in general be non-local. In the case of supersymmetric gauge theories, this goes hand in hand with non-holomorphic contributions to gauge couplings, which, in $N = 2$ theories, cannot be expressed in terms of a holomorphic prepotential [46, 48, 47]. Symmetries, such as S- and T-duality provide an efficient way of controlling these non-holomorphic terms. While the holomorphic couplings derived from the holomorphic prepotential are not consistent with S- and T-duality, the additional non-holomorphic contributions transform in such a way that they restore these symmetries. The same remark applies to the black hole entropy, as has been shown for the particular case of supersymmetric black holes in string compactifications with $N = 4$ supersymmetry. There, S-duality is supposed to be an exact symmetry, and therefore physical quantities like gauge couplings, gravitational couplings and the entropy must be S-duality invariant. But this is only possible if non-holomorphic contributions are taken into account [49, 47]. In the notation used here this amounts to modifying the black hole free energy by adding a real-valued function Ω, which is homogeneous of degree 2 and not harmonic [62, 31]:

$$\mathcal{F}_{\text{BH}} \rightarrow \widehat{\mathcal{F}}_{\text{BH}} = 4\pi\mathrm{Im}F(Y^I, \Upsilon) + 4\pi\Omega(Y^I, \overline{Y}^I, \Upsilon, \overline{\Upsilon}) . \qquad (45)$$

Non-holomorphic terms can also be studied in the framework of topological string theory and are then encoded in a holomorphic anomaly equation [23]. The role of non-holomorphic contributions has recently received considerable attention [50, 4, 5] (see also [51]). It appears that these proposals do not fully agree with the one of [31], which was explained above. One way to clarify the role of non-holomorphic corrections is the study of subleading terms in explicit examples, which will be discussed in the next sections.

In the last section we briefly explained how black hole solutions are related to flux compactifications. It is interesting to note that variational principles are another feature that they share. Over the last years it has been realized that the geometries featuring in flux compactifications are calibrated geometries, i.e., one can compute volumes of submanifolds by integrating suitable calibrating forms over them, without knowing the metric explicitly (see for example [51, 43]). Such geometries can be characterized in terms of variational principles, such as Hitchin's [44]. In physical terms the idea is to write down an abelian gauge theory for higher rank gauge fields (aka differential forms) such that the equations of motion are the equations characterising the geometry. The topological partition function Z_{top} should then be interpreted as a wave function of the quantized version of this theory. Conversely the variational principle provides the semiclassical approximation of the quantum mechanics on the moduli space.

6. Fundamental strings and 'small' black holes

So far our discussion was quite abstract and in parts formal. Therefore we now want to test these ideas in concrete models. As was first realized in [52], the $\frac{1}{2}$ BPS states of the toroidally compactified heterotic string provide an ideal test ground for the idea that there is an exact relation between black hole microstates and the string partition function. Since this compactification has $N = 4$ rather than $N = 2$ supersymmetry, one gets an enhanced control over both $\mathcal{S}_{\text{macro}}$ and $\mathcal{S}_{\text{micro}}$. For generic moduli, the massless spectrum of the heterotic string compactified on T^6 consists of the $N = 4$ supergravity multiplet together with 22 $N = 4$ vector multiplets. Since the gravity multiplet contains 4 graviphotons, the gauge group is $G = U(1)^{28}$. There are 28 electric charges q and 28 magnetic charges p which take values in the Narain lattice $\Gamma_{6,22}$. This lattice is even and selfdual with respect to the bilinear form of signature $(6, 22)$, and hence it is unique up to isometries. The T-duality group $O(6, 22, \mathbb{Z})$ group consists of those isometries which are lattice automorphisms. The S-duality group $SL(2, \mathbb{Z})$ acts as $\mathbf{2} \otimes \mathbb{1}$ on the $(28+28)$-component vector $(q, p) \in \Gamma_{6,22} \oplus \Gamma_{6,22} \simeq \mathbb{Z}^2 \otimes \Gamma_{6,22}$, where $\mathbf{2}$ denotes the fundamental representation of $SL(2, \mathbb{Z})$ and $\mathbb{1}$ the identity map on $\Gamma_{6,22}$.

It turns out that the $N = 2$ formalism described earlier can be used to construct supersymmetric black hole solutions of the $N = 4$ theory [47]. If one uses the up-to-two-derivatives part of the effective Lagrangian, the entropy is given by

[53, 47]

$$\mathcal{S}_{\text{macro}} = \frac{A}{4} = \pi\sqrt{q^2p^2 - (q \cdot p)^2} \ . \tag{46}$$

Here $a \cdot b$ denotes the scalar product with signature $(6, 22)$, so that the above formula is manifestly invariant under T-duality. It can be shown that $(q^2, p^2, q \cdot p)$ transforms in the **3**-representation under S-duality, and that the quadratic form $q^2p^2 - (q \cdot p)^2$ is invariant.[22] Therefore $\mathcal{S}_{\text{macro}}$ is manifestly S-duality invariant as well. The supersymmetric black hole solutions form two classes, corresponding to the two possible types of BPS multiplets [54, 55] (see also [56] for a review). The $\frac{1}{2}$ BPS solutions with 8 (out of a maximum of 16) Killing spinors are characterized by

$$q^2p^2 - (q \cdot p)^2 = 0 \tag{47}$$

and therefore have a degenerate horizon, at least in the lowest order approximation. Particular solutions of (47) are $p = 0$ ('electric black holes') and $q = 0$ ('magnetic black holes').

The $N = 4$ theory also has $\frac{1}{4}$ BPS solutions with only 4 Killing spinors. They satisfy

$$q^2p^2 - (q \cdot p)^2 \neq 0 \tag{48}$$

and therefore have a non-vanishing horizon. They are 'genuinely dyonic' in the sense that it is not possible to set all electric (or all magnetic) charges to zero by an S-duality transformation. Thus they are referred to as dyonic black holes. We will discuss them in the next section.

Let us return to the electric black holes. We saw that the horizon area is zero, $A = 0$, and so is the black hole entropy $\mathcal{S}_{\text{macro}} = 0$. Geometrically, the solution has a null singularity, i.e., the curvature singularity coincides with the horizon. One might wonder whether stringy or quantum corrections resolve this singularity. Moreover, a vanishing black hole entropy means that there is only one microstate, and one should check whether this is true.

The candidate microstates for the supersymmetric electric black hole are fundamental strings sitting in $\frac{1}{2}$ BPS multiplets [57]. These are precisely the states where the left-moving, supersymmetric sector is put into its ground state while exciting oscillations in the right-moving, bosonic sector. Such states take the following form:

$$\prod_l \alpha^{i_l}_{-m_l} |(P_L, P_R)\rangle \otimes |8 \oplus 8\rangle \tag{49}$$

Here the α's are creation operators for the right-moving oscillation mode of level $m_l = 1, 2, \ldots$, which can be aligned along the two transverse space directions, $i_l = 1, 2$, along the six directions of the torus, $i_l = 3, \ldots 8$, or along the maximal torus of the rank 16 gauge group of the ten-dimensional theory, $i_l = 9, \ldots, 24$. P_L and P_R are the left- and right-moving momenta, or, in other words, $q = (P_L, P_R) \in \Gamma_{6,22}$ are the electric charges. Finally $|8 \oplus 8\rangle$ is the left-moving ground state, which is a four-dimensional $N = 4$ vector supermultiplet with eight bosonic and eight

---

[22]Note that $SL(2, \mathbb{R}) \simeq SO(1, 2)$ and that the quadratic form $q^2p^2 - (q \cdot p)^2$ has signature $(1, 2)$.

fermionic degrees of freedom. Since the space-time supercharges are constructed out of the left-moving oscillators it is clear that this state transforms in the same way as the left-moving ground state, and therefore is an $\frac{1}{2}$ BPS state.

Physical states satisfy the mass formula

$$\alpha' M^2 = N - 1 + \tfrac{1}{2} P_R^2 + \tilde{N} + \tfrac{1}{2} P_L^2 , \tag{50}$$

where $N, \tilde{N}$ is the total right- and left-moving excitation level, respectively. Moreover physical states satisfy the level matching condition

$$N - 1 + \tfrac{1}{2} P_R^2 = \tilde{N} + \tfrac{1}{2} P_L^2 . \tag{51}$$

For $\frac{1}{2}$ BPS states we have $\tilde{N} = 0$ and level matching fixes the level N and the mass M in terms of the charges:[23]

$$N = \tfrac{1}{2}(P_L^2 - P_R^2) + 1 = -\tfrac{1}{2}q^2 + 1 = \tfrac{1}{2}|q^2| + 1 . \tag{52}$$

The problem of counting the number of $\frac{1}{2}$ BPS states amounts to counting partitions of an integer N (modulo the 24-fold extra degeneracy introduced by the additional labels i_l). This is a classical problem which has been studied by Hardy and Ramanujan [58]. The number $d(q)$ of states at of given charge admits the integral representation

$$d(q) = \oint_C d\tau \, \frac{\exp(i\pi\tau q^2)}{\eta^{24}(\tau)} . \tag{53}$$

Here τ take values in the upper half plane, $\eta(\tau)$ is the Dedekind η-function and C is a suitable integration contour. For large charges ($|q^2| \gg 1$), the asymptotic number of states is governed by the Hardy-Ramanujan formula

$$d(q) = \exp\left(4\pi\sqrt{\tfrac{1}{2}|q^2|} - \tfrac{27}{4}\log|q^2| + \cdots\right) . \tag{54}$$

The statistical entropy of string states therefore is

$$\mathcal{S}_{\mathrm{micro}}(q) = 4\pi\sqrt{\tfrac{1}{2}|q^2|} - \tfrac{27}{4}\log|q^2| + \cdots . \tag{55}$$

Comparing to the black hole entropy of electric supersymmetric black holes, we realize that there is a discrepancy, because $\mathcal{S}_{\mathrm{macro}}(q) = 0$. One can now reanalyse the black hole solutions and take into account higher derivative terms. As a first step one takes those terms which occur at tree level in the heterotic string theory. These are the same terms that one gets by dimensional reduction of higher derivative terms ('R^4-terms') of the ten-dimensional effective theory. Already this leading order higher derivative term is sufficient to resolve the null singularity and to give the electric black hole a finite horizon with area

$$\frac{A}{4} = 2\pi\sqrt{\tfrac{1}{2}|q^2|} = \tfrac{1}{2}\mathcal{S}_{\mathrm{micro}} + \cdots , \tag{56}$$

as was shown in [59] using the results of [15, 47]. The resulting area is large in Planckian but small in string units, reflecting that the resolution is a stringy effect.

---

[23] With our conventions q^2 is negative for physical BPS states with large excitation number N.

Hence these black holes are called 'small black holes', in contrast to the 'large' dyonic black holes, which already have a finite area in the classical approximation.

Again it is crucial to deviate from the area law and to use the generalized definition of black hole entropy (21), which results in [59]:

$$\mathcal{S}_{\text{macro}} = \frac{A}{4} + \text{Correction} = \frac{A}{4} + \frac{A}{4} = \frac{A}{2} = 4\pi\sqrt{\tfrac{1}{2}|q^2|} = \mathcal{S}_{\text{micro}} + \cdots . \qquad (57)$$

Thus we find that $\mathcal{S}_{\text{macro}}$ and $\mathcal{S}_{\text{micro}}$ are equal up to subleading contributions in the charges.

We can try to improve on this result by including further subleading contributions to $\mathcal{S}_{\text{macro}}$. According to [47, 62] the next relevant term is a non-holomorphic correction to the entropy which gives rise to a term logarithmic in the charges:

$$\mathcal{S}_{\text{macro}} = 4\pi\sqrt{\tfrac{1}{2}|q^2|} - 6\log|q^2| . \qquad (58)$$

This has the same form as $\mathcal{S}_{\text{micro}}$, but the coefficient of the subleading term is different. This is, however, to be expected, if $\mathcal{S}_{\text{macro}}$ and $\mathcal{S}_{\text{micro}}$ correspond to different ensembles. The actual test consists of the following: take the black hole free energy corresponding to the above $\mathcal{S}_{\text{macro}}$, and evaluate the integral (30), or any candidate modification thereof like (42), in a saddle point approximation and compare the result to $\mathcal{S}_{\text{micro}}$. This is, however not completely straightforward, because the measure in (42) vanishes identically when neglecting non-holomorphic and non-perturbative corrections. This reflects that the attractor points for electric black holes sit at the boundary of the classical moduli space (the Kähler cone). This boundary disappears in the quantum theory, and non-holomorphic and non-perturbative corrections make the measure finite. But still, the point around which one tries to expand does not correspond to a classical limit. This might explain why there is still disagreement for the term of the form $\log|q^2|$ when taking into account the leading non-holomorphic contribution to the measure ([31]). Moreover, it appears that [4, 5], who take a different attitude towards non-holomorphic contributions than [47, 31], also find a mismatch for the logarithmic term. It is tempting to conclude that the conjecture (29) just does not apply to small black holes. However, it was shown in [4, 5] that there is an infinite series of sub-subleading contributions, involving inverse powers of the charges, which matches perfectly! This suggest that there is a more refined version of the conjecture which applies to small black holes.

7. Dyonic strings and 'large' black holes

Let us finally briefly discuss $\frac{1}{4}$ BPS black holes. While the leading order black hole entropy is (46), one can also derive a formula which takes into account the non-perturbative and non-holomorphic corrections [47, 31]:

$$S_{\text{macro}}(q, p) = -\pi \left[\frac{q^2 - i(S - \overline{S})q \cdot p + |S|^2 p^2}{S + \overline{S}} - 2\log[(S + \overline{S})^6 |\eta(S)|^{24}] \right]_{\text{horizon}}$$

(59)

Here S denotes the dilaton and $\eta(S)$ is the Dedekind η-function. Recalling that the dilaton is related to the string coupling g_S by $S = \frac{1}{g_S^2} + i\theta$, and using the expansion of the η-function,

$$\eta(S) = -\tfrac{1}{12}\pi S - e^{-2\pi S} + \mathcal{O}(e^{-4\pi S})$$

(60)

we see that (59) includes an infinite series of instanton corrections.

In order to show that (59) is invariant under T-duality and S-duality, note that $q^2, p^2, p \cdot q$ and S are invariant under T-duality. Under S-duality $(q^2, p^2, p \cdot q)$ transforms in the **3** of $SL(2, \mathbb{Z})$, while

$$S \to \frac{aS - ib}{icS + d}, \quad \begin{pmatrix} a & b \\ c & d \end{pmatrix} \in SL(2, \mathbb{Z}) .$$

(61)

It is straightforward to see that $(S + \overline{S})^{-1}(1, -i(S - \overline{S}), |S|^2)$ transforms in the **3** of $SL(2, \mathbb{Z})$ so that the first term of (59) is S-duality invariant. S-duality invariance of the second term follows from the fact that $\eta^{24}(S)$ is a modular form of degree 12. Observe that the non-holomorphic term $\sim \log(S + \overline{S})$ is needed to make the second term of (59) S-duality invariant.

The entropy formula (59) is not fully explicit, since one cannot solve the attractor equations explicitly for the dilaton as a function of the charges. However, the dilaton attractor equations take the suggestive form

$$\frac{\partial S_{\text{macro}}(q, p, S, \overline{S})}{\partial S} = 0 .$$

(62)

Now we need to look for candidate microstates [60]. Since these must carry electric and magnetic charge and must sit in $\frac{1}{4}$ BPS multiplets, they cannot be fundamental string states. However, the underlying ten-dimensional string theory contains besides fundamental strings also solitonic five-branes, which carry magnetic charge. Upon double dimensional reduction to six dimension these can become magnetic strings, which sit in $\frac{1}{2}$ BPS multiplets. By forming bound states with fundamental strings one can obtain dyonic strings forming $\frac{1}{4}$ BPS multiplets. Further double dimensional reduction to four dimensions gives dyonic zero-branes, which at finite coupling should correspond to dyonic $\frac{1}{4}$ BPS black holes.

Based on the conjecture that the world volume theory of the heterotic five-brane is a six-dimensional string theory, one can derive a formula for the degeneracy of dyonic states [60]:

$$d_{DVV}(q, p) = \oint_{C_3} d\Omega \frac{\exp i\pi(Q, \Omega Q)}{\Phi_{10}(\Omega)} ,$$

(63)

where Ω is an element of the rank 2 Siegel upper half space (i.e., a symmetric complex two-by-two matrix with positive definite imaginary part). The vector

$Q = (q, p) \in \Gamma_{6,22} \oplus \Gamma_{6,22}$ combines the electric and magnetic charges, Φ_{10} is the degree 2, weight 10 Siegel cusp form, and C_3 is a three-dimensional integration contour in the Siegel upper half space. This formula is a natural generalization of the degeneracy formula (53) for electric $\frac{1}{2}$ BPS states. Recently, an alternative derivation has been given [61], which uses the known microscopic degeneracy of the five-dimensional D5-D1-brane bound state and the relation between five-dimensional and four-dimensional black holes [42].

It has been shown in [60] that the saddle point value of the integral (63) gives the leading order black hole entropy (46). More recently it has been shown in [62] that a saddle point evaluation of (63) yields precisely the full macroscopic entropy (59). In particular, the conditions for a saddle point of the integrand of (63) are precisely the dilaton attractor equations (62).

The natural next step is to investigate whether the microscopic state degeneracy (63) is consistent with the symplectically covariant version (42) of (30). This can indeed be shown [31] using the recent result of [63], who have evaluated the mixed partition functions of BPS black holes in $N = 4$ and $N = 8$ string compactifications. In particular, the non-trivial measure found in [63] can be obtained from (42) by taking the limit of large charges [31].

8. Discussion

We have seen that there is an impressive agreement between the counting of supersymmetric string states and the entropy of supersymmetric black holes. In particular, these comparisons are sensitive to the distinction between the naive area law and Wald's generalized definition. Moreover, there appears to be a direct link between the black hole entropy and the string partition function. This is a big leap forward towards a conceptual understanding of black hole microstates. The stringy approach to black hole entropy also has its limitations. One is that in order to identify the black hole microstates one has to extrapolate to asymptotically small coupling. In particular, one does not really get a good understanding of the black hole microstates 'as such,' but only how they look like in a different regime of the theory. A second, related problem is that one needs supersymmetry to have sufficient control over the extrapolation, the state counting and the construction of the corresponding black hole solutions. Therefore, quantitative agreement has only been established for supersymmetric black holes, which are charged, extremal black holes, and therefore not quite relevant for astrophysics. It should be stressed, however, that the proportionality between microscopic entropy and area can be established by a variety of methods, including the string-black hole correspondence reviewed at the beginning. Moreover, the work of [36, 37] and [40] suggests that the relation between black hole entropy and string theory states holds without supersymmetry.

The main limitation of string theory concerning quantum gravity in general and black holes in particular is that its core formalism, string perturbation theory around on-shell backgrounds, cannot be used directly to investigate the dynamics of generic, curved, non-stationary space-times. But we have seen that even within these limitations one can obtain remarkable results, which define benchmarks for other candidate theories of quantum gravity. Historically, one important guideline for finding new physical theories has always been that one should be able to reproduce established theories in a certain limit. This is reflected by the prominent role that the renormalization group and effective field theories play in contemporary quantum field theory. Concerning quantum gravity, it appears to be important to keep in touch with classical gravity in terms of a controlled semi-classical limit. Specifically, a theory of quantum gravity should allow one to construct a low energy effective field theory, which contains the Einstein-Hilbert term plus computable higher derivative corrections. Moreover, if microscopic black hole states can be identified and counted, the question as to whether they obey the area law or Wald's generalized formula should be answered.

The study of black holes has also lead to new ideas which will hopefully improve our understanding of string theory. Notably we have seen that there is a lot in common between supersymmetric black holes and flux compactifications, in particular the role of variational principles. The relation between the black hole partition function and the topological string partition function, and the interpretation of the latter as a wave function shows that there is a kind of minisuperspace approximation, which can be used to investigate the dynamics of flux compactifications and quantum cosmology using a stringy version of the Wheeler de Witt equation. This could not only improve the conceptual understanding of string theory, but would also increase the overlap between string theory and canonical approaches to quantum gravity.

More work needs to be done in order to further develop the proposal made by [3]. One key question is the relation between the macroscopic and microscopic entropies for small black holes, another one is the role of non-perturbative corrections in general [4, 5, 63, 31]. Future work will have to decide whether the relation discovered by [3] is an exact or only an asymptotic statement. Besides non-perturbative corrections also non-holomorphic corrections are important. We have discussed a concrete proposal for treating non-holomorphic corrections based on [31], but it is not obvious how this relates in detail to the microscopic side, i.e., to the topological string.

Acknowledgements

I would like to thank Bertfried Fauser and Jürgen Tolkdorf for organising a very diverse and stimulating workshop. The original results reviewed in this contribution were obtained in collaboration with Gabriel Lopes Cardoso, Bernard de Wit and Jürg Käppeli, who also gave valuable comments on the manuscript.

References

[1] L. Susskind, *Some Speculations about Black Hole Entropy in String Theory*. In C. Teitelboim (ed.), The Black Hole, 118. hep-th/9309145.

[2] G. Horowitz and J. Polchinski, Phys. Rev. D 55 (1997) 6189, hep-th/9612146.

[3] H. Ooguri, A. Strominger and C. Vafa, Phys. Rev. D 70 (2004) 106007, hep-th/0405146.

[4] A. Dabholkar, F. Denef, G. W. Moore, B. Pioline, JHEP 08 (2005) 021, JHEP 08 (2005) 021, hep-th/0502157.

[5] A. Dabholkar, F. Denef, G. W. Moore, B. Pioline, hep-th/0507014.

[6] A. Strominger and C. Vafa, Phys. Lett. B 379 (1996) 99, hep-th/9601029.

[7] G. W. Gibbons, *Supersymmetric Soliton States in Extended Supergravity Theories*. In P. Breitenlohner and H. P Dürr (eds.), Unified Theories of Elementary Particles, Springer, 1982.

[8] G. W. Gibbons and C. M. Hull, Phys. Lett. B 109 (1982) 190.

[9] B. de Wit, P. G. Lauwers, R. Philippe, S.-Q. Su and A. Van Proeyen, Phys. Lett. B 134 (1984) 37.

[10] B. de Wit and A. Van Proeyen, Nucl. Phys. B 245 (1984) 89.

[11] D. V. Alekseevsky, V. Cortes and C. Devchand, J. Geom. Phys. 42 (2002) 85, math.dg/9910091.

[12] J. Kowalski-Glikman, Phys. Lett. B 150 (1985) 125.

[13] G. L. Cardoso, B. de Wit, J. Käppeli and T. Mohaupt, JHEP 0012 (2000) 019, hep-th/0009234.

[14] S. Ferrara, R. Kallosh and A. Strominger, Phys. Rev. D 52 (1995) 5412, hep-th/9508072. A. Strominger, Phys. Lett. B 383 (1996) 39, hep-th/9602111. S. Ferrara and R. Kallosh, Phys. Rev. D 54 (1996) 1514, hep-th/9602136. S. Ferrara and R. Kallosh, Phys. Rev. D 54 (1996) 1525, hep-th/9603090.

[15] G. L. Cardoso, B. de Wit and T. Mohaupt, Phys. Lett. B 451 (1999) 309, hep-th/9812082.

[16] K. Behrndt, G. L. Cardoso, B. de Wit, R. Kallosh, D. Lüst and T. Mohaupt, Nucl. Phys. B 488 (1997) 236, hep-th/9610105.

[17] M. Shmakova, Phys. Rev. D 56 (1977) 540, hep-th/9612076.

[18] K. Behrndt, D. Lüst and W. A. Sabra, Nucl. Phys. B 510 (1998) 264,hep-th/9705169.

[19] J. Maldacena, A. Strominger and E. Witten, JHEP 12 (1997) 002, hep-th/9711053.

[20] C. Vafa, Adv. Theor. Math. Phys. 2 (1998) 207, hep-th/9711067.

[21] B. de Wit, Nucl. Phys. Proc. Suppl. 49 (1996) 191, hep-th/9602060.

[22] T. Mohaupt, Fortsch. Phys. 49 (2001) 3, hep-th/0007195.

[23] M. Bershadsky, S. Cecotti, H. Ooguri and C. Vafa, Comm. Math. Phys. 165 (1993) 311, hep-th/9309140.

[24] I. Antoniadis, E. Gava, N. S. Narain and T. R. Taylor, Nucl. Phys. B 413 (1994) 162, hep-th/9307158.

[25] T. Mohaupt, Class. Quant. Grav. 17 (2000) 3429, hep-th/0004098.

[26] R. Wald, Phys. Rev. D 48 (1993) 3427, gr-qc/9307038.

[27] V. Iyer and R. Wald, Phys. Rev. D 50 (1994) 846, gr-qc/9403028.

[28] E. Witten, hep-th/9306122.

[29] H. Ooguri, C. Vafa and E. Verlinde, hep-th/0502211.

[30] S. Gukov, K. Saraikin and C. Vafa, hep-th/0509109.

[31] G. L. Cardoso, B. de Wit, J. Käppeli and T. Mohaupt, *in preparation*.

[32] B. de Wit, hep-th/0503211.

[33] B. de Wit, hep-th/0511261.

[34] J. Käppeli, hep-th/0511221.

[35] G. L. Cardoso, B. de Wit and T. Mohaupt, Fortsch. Phys. 48 (2000) 49, hep-th/9904005.

[36] A. Sen, hep-th/0506177.

[37] A. Sen, hep-th/0508042.

[38] K. Behrndt, G. L. Cardoso and S. Mahapatra, hep-th/0506251.

[39] D. Gaiotto, A. Strominger and X. Yin, hep-th/0503217, hep-th/0504126.

[40] P. Kraus and F. Larsen, JHEP 09 (2005) 034, hep-th/0506176.

[41] J. A. Harvey, R. Minasian and G. Moore, JHEP 09 (1998) 004. hep-th/9808060.

[42] M. Guica, L. Huang, W. Li and A. Strominger, hep-th/0505188.

[43] R. Dijkgraaf, S. Gukov, A. Neitzke and C. Vafa, hept-th/0411073.

[44] N. Hitchin, math.dg/0107101, math.dg/0010054.

[45] V. Kaplunovsky and J. Louis, Nucl. Phys. B 444 (1995) 191, hep-th/9502077.

[46] B. de Wit, V. Kaplunovsky, J. Louis and D. Lüst, Nucl. Phys. B 451 (1995) 53, hep-th/9504006.

[47] G. L. Cardoso, B. de Wit and T. Mohaupt, Nucl.Phys. B567 (2000) 87, hep-th/9906094.

[48] B. de Wit, G. L. Cardoso, D. Lüst, T. Mohaupt and S.-J. Rey, Nucl. Phys. B 481 (1996) 353, hep-th/9607184.

[49] J. A. Harvey and G. Moore, Phys. Rev. D 57 (1998) 2323, hep-th/9610237.

[50] E. Verlinde, hep-th/0412139.

[51] A. A. Gerasimov and S. L. Shatashvili, JHEP 0411 (2004) 074, hep-th/0409238.

[52] A. Dabholkar, Phys. Rev. Lett. 94 (2005) 241301, hep-th/0409148.

[53] M. Cvetic and D. Youm, Phys. Rev. D 53 (1996) 584, hep-th/9507090. M. Cvetic and A. Tseytlin, Phys. Rev. D53 (1996) 5619, Erratum-ibid. D55 (1997) 3907, hep-th/9512031.

[54] J. Schwarz and A. Sen, Phys. Lett. B 312 (1993) 105, hep-th/9305185.

[55] M. Duff, J. T. Liu and J. Rahmfeld, Nucl.Phys. B 459 (1996) 125, hep-th/9508094.

[56] G. L. Cardoso, G. Curio, D. Lüst, T. Mohaupt and S.-J. Rey, Nucl. Phys. B 464 (1996) 18, hep-th/9512129.

[57] A. Dabholkar and J. Harvey, Phys. Rev. Lett. 63 (1989) 478. A. Dabholkar, G. W. Gibbons, J. A. Harvey and F. Ruiz Ruiz, Nucl. Phys. B 340 (1990) 33.

[58] G. H. Hardy and S. Ramanujan, Proc. Lond. Math. Soc. 2 (1918) 75.

[59] A. Dabholkar, R. Kallosh and A. Maloney, JHEP 0412 (2004) 059 hep-th/0410076.

[60] R. Dijkgraaf, E. Verlinde and H. Verlinde, Nucl.Phys. B 484 (1997) 543, hep-th/9607026.

[61] D. Shih, A. Strominger and X. Yin, hep-th/0505094.

[62] G. L. Cardoso, B. de Wit, J. Käppeli and T. Mohaupt, JHEP 12 (2004) 075, hep-th/0412287.

[63] D. Shih and X. Yin, hep-th/0508174.

[64] G. L. Cardoso, B. de Wit, J. Käppeli and T. Mohaupt, Forschr. Phys. 46 (2001) 557, hep-th/0012232.

Thomas Mohaupt
Theoretical Physics Divison
Department of Mathematical Sciences
University of Liverpool
Liverpool L69 3BX, UK
e-mail: Thomas.Mohaupt@liv.ac.uk

Quantum Gravity
B. Fauser, J. Tolksdorf and E. Zeidler, Eds., 263–281
© 2006 Birkhäuser Verlag Basel/Switzerland

The Principle of the Fermionic Projector: An Approach for Quantum Gravity?

Felix Finster

Abstract. In this chapter we introduce the mathematical framework of the principle of the fermionic projector and set up a variational principle in discrete space-time. The underlying physical principles are discussed. We outline the connection to the continuum theory and state recent results. In the last two sections, we speculate on how it might be possible to describe quantum gravity within this framework.

Mathematics Subject Classification (2000). Primary 81T75; Secondary 46L87.

Keywords. Fermionic projector, discrete space-time, variational principle, non-locality, non-causality, continuum limit.

The principle of the fermionic projector [3] provides a new model of space-time together with the mathematical framework for the formulation of physical theories. It was proposed to formulate physics in this framework based on a particular variational principle. Here we explain a few basic ideas of the approach, report on recent results and explain the possible connection to quantum gravity.

It is generally believed that the concept of a space-time continuum (like Minkowski space or a Lorentzian manifold) should be modified for distances as small as the Planck length. We here assume that space-time is discrete on the Planck scale. Our notion of "discrete space-time" differs from other discrete approaches (like for example lattice gauge theories or quantum foam models) in that we do not assume any structures or relations between the space-time points (like for example the nearest-neighbor relation on a space-time lattice). Instead, we set up a variational principle for an ensemble of quantum mechanical wave functions. The idea is that for mimimizers of our variational principle, these wave functions should induce relations between the discrete space-time points, which, in a suitable limit, should go over to the topological and causal structure of a Lorentzian manifold. More specifically, in this limit the wave functions should group to a configuration of Dirac seas.

I would like to thank the Erwin Schrödinger Institute, Wien, for its hospitality.

For clarity, we first introduce the mathematical framework (Section 1) and discuss it afterwards, working out the underlying physical principles (Section 2). Then we outline the connection to the continuum theory (Sections 3 and 4) and state some results (Sections 5). Finally, we give an outlook on classical gravity (Section 6) and the field quantization (Section 7).

1. A variational principle in discrete space-time

We let $(H, <.|.>)$ be a complex inner product space of signature (N, N). Thus $<.|.>$ is linear in its second and antilinear in its first argument, and it is symmetric,

$$\overline{<\Psi \,|\, \Phi>} \;=\; <\Phi \,|\, \Psi> \qquad \text{for all } \Psi, \Phi \in H \,,$$

and non-degenerate,

$$<\Psi \,|\, \Phi> \;=\; 0 \;\; \text{for all } \Phi \in H \quad \Longrightarrow \quad \Psi \;=\; 0 \,.$$

In contrast to a scalar product, $<.|.>$ is *not* positive. Instead, we can choose an orthogonal basis $(e_i)_{i=1,\ldots,2N}$ of H such that the inner product $<e_i \,|\, e_i>$ equals $+1$ if $i = 1, \ldots, N$ and equals -1 if $i = N + 1, \ldots, 2N$.

A *projector* A in H is defined just as in Hilbert spaces as a linear operator which is idempotent and self-adjoint,

$$A^2 = A \qquad \text{and} \qquad <A\Psi \,|\, \Phi> = <\Psi \,|\, A\Phi> \quad \text{for all } \Psi, \Phi \in H \,.$$

Let M be a finite set. To every point $x \in M$ we associate a projector E_x. We assume that these projectors are orthogonal and complete in the sense that

$$E_x \, E_y \;=\; \delta_{xy} \, E_x \qquad \text{and} \qquad \sum_{x \in M} E_x \;=\; \mathbb{1} \,. \tag{1}$$

Furthermore, we assume that the images $E_x(H) \subset H$ of these projectors are non-degenerate subspaces of H, which all have the same signature (n, n). We refer to (n, n) as the *spin dimension*. The points $x \in M$ are called *discrete space-time points*, and the corresponding projectors E_x are the *space-time projectors*. The structure $(H, <.|.>, (E_x)_{x \in M})$ is called *discrete space-time*.

We introduce one more projector P on H, the so-called *fermionic projector*, which has the additional property that its image $P(H)$ is a *negative definite* subspace of H. We refer to the rank of P as the *number of particles* $f := \dim P(H)$.

A space-time projector E_x can be used to restrict an operator to the subspace $E_x(H) \subset H$. Using a more graphic notion, we also refer to this restriction as the *localization* at the space-time point x. For example, using the completeness of the space-time projectors (1), we readily see that

$$f \;=\; \mathrm{Tr}\, P \;=\; \sum_{x \in M} \mathrm{Tr}(E_x P) \,. \tag{2}$$

The expression $\mathrm{Tr}(E_x P)$ can be understood as the localization of the trace at the space-time point x, and summing over all space-time points gives the total trace.

When forming more complicated composite expressions, it is convenient to use the short notations

$$P(x, y) = E_x P E_y \qquad \text{and} \qquad \Psi(x) = E_x \Psi. \tag{3}$$

The operator $P(x, y)$ maps $E_y(H) \subset H$ to $E_x(H)$, and it is often useful to regard it as a mapping only between these subspaces,

$$P(x, y) : E_y(H) \rightarrow E_x(H).$$

Using (1), we can write the vector $P\Psi$ as follows,

$$(P\Psi)(x) = E_x P\Psi = \sum_{y \in M} E_x P E_y \Psi = \sum_{y \in M} (E_x P E_y)(E_y \Psi),$$

and thus

$$(P\Psi)(x) = \sum_{y \in M} P(x, y) \Psi(y). \tag{4}$$

This relation resembles the representation of an operator with an integral kernel. Therefore, we call $P(x, y)$ the *discrete kernel* of the fermionic projector.

We can now set up our variational principle. We define the *closed chain* A_{xy} by

$$A_{xy} = P(x, y) P(y, x) = E_x P E_y P E_x; \tag{5}$$

it maps $E_x(H)$ to itself. Let $\lambda_1, \ldots, \lambda_{2n}$ be the zeros of the characteristic polynomial of A_{xy}, counted with multiplicities. We define the *spectral weight* $|A_{xy}|$ by

$$|A_{xy}| = \sum_{j=1}^{2n} |\lambda_j|.$$

Similarly, one can take the spectral weight of powers of A_{xy}, and by summing over the space-time points we get positive numbers depending only on the form of the fermionic projector relative to the space-time projectors. Our variational principle is to

$$\text{minimize} \quad \sum_{x, y \in M} |A_{xy}^2| \tag{6}$$

by considering variations of the fermionic projector which satisfy the constraint

$$\sum_{x, y \in M} |A_{xy}|^2 = \text{const}. \tag{7}$$

In the variation we also keep the number of particles f as well as discrete space-time fixed. Using the method of Lagrangian multipliers, for every minimizer P there is a real parameter μ such that P is a stationary point of the *action*

$$S_\mu[P] = \sum_{x, y \in M} \mathcal{L}_\mu[A_{xy}] \tag{8}$$

with the *Lagrangian*

$$\mathcal{L}_\mu[A] = |A^2| - \mu |A|^2. \tag{9}$$

This variational principle was first introduced in [3]. In [4] it is analyzed mathematically, and it is shown in particular that minimizers exist:

Theorem 1.1. *The variational principle (6, 7) attains its minimum.*

2. Discussion of the underlying physical principles

We come to the physical discussion. Obviously, our mathematical framework does not refer to an underlying space-time continuum, and our variational principle is set up intrinsically in discrete space-time. In other words, our approach is *background free*. Furthermore, the following physical principles are respected, in a sense we briefly explain.

- The **Pauli Exclusion Principle**: We interpret the vectors in the image of P as the quantum mechanical states of the particles of our system. Thus, choosing a basis $\Psi_1, \ldots, \Psi_f \in P(H)$, the Ψ_i can be thought of as the wave functions of the occupied states of the system. Every vector $\Psi \in H$ either lies in the image of P or it does not. Via these two conditions, the fermionic projector encodes for every state Ψ the occupation numbers 1 and 0, respectively, but it is impossible to describe higher occupation numbers. More technically, we can form the anti-symmetric many-particle wave function

$$\Psi = \Psi_1 \wedge \cdots \wedge \Psi_f .$$

  Due to the anti-symmetrization, this definition of Ψ is (up to a phase) independent of the choice of the basis $\Psi_1, \ldots, \Psi_f$. In this way, we can associate to every fermionic projector a fermionic many-particle wave function which obeys the Pauli Exclusion Principle. For a detailed discussion we refer to [3, §3.2].

- A **local gauge principle**: Exactly as in Hilbert spaces, a linear operator U in H is called *unitary* if

$$<U\Psi \,|\, U\Phi> \; = \; <\Psi \,|\, \Phi> \qquad \text{for all } \Psi, \Phi \in H.$$

It is a simple observation that a joint unitary transformation of all projectors,

$$E_x \;\rightarrow\; U E_x U^{-1}, \qquad P \;\rightarrow\; U P U^{-1} \qquad \text{with } U \text{ unitary} \qquad (10)$$

keeps our action (6) as well as the constraint (7) unchanged, because

$$P(x,y) \;\rightarrow\; U\, P(x,y)\, U^{-1}, \qquad A_{xy} \;\rightarrow\; U A_{xy} U^{-1}$$
$$\det(A_{xy} - \lambda \mathbb{1}) \;\rightarrow\; \det\big(U(A_{xy} - \lambda\mathbb{1})\, U^{-1}\big) \;=\; \det(A_{xy} - \lambda\mathbb{1}),$$

and so the λ_j stay the same. Such unitary transformations can be used to vary the fermionic projector. However, since we want to keep discrete space-time fixed, we are only allowed to consider unitary transformations which do not change the space-time projectors,

$$E_x \;=\; U E_x U^{-1} \qquad \text{for all } x \in M . \qquad (11)$$

Then (10) reduces to the transformation of the fermionic projector

$$P \rightarrow UPU^{-1}.\tag{12}$$

The conditions (11) mean that U maps every subspace $E_x(H)$ into itself. Hence U splits into a direct sum of unitary transformations

$$U(x) := UE_x : E_x(H) \rightarrow E_x(H),\tag{13}$$

which act "locally" on the subspaces associated to the individual space-time points.

Unitary transformations of the form (11, 12) can be identified with local gauge transformations. Namely, using the notation (3), such a unitary transformation U acts on a vector $\Psi \in H$ as

$$\Psi(x) \longrightarrow U(x)\,\Psi(x).$$

This formula coincides with the well-known transformation law of wave functions under local gauge transformations (for more details see [3, §1.5 and §3.1]). We refer to the group of all unitary transformations of the form (11, 12) as the *gauge group*. The above argument shows that our variational principle is *gauge invariant*. Localizing the gauge transformations according to (13), we obtain at any space-time point x the so-called *local gauge group*. The local gauge group is the group of isometries of $E_x(H)$ and can thus be identified with the group $U(n, n)$. Note that in our setting the local gauge group cannot be chosen arbitrarily, but it is completely determined by the spin dimension.

- The **equivalence principle**: At first sight it might seem impossible to speak of the equivalence principle without having the usual space-time continuum. What we mean is the following more general notion. The equivalence principle can be expressed by the invariance of the physical equations under general coordinate transformations. In our setting, it makes no sense to speak of co-ordinate transformations nor of the diffeomorphism group because we have no topology on the space-time points. But instead, we can take the largest group which can act on the space-time points: the group of all permutations of M. Our variational principle is obviously *invariant under the permutation group* because permuting the space-time points merely corresponds to reordering the summands in (6, 7). Since on a Lorentzian manifold, every diffeomorphism is bijective and can thus be regarded as a permutation of the space-time points, the invariance of our variational principle under the permutation group can be considered as a generalization of the equivalence principle.

An immediate objection to the last paragraph is that the symmetry under permutations of the space-time points is not compatible with the topological and causal structure of a Lorentzian manifold, and this leads us to the discussion of the physical principles which are *not* taken into account in our framework. Our definitions involve **no locality** and **no causality**. We do not consider these principles as being fundamental. Instead, our concept is that the causal structure is induced on the space-time points by the minimizer P of our variational principle. In particular, minimizers should spontaneously break the above permutation symmetry to a smaller symmetry group, which, in a certain limiting case describing the vacuum, should reduce to Poincaré invariance. Explaining in detail how this is supposed to work goes beyond the scope of this introductory article (for a first step in the mathematical analysis of spontaneous symmetry breaking see [5]). In order to tell the reader right away what we have in mind, we shall first simply assume the causal structure of Minkowski space and consider our action in the setting of relativistic quantum mechanics (Section 3). This naive procedure will *not* work, but it will nevertheless illustrate our variational principle and reveal a basic difficulty. In Section 4 we will then outline the connection to the continuum theory as worked out in [3].

3. Naive correspondence to a continuum theory

Let us see what happens if we try to get a connection between the framework of Section 1 and relativistic quantum mechanics in the simplest possible way. To this end, we just replace M by the space-time continuum $\mathbb{R}^4$ and the sums over M by space-time integrals. For a vector $\Psi \in H$, the corresponding $\Psi(x) \in E_x(H)$ as defined by (3) should be a 4-component Dirac wave function, and the scalar product $<\Psi(x) \,|\, \Phi(x)>$ on $E_x(H)$ should correspond to the usual Lorentz invariant scalar product on Dirac spinors $\overline{\Psi}\Phi$ with $\overline{\Psi} = \Psi^\dagger \gamma^0$ the adjoint spinor. Since this last scalar product is indefinite of signature $(2, 2)$, we are led to choosing $n = 2$, so that the spin dimension is $(2, 2)$.

In view of (4), the discrete kernel should in the continuum go over to the integral kernel of an operator P on the Dirac wave functions,

$$(P\Psi)(x) = \int_M P(x, y)\, \Psi(y)\, d^4y \,.$$

The image of P should be spanned by the occupied fermionic states. We take Dirac's concept literally that in the vacuum all negative-energy states are occupied by fermions forming the so-called *Dirac sea*. This leads us to describe the vacuum by the integral over the lower mass shell

$$P(x, y) = \int \frac{d^4k}{(2\pi)^4}\, (\slashed{k} + m)\, \delta(k^2 - m^2)\, \Theta(-k^0)\, e^{-ik(x-y)} \qquad (14)$$

(we consider for simplicity only one Dirac sea of mass m; the factor $(\slashed{k} + m)$ is needed in order to satisfy the Dirac equation $(i\slashed{\partial}_x - m)\, P(x, y) = 0$).

We now consider our action for the above fermionic projector. Since we do not want to compute the Fourier integral (14) in detail, we simply choose x and y for which the integrals in (14) exist (for details see below) and see what we get using only the *Lorentz symmetry* of P. We can clearly write $P(x, y)$ as

$$P(x, y) = \alpha\, (y - x)_j \gamma^j + \beta\, \mathbb{1}$$

with two complex parameters α and β. Taking the complex conjugate of (14), we see that

$$P(y, x) = \overline{\alpha}\,(y - x)_j \gamma^j + \overline{\beta}\,\mathbb{1}\,.$$

As a consequence,

$$A_{xy} = P(x, y)\,P(y, x) = a\,(y - x)_j \gamma^j + b\,\mathbb{1} \tag{15}$$

with real parameters a and b given by

$$a = \alpha\overline{\beta} + \beta\overline{\alpha}\,, \qquad b = |\alpha|^2\,(y - x)^2 + |\beta|^2\,. \tag{16}$$

Using the formula $(A_{xy} - b\mathbb{1})^2 = a^2\,(y - x)^2$, one can easily compute the zeros of the characteristic polynomial of A_{xy},

$$\lambda_1 = \lambda_2 = b + \sqrt{a^2\,(y - x)^2}\,, \qquad \lambda_3 = \lambda_4 = b - \sqrt{a^2\,(y - x)^2}\,.$$

If the vector $(y - x)$ is spacelike, we conclude from the inequality $(y - x)^2 < 0$ that the argument of the above square root is negative. As a consequence, the λ_j appear in *complex conjugate pairs*,

$$\overline{\lambda_1} = \lambda_3\,, \qquad \overline{\lambda_2} = \lambda_4\,.$$

Furthermore, the λ_j all have the same absolute value $|\lambda_j| =: |\lambda|$, and thus the action (6) reduces to

$$\mathcal{S}_\mu[A] = |\lambda|^2\,(4 - 16\,\mu)\,.$$

This simplifies further if we choose the Lagrangian multiplier μ equal to $\frac{1}{4}$, because then the action vanishes identically. If conversely $(y - x)$ is timelike, the λ_i are all real. Using (16), one easily verifies that they are all *positive* and thus $\mathcal{S}_{\frac{1}{4}}[A] = (\lambda_1 - \lambda_3)^2$. We conclude that

$$\mathcal{S}_{\frac{1}{4}}[A_{xy}] = \begin{cases} 4a^2\,(y - x)^2 & \text{if } (y - x) \text{ is timelike} \\ 0 & \text{if } (y - x) \text{ is spacelike}\,. \end{cases} \tag{17}$$

This consideration gives a simple *connection to causality*: In the two cases where $(y - x)$ is timelike or spacelike, the spectral properties of the matrix A_{xy} are completely different (namely, the λ_j are real or appear in complex conjugate pairs, respectively), and this leads to a completely different form of the action (17). More specifically, if the λ_j are non-real, this property is (by continuity) preserved under small perturbations of A_{xy}. Thinking of a dynamical situation, this suggests that perturbations of $P(x, y)$ for spacelike $(y - x)$ should not effect the action or, in other words, that events at points x and y with spacelike separation should not be related to each other by our variational principle. We remark that choosing $\mu = \frac{1}{4}$ is justified by considering the Euler-Lagrange equations corresponding to our variational principle, and this also makes the connection to causality clearer (see [3, §3.5 and §5]).

Apart from the oversimplifications and many special assumptions, the main flaw of this section is that the Fourier integral (14) does not exist for all x and y. More precisely, $P(x, y)$ is a well-defined distribution, which is even a smooth function if $(y - x)^2 \neq 0$. But *on the light cone* $(y - x)^2 = 0$, this distribution is *singular* (for more details see [3, §2.5]). Thus on the light cone, the pointwise product in (15) is ill-defined and our above arguments fail. The resolution of this problem will be outlined in the next section.

4. The continuum limit

We now return to the discrete setting of Section 1 and shall explain how to get a rigorous connection to the continuum theory. One approach is to study the minimizers in discrete space-time and to try to recover structures known from the continuum. For example, in view of the spectral properties of A_{xy} in Minkowski space as discussed in the previous section, it is tempting to introduce in discrete space-time the following notion (this definition is indeed symmetric in x and y, see [3, §3.5]).

Def. 4.1. *Two discrete space-time points $x, y \in M$ are called* **timelike** *separated if the zeros λ_j of the characteristic polynomial of A_{xy} are all real and not all equal. They are said to be* **spacelike** *separated if the λ_j are all non-real and have the same absolute value.*

The conjecture is that if the number of space-time points and the number of particles both tend to infinity at a certain relative rate, the above "discrete causal structure" should go over to the causal structure of a Lorentzian manifold. Proving this conjecture under suitable assumptions is certainly a challenge. But since we have a precise mathematical framework in discrete space-time, this seems an interesting research program.

Unfortunately, so far not much work has been done on the discrete models, and at present almost nothing is known about the minimizers in discrete space-time. For this reason, there seems no better method at the moment than to impose that the fermionic projector of the *vacuum* is obtained from a Dirac sea configuration by a suitable regularization process on the Planck scale [3, Chapter 4]. Since we do not know how the physical fermionic projector looks like on the Planck scale, we use the *method of variable regularization* and consider a large class of regularizations [3, §4.1].

When introducing the fermionic projector of the vacuum, we clearly put in the causal structure of Minkowski space as well as the free Dirac equation ad hoc. What makes the method interesting is that we then introduce a *general interaction* by inserting a general (possibly nonlocal) perturbation operator into the Dirac equation. Using methods of hyperbolic PDEs (the so-called *light-cone expansion*), one can describe the fermionic projector with interaction in detail [3, §2.5]. It turns out that the regularization of the fermionic projector with interaction is completely determined by the regularization of the vacuum (see [3, §4.5 and Appendix D]). Due to the regularization, the singularities of the fermionic projector have disappeared, and one can consider the *Euler-Lagrange equations* corresponding to our variational principle (see [3, §4.5 and Appendix F]). Analyzing the dependence on the regularization in detail, we can perform an expansion in powers of the Planck length. This gives differential equations involving Dirac and gauge fields, which involve a small number of so-called *regularization parameters*, which depend on the regularization and which we treat as free parameters (see [3, §4.5 and Appendix E]). This procedure for analyzing the Euler-Lagrange equations in the continuum is called *continuum limit*. We point out that only the singular behavior of $P(x, y)$ on the light cone enters the continuum limit, and this gives causality.

5. Obtained results

In [3, Chapters 6-8] the continuum limit is analyzed in spin dimension $(16, 16)$ for a fermionic projector of the vacuum, which is the direct sum of seven identical massive sectors and one massless left-handed sector, each of which is composed of three Dirac seas. Considering general chiral and (pseudo)scalar potentials, we find that the sectors spontaneously form pairs, which are referred to as *blocks*. The resulting *effective interaction* can be described by chiral potentials corresponding to the effective gauge group

$$SU(2) \otimes SU(3) \otimes U(1)^3 .$$

This model has striking similarity to the standard model if the block containing the left-handed sector is identified with the leptons and the three other blocks with the quarks. Namely, the effective gauge fields have the following properties.

- The $SU(3)$ corresponds to an unbroken gauge symmetry. The $SU(3)$ gauge fields couple to the quarks exactly as the strong gauge fields in the standard model.
- The $SU(2)$ potentials are left-handed and couple to the leptons and quarks exactly as the weak gauge potentials in the standard model. Similar to the CKM mixing in the standard model, the off-diagonal components of these potentials must involve a non-trivial mixing of the generations. The $SU(2)$ gauge symmetry is spontaneously broken.
- The $U(1)$ of electrodynamics can be identified with an Abelian subgroup of the effective gauge group.

The effective gauge group is larger than the gauge group of the standard model, but this is not inconsistent because a more detailed analysis of our variational principle should give further constraints for the Abelian gauge potentials. Moreover, there are the following differences to the standard model, which we derive mathematically without working out their physical implications.

- The $SU(2)$ gauge field tensor F must be simple in the sense that $F = \Lambda s$ for a real 2-form Λ and an $su(2)$-valued function s.
- In the lepton block, the off-diagonal $SU(2)$ gauge potentials are associated with a new type of potential, called nil potential, which couples to the right-handed component.

6. Outlook: The classical gravitational field

The permutation symmetry of our variational principle as discussed in Section 2 guarantees that the equations obtained in the continuum limit are invariant under diffeomorphisms. This gives us the hope that classical gravity might already be taken into account, and that even quantum gravity might be incorporated in our framework if our variational principle is studied beyond the continuum limit. Unfortunately, so far these questions have hardly been investigated. Therefore, at this point we leave rigorous mathematics and must enter the realm of what a cricial scientist might call pure speculation. Nevertheless, the following discussion might be helpful to give an idea of what our approach is about, and it might also give inspiration for future work in this area.

The only calculations for gravitational fields carried out so far are the calculations for linearized gravity [6, Appendix B]. The following discussion of classical gravity is based on these calculations. For the metric, we consider a linear perturbation h_{jk} of the Minkowski metric $\eta_{jk} = \mathrm{diag}(1, -1, -1, -1)$,

$$g_{jk}(x) = \eta_{jk} + h_{jk}(x) .$$

In linearized gravity, the diffeomorphism invariance corresponds to a large freedom to transform the h_{jk} without changing the space-time geometry (this freedom is usually referred to as "gauge freedom", but we point out for clarity that it is not related to the "local gauge freedom" as discussed in Section 2). This freedom can be used to arrange that (see e.g. [7])

$$\partial^k h_{jk} = \frac{1}{2} \partial_j h_{kl} \, \eta^{kl} .$$

Computing the corresponding Dirac operator and performing the light-cone expansion, the first-order perturbation of the fermionic projector takes the form

$$\Delta P(x,y) = \mathcal{O}(\slashed{\xi} \, z^{-1}) + \mathcal{O}(\xi^{[k} \gamma^{l]} \, z^{-1}) + \mathcal{O}(m) + \mathcal{O}((h_{ij})^2) \tag{18}$$

$$+ \frac{1}{2} \left(\int_x^y h_j^k \right) \xi^j \frac{\partial}{\partial y^k} \, P(x,y)$$

$$\left. - \frac{i}{16\pi^3} \left(\int_x^y (2\alpha - 1) \, \gamma^i \, \xi^j \, \xi^k \, (h_{jk,i} - h_{ik,j}) \right) d\alpha \, z^{-2} \right\} \tag{19}$$

$$\left. - \frac{1}{32\pi^3} \left(\int_x^y \varepsilon^{ijlm} \, (h_{jk,i} - h_{ik,j}) \, \xi^k \, \xi_l \, \rho \gamma_m \right) d\alpha \, z^{-2} \right.$$

$$+ \frac{i}{32\pi^3} \left(\int_x^y (\alpha^2 - \alpha) \, \xi^j \, \gamma^k \, G_{jk} \right) d\alpha \, z^{-1} , \tag{20}$$

where we set $\xi \equiv y - x$, the integrals go along straight lines joining the points x and y,

$$\int_x^y f \, d\alpha = \int_0^1 f(\alpha y + (1-\alpha)x) \, d\alpha ,$$

and z^{-1}, z^{-2} are distributions which are singular on the light cone,

$$z^{-1} = \frac{\text{PP}}{\xi^2} + i\pi\delta(\xi^2)\,\epsilon(\xi^0)\,, \quad z^{-2} = \frac{\text{PP}}{\xi^4} - i\pi\delta'(\xi^2)\,\epsilon(\xi^0)$$

(where PP denotes the principal part, and ϵ is the step function $\epsilon(x) = 1$ if $x > 0$ and $\epsilon(x) = -1$ otherwise). In this formula we only considered the most singular contributions on the light cone and did no take into account the higher orders in the rest mass m of the Dirac particles (for details see [6]). Nevertheless, the above formula gives us some general information on how the fermionic projector depends on a classical gravitational field. The contribution (19) describes an "infinitesimal deformation" of the light cone corresponding to the fact that the gravitational field affects the causal structure. Since it involves at most first derivatives of the metric, the curvature does not enter, and thus (19) can be compensated by a gauge and an infinitesimal coordinate tranformation. The diffeomorphism invariance of the equations of the continuum limit ensures that the contribution (19) drops out of these equations (and this can also be verified by a direct computation of the closed chain). We conclude that the equations of the continnum limit will be governed by the contribution (20). It is remarkable that the *Einstein tensor* G_{jk} appears. Thus, provided that the equations of the continuum limit give sufficiently strong constraints, we obtain the vacuum Einstein equations.

The situation becomes even more interesting if fermionic matter is involved. In this case, the wave function Ψ of a particle (or similarly anti-particle) will lead to a perturbation of the fermionic projector of the form

$$\Delta P(x, y) = -\Psi(x)\,\overline{\Psi(y)}\,. \tag{21}$$

Performing a multi-pole expansion around x, the zeroth moment $-\Psi(x)\overline{\Psi(x)}$ corresponds to the electromagnetic current and should be taken care of by the Maxwell equations. The first moment

$$-(y - x)^j\,\Psi(x)\,\partial_j\overline{\Psi(x)}$$

is proportional to the energy-momentum tensor of the Dirac wave function. Imposing that this contribution should be compensated by the first moment of (20), we obtain a relation of the form

$$\frac{i}{32\pi^3}\,\frac{1}{6}\,\xi^j\,G_{jk}\,z^{-1} = \xi^j\,T_{jk}[\Psi]\,. \tag{22}$$

This calculation was too naive, because the left side of the equation involves the singular distribution z^{-1}, whereas the right side is smooth. This is also the reason why the method of the continuum limit as developed in [3] cannot be applied directly to the gravitational field. On the other hand, this is not to be expected, because the formalism of the continuum limit only gives dimensionless constants, whereas the gravitational constant has the dimension of length squared. These extra length dimensions enter (22) by the factor z^{-1}. The simplest method to convert (22) into a reasonable differential equation is to argue that the concept of the space-time continuum should be valid only down to the Planck scale, where the discreteness of space-time should lead to some kind of "ultraviolet regularization." Thus it seems natural to replace the singular factor z^{-1} by the value of this factor on the Planck scale. This leads to an equation of the form

$$G_{jk} \frac{1}{l_P^2} \sim T_{jk} , \tag{23}$$

where l_P denotes the Planck length. These are precisely the Einstein equations. We point out that the above argument is not rigorous, in particular because the transition from (22) to (23) would require a methods which go beyond the formalism of the continuum limit. Nevertheless, our consideration seems to explain why the Planck length enters the Einstein equations and in particular why the coupling of matter to the gravitational field is so extremely weak. Also, we get some understanding for how the Einstein equations could be derived from our variational principle.

7. Outlook: The field quantization

We hope that in our approach, the field quantization is taken into account as soon as one goes beyond the continuum limit and takes into account the discreteness of space-time. Since the basic mechanism should be the same for the gravitational field as for any other bosonic field, we can just as well consider here for simplicity an electromagnetic field. The basic ideas are quite old and were one of the motivations for thinking of the principle of the fermionic projector [2]. Nevertheless, the details have not been worked out in the meantime, simply because it seemed more important to first get a rigorous connection to the continuum theory by analyzing the continuum limit. Thus the following considerations are still on the same speculative level as nine years ago. In order to convey the reader some of the spontaneity of the early text, we here simply give a slightly revised English translation of [2, Section 1.4]:

In preparation of the discussion of field quantization, we want to work out why quantized bosonic fields are needed, i.e. what the essence of a "quantization" of these fields is. To this aim, we shall analyze to which extent we can get a connection to quantum field theory by considering *classical* gauge fields. For simplicity, we restrict attention to one type of particles and an electromagnetic interaction, but the considerations apply just as well to a general interaction including gravitational fields. Suppose that when describing the interacting system of fermions in the continuum limit we get the system of coupled differential equations

$$(i\partial\!\!\!/ + e A\!\!\!/ - m)\, \Psi = 0, \qquad F^{ij}_{,j} = e\, \overline{\Psi} \gamma^i \Psi . \tag{24}$$

These equations are no longer valid at energies as high as the Planck energy, because the approximations used in the formalism of the continuum limit are no longer valid. Our variational principle in discrete space-time should then still describe our system, but at the moment we do not know how the corresponding interaction looks like. For simplicity, we will assume in what follows that the fermions do not interact at such high energies. In this way, we get in the classical Maxwell equations a natural cutoff for very large momenta.

When describing (24) perturbatively, one gets Feynman diagrams. To this end we can proceed just as in [1]: We expand Ψ and A in powers of e,

$$\Psi = \sum_{j=0}^{\infty} e^j \, \Psi^{(j)}, \qquad A = \sum_{j=0}^{\infty} e^j \, A^{(j)}$$

and substitute these expansions in the differential equations (24). In these equations, the contributions to every order in e must vanish, and thus one solves for the highest appearing index $^{(j)}$. In the Lorentz gauge, we thus obtain the formal relations

$$\Psi^{(j)} = - \sum_{k+l=j-1} (i\partial\!\!\!/-m)^{-1} \left(A\!\!\!/^{(k)} \, \Psi^{(l)} \right), \quad A_i^{(j)} = - \sum_{k+l=j-1} \Box^{-1} \left(\overline{\Psi}^{(k)} \gamma_i \Psi^{(l)} \right),$$

$$(25)$$

which by iterative substitutions can be brought into a more explicit form. Taking into account the pair creation, we obtain additional diagrams which contain closed fermion lines, due to the Pauli Exclusion Principle with the correct relative signs. In this way we get all Feynman diagrams.

We come to the renormalization. Since we obtain all the Feynman diagrams of quantum field theory, the only difference of our approach to standard quantum field theory is the natural cutoff for large momenta. In this way all ultraviolet divergences disappear, and the difference between naked and effective coupling constants becomes finite. One can (at least in principle) express the effective coupling constants in terms of the naked coupling constants by adding up all the contributions by the self-interaction. Computations using the renormalization group show that the effective masses and coupling constants depend on the energy. The effective constants at the Planck scale should be considered as our naked coupling constants.

The fact that the theory can be renormalized is important for us, because this ensures that the self-interaction can be described merely by a change of the masses and coupling constants. But renormalizability is not absolutely necessary for a meaningful theory. For example, the renormalizability of diagrams is irrelevant for classes of diagrams which (with our cutoff) are so small that they are negligible. Furthermore, one should be aware that the introduction of a cutoff is an approximation which has no justification. In order to understand the self-interaction at high energies one would have to analyze our variational principle without using the formalism of the continuum limit.

We explained the connection to the Feynman diagrams and the renormalization in order to point out that perturbative quantum field theory is obtained already with *classical* bosonic fields if one studies the coupled interaction between the classical field and the fermions. With second quantization of the gauge fields one can obtain the Feyman diagrams using Wick's theorem in a more concise way, but at this point it is unnecessary both from the mathematical and physical point of view to go over from classical to quantized bosonic fields. In particular, one should be aware of the fact that all the high precision tests of quantum field theory (like the Lamb shift and the anomalous g factor) are actually no test of the field quantization. One does not need to think of a photon line as an "exchange of a virtual photon"; the photon propagator can just as well be considered simply as the operator $\Box^{-1}$ in (25), which appears in the perturbation expansion of the coupled differential equations (24). Also the equation $E = \hbar\omega$, which in a graphic language tells us about the "energy of one photon," does not make a statement on the field quantization. This can be seen as follows: In physics, the energy appears in two different contexts. In classical field theory, the energy is a conserved quantity following from the time translation invariance of the Lagrangian. In quantum theory, on the other hand, the sum of the frequencies of the wave functions and potentials is conserved in any interaction, simply because in the perturbation expansion plane waves of different frequencies are orthogonal. These "classical" and "quantum mechanical" energies are related to each other via the equation $E = \hbar\omega$. Planck's constant can be determined without referring to the electromagnetic field (for example via the Compton wavelength of the electron). Since the classical and quantum mechanical energies are both conserved, it is clear that the relation $E = \hbar\omega$ must hold in general. (Thus the energy transmitted by a photon line of frequency ω really is $\hbar\omega$.)

After these considerations there remain only a few effects which are real tests of the field quantization. More precisely, these are the following observations,

1. Planck's radiation law
2. the Casimir effect
3. the wave-particle duality of the electromagnetic field, thus for example the double-slid experiment

For the derivation of Planck's radiation law, one uses that the energy of an electromagnetic radiation mode cannot take continuous values, but that its energy is quantized in steps of $\hbar\omega$. The Casimir effect measures the zero point energy of the radiation mode. In order to understand field quantization, one needs to find a convincing explanation for the above observations. However, the formalism of quantum field theory does not immediately follow from the above observations. For example, when performing canonical quantization one assumes that each radiation mode can be described by a quantum mechanical oscillator. This is a possible explanation, but it is not a compelling consequence of the discreteness of the energy levels.

We shall now explain how the above observations could be explained in the framework of the principle of the fermionic projector. In order to work out the difference between the continuum limit and the situation in discrete space-time, we will discuss several examples. It will always be sufficient to work also in discrete space-time with the classical notions. For example, by an electromagnetic wave in discrete space-time we mean a variation of the fermionic projector which in the continuum limit can be described via a perturbation of the Dirac operator by a classical electromagnetic field.

We begin with a simple model in discrete space-time, namely a completely filled Dirac sea and an electromagnetic field in the form of a radiation mode. We want to analyze the effect of a variation of the amplitude of the electromagnetic wave. In the continuum limit, we can choose the amplitude arbitrarily, because the Maxwell equations will in any case be satisfied. However, the situation is more difficult in discrete space-time. Then the variation of the amplitude corresponds to a variation of the fermionic projector. However, when performing the perturbation expansion for P in discrete space-time, we need to take into account several contributions which could be left out in the continuum limit. These additional contributions do not drop out of the Euler-Lagrange equations corresponding to our variational principle. If these equations are satisfied for a given fermionic projector P, we cannot expect that they will still hold after changing the amplitude of the electromagnetic wave. More generally, in discrete space-time there seems to be no continuous family $P(\tau)$ of solutions of the Euler-Lagrange equations. This means in particular that the amplitude of the electromagnetic wave can take only discrete values.

Alternatively, the difference between the continuum limit and the description in discrete space-time can be understood as follows: In discrete space-time, the number f of particles is an integer. If for different values of f we construct a fermionic projector of the above form, the amplitude of the corresponding electromagnetic wave will in general be different. Let us assume for simplicity that for each f (in a reasonable range) there is exactly one such projector P_f with corresponding amplitude A_f. Since f is not known, we can choose f arbitrarily. Thus the amplitude of the wave can take values in the discrete set $\{A_f\}$. In the continuum limit, however, the fermionic projector is an operator of infinite rank. Thus it is clear that now we do not get a restriction for the amplitude of the electromagnetic wave, and the amplitude can be varied continuously.

We conclude that in discrete space-time a natural "quantization" of the amplitude of the electromagnetic wave should appear. Before we can get a connection to the Planck radiation and the Casimir effect, we need to refine our consideration. Namely, it seems unrealistic to consider an electromagnetic wave which is spread over the whole of space-time. Thus we now consider a wave in a four-dimensional box (for example with fixed boundary values). Let us assume that the box has length L in the spatial directions and T in the time direction. In this case, again only discrete values for the amplitude of the wave should be admissible. But now the quantization levels should depend on the size of the box, in particular on the parameter T. Qualitatively, one can expect that for smaller T the amplitude of the wave must be larger in order to perturb the fermionic projector in a comparable way. This means that the quantization levels become finer if T becomes larger. Via the classical energy density of the electromagnetic field, the admissible amplitudes $\{A_j\}$ can be translated into field energies of the wave. Physically speaking, we create a wave at time t and annihilate it a later time $t + T$. Since, according to our above consideration, the relation $E = \hbar\omega$ should hold in any interacting system, we find that the field energy must be "quantized" in steps of $\hbar\omega$. On the other hand, we just saw that the quantization levels depend on T. In order to avoid inconsistencies, we must choose T such that the quantization steps for the field energy are just $\hbar\omega$.

In this way we obtain a condition which at first sight seems very strange: If we generate an electromagnetic wave at some time t, we must annihilate it at some later time $t + T$. Such an additional condition which has no correspondence in the continuum limit, is called a *non-local quantum condition*. We derived it under the assumption of a "quantization" of the amplitude from the equations of the continuum limit (classical field equations, description of the interaction by Feyman diagrams). Since the Euler-Lagrange equations of discrete space-time should in the continuum limit go over to the classical equations, a solution in discrete space-time should automatically satisfy the non-local quantum condition.

Of course, the just-derived condition makes no physical sense. But our system of one radiation mode is also oversimplified. Thus before drawing further conclusions, let us consider the situation in more realistic settings: In a system with several radiation modes, we cannot (in contrast to the situation with canonical quantization) treat the different modes as being independent, because the variation of the amplitude of one mode will influence the quantization levels of all the other radiation modes. This mutual influence is non-local. Thus an electromagnetic wave also changes the energy levels of waves which are in large spacelike distance. The situation becomes even more complicated if fermions are brought into the system, because then the corresponding Dirac currents will also affect the energy levels of the radiation modes. The complexity of this situation has two consequences: First, we can make practically no stament on the energy levels, we only know that the quantization steps are $\hbar\omega$. Thus we can describe the energy of the lowest level only statistically. It seems reasonabl to assume that they are evenly distributed in the interval $[0, \hbar\omega)$. Then we obtain for the possible energy levels of each radiation mode on average the values $(n + \frac{1}{2})\hbar\omega$. Secondly, the non-local quantum conditions are now so complicated that we can no longer specify them. But it seems well possible that they can be satisfied in a realistic physical situation. We have the conception that such non-local quantum conditions determine all what in the usual statistical interpretation of quantum mechanics is said to be "undetermined" or "happens by chance". We will soon come back to this point when discussing the wave-particle dualism.

After these consideration we can explain the above observations 1. and 2.: Since the energy of each radiation mode is quantized in steps of $\hbar\omega$, we obtain Planck's radiation law, whereas the average energy of $\frac{1}{2}\hbar\omega$ of the ground state energy explains the Casimir effect. We conclude that under the assumption of a "quantization" of the amplitude of the electromagnetic wave we come to the same conclusions as with canonical quantization. The reason is that with the Feynman diagrams and the equation $E = \hbar\omega$ we had all the formulas for the quantitative description at our disposal, and therefore it was sufficient to work with a very general "discreteness" of the energy levels.

We come to the wave-particle dualism. Since this is a basic effect in quantum mechanics, which appears similarly for bosons and fermions, we want to discuss this point in detail. First we want to compare our concept of bosons and fermions. Obviously, we describe bosons and fermions in a very different way: the wave functions of the fermions span the image of the projector P, whereas the bosons correspond (as described above) to the discrete excitation levels of the classical bosonic fields. In our description, the Fock space or an equivalent formalism does not appear. It might not seem satisfying that in this way the analogy in the usual description of bosons and fermions, namely the mere replacements of commutators by anti-commutators, gets lost. However, we point out that the *elementary* bosons and fermions differ not only by their statistics but also in the following important point. For the fermions (leptons, quarks) we have a conservation law (lepton number, baryon number), not so for the gauge bosons. This difference is taken into account in our formalism: Every fermion corresponds to a vector in $P(H)$. We can transform fermions into each other and can create/annihilate them in pairs. But we cannot change the total number f of particles. In particular, we cannot annihilate a single fermion. In contrast, since the gauge bosons merely correspond to discrete values of the bosonic fields. They can be generated or annihilated arbitrarily in the interaction, provided that the conservation law for energy and momentum is satisfied.

In order to clarify the connection to the Fock space, we briefly mention how we describe composite particles (for example mesons or baryons). They are all composed of the elementary fermions. Thus a particle composed of p components corresponds to a vector of $(P(H))^p$. This representation is not suitable for practical purposes. It is more convenient to use for the elemenatry fermions the Fock space formalism. Then the creation/annihlation operators for the composite particle are a product of p fermionic creation/annihilation operators. If p is even (or odd), we can generate with these creation/annihilation operators the whole bosonic (or fermionic) Fock space. In this way, we obtain for composite particles the usual formalism. However, we point out that in our description this formalism has no fundamental significance.

Due to the different treatment of elementary fermions and bosons, we need to find an explanation for the wave-particle dualism which is independent of the particular description of these particles. For a fermion, this is a vector $\Psi \in P(H)$, for a boson the gauge field. Thus in any case, the physical object is not the point-like particle, but the wave. At first sight this does not seem reasonable, because we have not at all taken into account the particle character. Our concept is that the particle character is a consequence of a "discreteness" of the interaction described by our variational principle. In order to specify what we mean by "discreteness" of the interaction, we consider the double slid experiment. We work with an electron, but the consideration applies just as well to a photon, if the wave function of the electron is replaced by the electromagnetic field. When it hits the photographic material on the screen, the electron interacts with the silver atoms, and the film is exposed. In the continuum limit we obtain the same situation as in wave mechanics: the waves originating at the two slids are superposed and generate on the screen an interference pattern. Similar to our discussion of the electromagnetic radiation mode, the continuum limit should describe the physical situation only approximately. But when considerung the variational principle in discrete space-time, the situation becomes much more complicated. Let us assume that the interaction in discrete space-time is "discrete" in the sense that the electron prefers to interact with only one atom of the screen. This assumption is already plausible in the continuum limit. Namely, if the electron interacts with a silver atom, one electron from the atom must be excited. Since this requires a certain minimal energy, the kinetic energy of the electron hitting the screen can excite only a small number of atoms. Thus the interaction between electron and the screen can take place only at individual silver atoms; the electron cannot pass its energy continuously onto the screen.

Under this assumption we get on the screen an exposed dot, and thus we get the impression of a pointlike particle. At which point of the screen the interactin takes place is determined by the detailed form of the fermionic projector P in discrete space-time. With the notion introduced above, we can also say that which silver atom is exposed is determined by non-local quantum conditions. At this point, the non-locality and non-causality of our variational principle in discrete space-time becomes important. Since the non-local quantum conditions are so complicated, we cannot predict at which point of the screen the interaction will take place. Even if we repeat the same experiment under seemingly identical conditions, the global situation will be different. As a consequence, we can only make statistical statements on the measurements. From the known continuum limit we know that the probabilities for the measurements are given by the usual quantum mechanical expectation values.

At this point we want to close the discussion. We conclude that the principle of the fermionic projector raises quite fundamental questions on the structure of space-time, the nature of field quantization and the interpretation of quantum mechanics. Besides working out the continuum limit in more detail, it will be a major goal of future work to give specific answers to these questions.

References

[1] J.D. Bjorken, S. Drell, "Relativistic Quantum Mechanics," McGraw Hill (1965)

[2] F. Finster, "Derivation of field equations from the principle of the fermionic projector," gr-qc/9606040, unpublished preprint

[3] F. Finster, "The Principle of the Fermionic Projector," *AMS/IP Studies in Advanced Mathematics Series* **35** (2006)

[4] F. Finster, "A variational principle in discrete space-time – existence of minimizers," math-ph/0503069 v2 (2005)

[5] F. Finster, "Fermion systems in discrete space-time – outer symmetries and sponta-neous symmetry breaking," math-ph/0601039 (2006)

[6] F. Finster, "Light-cone expansion of the Dirac sea to first order in the external potential," hep-th/9707128, *Michigan Math.J.* **46** (1999) 377-408

[7] Landau, L.D., Lifshitz, E.M., "Classical Field Theory," Pergamon Press (1977)

Felix Finster
NWF I – Mathematik
Universität Regensburg
D-93040 Regensburg, Germany
e-mail: `Felix.Finster@mathematik.uni-regensburg.de`

Quantum Gravity
B. Fauser, J. Tolksdorf and E. Zeidler, Eds., 283–292
© 2006 Birkhäuser Verlag Basel/Switzerland

Gravitational Waves and Energy Momentum Quanta

T. Dereli and R. W. Tucker

Abstract. We review the role of the classical stress-energy tensor in defining the concept of energy and its conservation in classical field theory. Conserved electromagnetic currents associated with spacetime Killing symmetries are discussed in an attempt to draw analogies with the concepts of photons and gravitons. By embedding Einstein's original formulation of General Relativity into a broader context we show that a dynamic covariant description of gravitational stress-energy emerges naturally from a variational principle. A tensor T^G is constructed from a contraction of the Bel tensor with a symmetric covariant second degree tensor field Φ that has a form analogous to the stress-energy tensor of the Maxwell field in an arbitrary space-time. For plane-fronted gravitational waves helicity-2 polarised (graviton) states can be identified carrying non-zero energy and momentum.

Mathematics Subject Classification (2000). Primary 83D05; Secondary 83C40; 83C35.

Keywords. Gravitational stress-energy, stress-energy pseudo-tensors, gravitational waves, gravitons, Bel-Robinson tensor.

1. Introduction

The development of quantum mechanics owed much to the recognition that experimental results involving light and matter were incomprehensible within the framework of classical physics. The observation that it was the frequency of radiation that initiated the release of electrons from certain irradiated metals rather than its intensity, was one of the results that established the concept of the quantum as the physical motivation behind the quantisation of energy. This, together with the propagation of energy by the electromagnetic field led ultimately to the identification of physical states with rays in Hilbert space and quantum algebras of linear operators as the algebra of observables, replacing the algebra of functions on classical phase space.

Once these notions had taken root the equivalence between matrix mechanics and wave mechanics for quantum systems became clear and their relation to classical mechanics was established in terms of relative scales determined by Planck's constant $\hbar$.

The discovery of the light-cone structure in spacetime and the subsequent recognition that the speed of light in vacuo c determined another scale for physical phenomena necessitated only minor modifications to accommodate quantum mechanics and quantum field theory within the framework of special relativity. With this assimilation new successful predictions, such as the existence of anti-particles and the magnetic moment of the electron, firmly established the paradigm for relativistic quantum field theory.

Today the wide application of the theory has led to unprecedented advances in technology. It also underpins the current picture of most basic interactions between the fundamental constituents of matter.

It is surprising then that the all pervasive presence of the gravitational interaction sits uneasily alongside these advances. Although there exist numerous contenders for a consistent theory of quantum gravitation a clear predictive formulation remains elusive. It is sometimes asserted that one reason for this is due to the natural scale of quantum gravitational phenomenal being set by the constants c, $\hbar$ and the Newtonian gravitational constant G.

The theory of classical gravitation (Einstein 1915) is most naturally formulated in terms of pseudo-Riemannian spacetime geometry and predicts the existence of a vacuum curvature tensor with Killing symmetries analogous to those possessed by plane electromagnetic waves. Such wavelike spacetimes are termed gravitational waves since they may induce spatial oscillations among material test particles (geodesic deviation) just as electromagnetic waves can cause electrically charged test particles to oscillate. A number of international experiments are seeking to detect such gravitational waves thought to have been produced by violent astrophysical processes.

In Einstein's classical theory variations of spacetime curvature require variations in the spacetime metric tensor. Hence spacetime itself is no longer the immutable background of special relativity and its variable properties in turn become linked with all the other fields in Nature that are coupled to gravitation via their stress-energy tensors.

In a quantum description of gravity it is natural to contemplate the quanta of gravitational wave spacetimes (gravitons) by analogy to the quanta of the electromagnetic waves (photons). Indeed many perturbation schemes exist that make such notions precise for "weak" gravitational waves on a classical background. Such weak gravitational quanta might be expected to induce processes by analogy with the photo-electric effect. However, Einstein's equations also admit as classical solutions, non-perturbative (exact) vacuum wave-like spacetimes. To date no quantum scheme for gravitation has shown how such classical spacetime geometries arise in any classical limit.

In this note we explore the role of the classical stress-energy tensor in electromagnetism and gravitation. In the presence of Killing symmetries the electromagnetic stress-energy tensor gives rise to conserved currents carrying physical attributes associated with the Killing symmetries. If the analogy between the photon and graviton is pursued then it is natural to seek some analogy with quantised gravitational "energy" that can be "exchanged" in some quantum process involving other quantised fields. However in Einstein's theory of gravitation the role of gravitational stress-energy is elusive. By embedding this theory into a wider context by introducing additional fields we explore the possibility of stress-energy being identified with purely local gravitational degrees of freedom. This model is analysed in the context of a classical gravitational wave spacetime.

2. Conserved quantities and electromagnetism

If U is a domain of spacetime with boundary

$$\partial U = \Sigma_1 + \Sigma_2 + \Pi$$

for spacelike hypersurfaces Σ_i and $\mathcal{J}$ a closed 3-form (i.e. $d\mathcal{J} = 0$) then

$$\int_{\partial U} \mathcal{J} = \int_U d\mathcal{J} = 0.$$

Thus

$$\int_{\Sigma_1} \mathcal{J} = \int_{-\Sigma_2} \mathcal{J} - \int_\Pi \mathcal{J}$$

and if U is chosen so that $\int_\Pi \mathcal{J} = 0$ the flux of $\mathcal{J}$ through Σ_1 equals the flux of $\mathcal{J}$ through Σ_2. We call such closed 3-forms conserved.

The Maxwell field system on spacetime is $dF = 0$ and $d \star F = j$ where $\star$ is the Hodge map associated with the spacetime metric tensor g and j is a source current 3-form for the 2-form F.

For any vector field V on spacetime and any Maxwell solution F define the "drive" 3-form

$$\tau_V = \frac{1}{2}\{i_V F \wedge \star F - i_V \star F \wedge F\}$$

where i_V denotes the interior (contraction) operator with respect to V. If V is a (conformal) Killing vector ($\mathcal{L}_V g = \lambda g$ for some scalar λ) then:

$$d\tau_V = -i_V F \wedge j$$

For each (conformal) Killing vector field these equations describe a "local conservation equation" ($d\tau_V = 0$) in a source free region ($j = 0$).

If V is a timelike unit vector ($g(V, V) = -1$) one may write uniquely

$$F = \tilde{E} \wedge \tilde{V} + B$$

where $i_V E = i_V B = 0$ and

$$\tau_V = -\tilde{E} \wedge \tilde{B} \wedge \tilde{V} + \frac{1}{2}\{g(E,E) + g(B,B)\}i_V(\star 1)$$

The form $\tilde{E} \wedge \tilde{B}$ was identified by Poynting (in a non-relativistic context) in a source free region as the local field energy transmitted normally across unit area per second (field energy current) and $\frac{1}{2}\{g(E,E) + g(B,B)\}$ the local field energy density.

More precisely $\int_\Sigma \tau_V$ is the field energy associated with the spacelike 3-chain Σ and $\int_{S^2} i_V \tau_V$ is the power flux across an oriented spacelike 2-chain S^2.

If X is a spacelike Killing vector generating spacelike translations along open integral curves then, with the split:

$$\tau_X = \mu_X \wedge \tilde{V} + \mathcal{G}_X,$$

the Maxwell stress 2-form μ_X may be used to identify mechanical forces produced by a flow of field momentum with density 3-form $\mathcal{G}_X$. In any local frame $\{X_a\}$ with dual coframe $\{e^b\}$ the 16 functions $T_{ab} = i_{X_b} \star \tau_{X_a}$ may be used to construct the second-rank stress-energy tensor:

$$T = T_{ab}\, e^a \otimes e^b$$

3. Conserved quantities and gravitation

The variational formulation of Einstein's theory of gravitation enables one to identify the contribution of all non-gravitational fields to the total stress-energy tensor T that enters the Einstein field equation

$$Ein = \frac{8\pi G}{c^4}T$$

where Ein is the Einstein tensor and G a universal coupling. Together with the field equations for matter one has a closed system of equations for interacting gravity and matter.

In space-times with Killing symmetries, T can be used to define global conserved quantities that generalise the Newtonian concepts of energy and momentum and their rates. In the absence of such symmetry, local densities of energy and power flux, momentum and stress follow in frames associated with time-like vector fields. Thus such local concepts are covariantly defined.

From these notions one sees that non-trivial gravitational fields that satisfy the Einstein vacuum equations have no covariantly defined mass-energy or momentum densities associated with them. This is usually regarded as being compatible with the *equivalence principle* which relies on notions of local "flatness". However if gravitation is associated with a non-zero curvature tensor such arguments for the absence of local gravitational stress-energy are not particularly persuasive.

An alternative approach is to associate *gravitational* and matter stress-energy with a class of closed 3-forms that can be derived from the Einstein tensor. With

$T = *\tau_a \otimes e^a$ in a local orthonormal coframe field $\{e^a\}$, Einstein's equation may be written

$$-d\Sigma_a = \frac{8\pi G}{c^4}\tau_a + \frac{1}{2}\left(\omega^b{}_c \wedge \omega^d{}_a \wedge *(e_d \wedge e_b \wedge e^c) - \omega^d{}_b \wedge \omega^b{}_c \wedge *(e_a \wedge e_d \wedge e^c)\right)$$

where

$$\Sigma_a \equiv \frac{1}{2}\omega^b{}_c \wedge *(e_a \wedge e_b \wedge e^c)$$

are the Sparling 2-forms with respect to this co-frame $\{e^a\}$ in which the connection 1-forms of the torsion free metric compatible (Levi-Civita) connection are written $\{\omega^a{}_b\}$.

A set of pseudo-stress-energy 3-forms $\{t_a\}$ for gravity may be chosen as

$$\frac{8\pi G}{c^4}t_a = \frac{1}{2}\left(\omega^b{}_c \wedge \omega^d{}_a \wedge *(e_d \wedge e_b \wedge e^c) - \omega^d{}_b \wedge \omega^b{}_c \wedge *(e_a \wedge e_d \wedge e^c)\right)$$

since the above gives the local conservation equation:

$$d(\tau_a + t_a) = 0.$$

A class of such "pseudo-tensors" with $t^a \sim t^a + dW^a$ for any set $\{W^a\}$ of 2-forms can be constructed. Members of this class satisfy $d(\tau_a + t_a) = 0$ but do not transform linearly under change of local frames and so cannot be associated with stress-energy *tensors*. Such a notion of gravitational stress-energy depends on the coordinates used to define the local frames needed for the construction of the components of a *pseudo-tensor*. Thus a *pseudo-tensor* that is zero in one frame may be non-zero in the neighbourhood of the same event in another.

4. The Bel-Robinson tensor

By analogy with the set of electromagnetic 3-forms

$$\tau_a = \frac{1}{2}\{i_{X_a} F \wedge \star F - i_{X_a} \star F \wedge F\}$$

it has long been suspected that the symmetric 3-index 3-forms [1]

$$\mathcal{B}_{abc} = \frac{1}{4}\{i_{X_a} R_{bd} \wedge \star R^d{}_c - i_{X_a} \star R^d{}_c \wedge R_{db} + b \leftrightarrow c\}$$

play some role in associating a covariant stress-energy tensor with the curvature of the gravitational field. We note that

$$\mathcal{B} = \mathcal{B}_{abc} \otimes e^a \otimes e^b \otimes e^c$$

is the rank 4 Bel-Robinson tensor [1], [2], [3]. It is well known that the Bel-Robinson tensor can be used to describe currents of "gravitational flux" of various kinds in different spacetime geometries (in both GR and Einstein-Cartan models). This was one of the motivations behind its genesis. The paper by Mashhoon et al [4]

---

[1] If $R^a{}_b = \frac{1}{2}R^a{}_{bcd} e^c \wedge e^d$ then $*R^a{}_b = \frac{1}{4}R^a{}_{bcd}\epsilon^{cd}{}_{pq} e^p \wedge e^q$ in terms of the alternating symbol $\epsilon^{cd}{}_{pq} = \eta^{ca}\eta^{db}\epsilon_{abpq}$

illustrates this in a linearised scheme. In the model to be discussed the structure of the Bel-Robinson tensor appears as a stress-energy tensor (just as the EM stress-energy tensor arises) from the variational equations of the theory rather that being postulated additionally.

To accommodate this notion one may seek variational equations from an action that incorporates these 3-forms into the field equations for gravitation. Standard General Relativity is based on a pseudo-Riemannian description of gravity in which the curvature of a torsion-free connection describes the gravitational field. A valid generalisation (made later by Brans and Dicke) includes an additional scalar field ϕ as a further gravitational degree of freedom. Here we contemplate instead an additional *symmetric second degree covariant tensor field* Φ that couples[2] directly to non-pseudo-Riemannian curvature.

The use of a non-pseudo-Riemannian geometry is not mandatory, particularly if one ignores fermionic matter. However the existence of spinors in Nature and the possibility of additional fermi-bose gravitational symmetries is most effectively dealt with in such a geometric framework. Even in the absence of such additional interactions the theory we describe can admit classical solutions with spacetime torsion. Whether such a theory can be reduced to a geometry without torsion in general is a dynamic question. Furthermore the variational formulation (with a dynamic connection) greatly facilitates the computations.

Consider then the gravitational action

$$\Lambda = \int_M \mathcal{L} * 1$$

where

$$\mathcal{L} = \frac{1}{\kappa} \mathcal{R} + R^a{}_{crs} R^{cbrs} \Phi_{ab} + \frac{\lambda}{2} \nabla_a \Phi_{bc} \nabla^a \Phi^{bc}$$

with $R^a{}_{bcd}$ denoting the components of the Riemann tensor, $\mathcal{R}$ the curvature scalar and κ, λ constants. Thus by analogy with the Brans-Dicke field we regard Φ as a gravitational field that along with the metric g and connection ∇ determines the geometry of spacetime in the absence of other *matter* fields. To derive the field equations by variation it is convenient to write a total action in terms of exterior forms and exterior covariant derivatives [5],[6],

$$\Lambda^T = \int_M L$$
$$= \int_M \frac{1}{\kappa} R_{ab} \wedge *(e^a \wedge e^b) + \frac{1}{2} \Phi^{ab} R_{ac} \wedge *R^c{}_b + \frac{\lambda}{2} D\Phi_{ab} \wedge *D\Phi^{ab} + \frac{1}{2} F \wedge *F$$

where the Maxwell action is included to emphasise the role of the gravitational contribution to the total stress-energy.

---

[2]Such couplings are sometimes referred to as "non-minimal".

It proves expedient to regard Λ^T as a functional of $\{e^a\}$, Φ, A and $\{\omega^a{}_b\}$ where $F = dA$ and $\omega^a{}_b$ denote the connection 1-forms. So

$$\delta\Lambda^T = \int_M \delta e^a \wedge \frac{\delta L}{\delta e^a} + \delta\omega^a{}_b \wedge \frac{\delta L}{\delta\omega^a{}_b} + \delta\Phi_{ab}\frac{\delta L}{\delta\Phi_{ab}} + \delta A \wedge \frac{\delta L}{\delta A}$$

where the partial variations have compact support on M. In the following $T^a \equiv de^a + \omega^a{}_b \wedge e^b$ denote the torsion 2-forms of ∇ in the local coframe $\{e^a\}$.

The variational field equations are:

$$-\frac{1}{2\kappa}R^{bc} \wedge *(e_a \wedge e_b \wedge e_c) = \tau_a + \lambda\tau_a^\Phi + \tau_a^{EM},$$

$$\frac{1}{2\kappa} * (e_a \wedge e_b \wedge e_c) \wedge T^c = \lambda(\Phi_{ac} * D\Phi^c{}_b - \Phi_{bc} * D\Phi^c{}_a)$$

$$+ \frac{1}{2}D(\Phi_{ac} * R^c{}_b) - \frac{1}{2}D(\Phi_{bc} * R^c{}_a),$$

$$\lambda D * D\Phi_{ab} = \frac{1}{2}R_{ac} \wedge *R^c{}_b,$$

$$d * F = 0 \tag{1}$$

where the stress-energy 3-forms are

$$\tau_a^\Phi = \frac{1}{2}(\iota_{X_a}D\Phi_{bc} * D\Phi^{bc} + D\Phi_{bc} \wedge \iota_{X_a} * D\Phi^{bc})$$

$$\tau_a^{EM} = \frac{1}{2}(\iota_{X_a}F \wedge *F - F \wedge \iota_{X_a} * F)$$

$$\tau_a = \Phi^{bc}\tau_{a,bc}$$

with

$$\tau_{a,bc} = \frac{1}{2}(\iota_{X_a}R_{bk} \wedge *R^k{}_c - R_{bk} \wedge \iota_{X_a} * R^k{}_c).$$

Since the left hand side of the first variational field equation above is proportional to the Einstein tensor Ein we identify $T^G \equiv *\tau_a \otimes e^a$ with the gravitational stress-energy tensor and can write

$$\frac{1}{2\kappa}Ein = T^{Total}$$

where

$$T^{Total} = T^G + T^\Phi + T^{EM}$$

is the total stress-energy tensor.

5. Wave solutions

Consider the field configuration described in coordinates u, v, x, y by the metric

$$g = du \otimes dv + dv \otimes du + dx \otimes dx + dy \otimes dy + 2H(u, x, y)du \otimes du$$

and

$$\Phi = 2\Phi_0 \, du \otimes du \,, \quad A = a(u, x, y) du,$$

for constant Φ_0 and a torsion-free connection. The symmetric tensor field Φ is both null and covariantly constant. It follows that the spacetime is wave-like i.e. $R^a{}_b \wedge * R^b{}_c = 0$ and for a covariantly constant Φ, i.e. $D\Phi = 0$, all but the Einstein and Maxwell equations are trivially satisfied. The non-vanishing Maxwell stress-energy 3-forms are

$$\tau_3^{EM} = \tau_0^{EM} = \left(\left(\frac{\partial a}{\partial x} \right)^2 + \left(\frac{\partial a}{\partial y} \right)^2 \right) du \wedge dx \wedge dy$$

while for the gravitational stress-energy 3-forms they are

$$\tau_3 = \tau_0 = \Phi_0 \left((\Delta H)^2 - 2\operatorname{Hess}(H) \right) du \wedge dx \wedge dy$$

with Laplacian and Hessian:

$$\Delta H = \frac{\partial^2 H}{\partial x^2} + \frac{\partial^2 H}{\partial y^2}$$

$$\operatorname{Hess}(H) = \frac{\partial^2 H}{\partial x^2} \frac{\partial^2 H}{\partial y^2} - \left(\frac{\partial^2 H}{\partial x \partial y} \right)^2$$

A class of plane wave solutions for g and F is obtained provided

$$\frac{1}{\kappa} \Delta H = \Phi_0 \left((\Delta H)^2 - 2\operatorname{Hess}(H) \right) + (\operatorname{grad}(a))^2$$
$$\Delta a = 0$$

A particular solution (with $a = 0$) describing gravitational waves without electromagnetic waves is given by

$$H = h_1(u)(x^2 - y^2) + 2h_2(u)xy + h_3(u)(x^2 + y^2) \tag{2}$$

with arbitrary real functions h_1, h_2, h_3 subject to the condition

$$h_1{}^2 + h_2{}^2 + \left(h_3 - \frac{1}{4\kappa\Phi_0} \right)^2 = \frac{1}{16\kappa^2\Phi_0{}^2}$$

Thus

$$\tau_0 = \tau_3 = 8\Phi_0(h_1{}^2 + h_2{}^2 + h_3{}^2) \, du \wedge dx \wedge dy.$$

Introducing the complex combinations $h(u) = h_1(u) + ih_2(u)$ and $z = x + iy = re^{i\theta}$ the metric tensor decomposes as

$$g = \eta + \mathcal{G}_+ + \mathcal{G}_- + \mathcal{G}_0$$

where η is the metric of Minkowski space-time and

$$\mathcal{G}_+ = \bar{h}(u)z^2 \, du \otimes du = \bar{\mathcal{G}}_-, \quad \mathcal{G}_0 = 2h_3(u)|z|^2 \, du \otimes du.$$

The $\mathcal{G}_\pm$ are null g-wave helicity eigen-tensors for linearised gravitation about $\eta + \mathcal{G}_0$:

$$\mathcal{L}_{\frac{1}{i}\frac{\partial}{\partial\theta}} \mathcal{G}_\pm = \pm 2\mathcal{G}_\pm$$

where $\mathcal{L}_X$ denotes the Lie derivative along the vector field X.

It is interesting to note that the components of the plane gravitational wave stress-energy tensor as defined above are bounded above and below. If one writes

$$\tau_0 = \mathcal{E}(u)\, du \wedge dx \wedge dy$$

then since

$$\mathcal{E}(u) = \frac{2}{\kappa} h_3(u)$$

one has

$$0 \leq \mathcal{E}(u) \leq \frac{1}{\kappa^2 \Phi_0}.$$

Thus causal observers will detect a bounded gravitational wave energy flux density.

The inclusion of a polarised electromagnetic plane wave propagating in the same direction as the gravitational wave is straightforward. The additional electromagnetic field specified by a with:

$$a(u, x, y) = \alpha(u)\, x + \beta(u)\, y$$

yields a coupled gravity and electromagnetic wave exact solution provided

$$h_1{}^2 + h_2{}^2 + \left(h_3 - \frac{1}{4\kappa\Phi_0} \right)^2 = \left(\frac{1}{16\kappa^2\Phi_0{}^2} - \frac{(\alpha^2 + \beta^2)}{4\Phi_0} \right) \quad .$$

With $2P(u) \equiv \alpha(u) - i\,\beta(u)$ one has

$$F = \frac{1}{2} Re\,(d(P(u)\, z)) \wedge du$$

with $d\,(P(u)dz) \wedge du$ an electromagnetic field of helicity 1. Again one finds that $\mathcal{E}(u)$ is bounded but in this case the lower bound is greater than zero and the upper bound less than $\frac{1}{\kappa^2\Phi_0}$.

6. Conclusions

Motivated by the structure of the classical stress-energy tensor for the electromagnetic field an extension of General Relativity has been developed in which the notion of a gravitational stress-energy tensor arises naturally. It is constructed from a contraction of the Bel-Robinson tensor with a symmetric covariant second degree tensor field and has a form analogous to the stress-energy tensor of the Maxwell field in an *arbitrary* space-time. This similarity is particularly apparent for the case of an exact plane-fronted gravitational wave solution in the presence of a covariantly constant null field. The latter serves to endow gravitational stress-energy tensor with appropriate physical dimensions.

For plane-fronted gravitational waves helicity-2 polarisation states can be identified carrying non-zero energy and momentum. Such states would be expected to be identified as "gravitons" in an quantum version based on this model. Similar states arise in GR but in the coordinates used above such states contribute zero to the classical pseudo-tensor constructed from t_a. The solution has been extended to include electromagnetic plane waves constructed from helicity 1 configurations.

We have argued that T^G can be used to define gravitational stress-energy in any causal frame and with this identification the exact wave spacetimes discussed here have tensorial gravitational stress-energy. However the properties of other $\{g, \nabla, \Phi\}$ configurations deserve further scrutiny.

Whether such considerations offer any new insights into the quantum nature of gravitation is an issue that must await some guidance from experiment.

References

[1] L. Bel, Introduction d'un tenseur du quatrième ordre, *C.R.Acad. Sci. Paris*, **247**, 1958:1297–1300

[2] I. Robinson, On the Bel-Robinson tensor, *Class. Q. Grav.*, **14**, no. 1A, 1997:A331–A333.

[3] S. Deser, *The immortal Bel-Robinson tensor*, Relativity and gravitation in general (Salamanca, 1998), 35–43, World Sci. Publishing, River Edge, NJ, 1999, gr-qc/9901007.

[4] B. Mashhoon, J. McClune, H. Quevedo, On the gravitoelectromagnetic stress-energy tensor, *Class. Quantum Grav.*, **16**, no. 4, 1999:1137–1148.

[5] T Dereli, R. W. Tucker, On the energy-momentum density of gravitational plane waves, *Class. Quantum Grav.*, **21**, no. 6, 2004:1459–1464.

[6] W. Thirring, A Course in Mathematical Physics, I and II, (2nd Ed.), Springer, (1993)

[7] N. Straumann, General Relativity, Springer, (2004)

T. Dereli
Department of Physics
Koç University
34450 Sarıyer, İstanbul
Turkey
e-mail: tdereli@ku.edu.tr

R. W. Tucker
Department of Physics
Lancaster University
Lancaster LA1 4YB
United Kingdom
e-mail: r.tucker@lancaster.ac.uk

Quantum Gravity
B. Fauser, J. Tolksdorf and E. Zeidler, Eds., 293–313
© 2006 Birkhäuser Verlag Basel/Switzerland

Asymptotic Safety in Quantum Einstein Gravity: Nonperturbative Renormalizability and Fractal Spacetime Structure

Oliver Lauscher and Martin Reuter

Abstract. The asymptotic safety scenario of Quantum Einstein Gravity, the quantum field theory of the spacetime metric, is reviewed and it is argued that the theory is likely to be nonperturbatively renormalizable. It is also shown that asymptotic safety implies that spacetime is a fractal in general, with a fractal dimension of 2 on sub-Planckian length scales.

Mathematics Subject Classification (2000). Primary 81T17; Secondary 81T16.

Keywords. Quantum Einstein gravity, renormalization group, non-Gaussian fixed points, fractal spacetime, asymptotic safety.

1. Introduction

Quantized General Relativity, based upon the Einstein-Hilbert action

$$S_{\text{EH}} = \frac{1}{16\pi G} \int d^4x \sqrt{-g} \left\{ -R + 2\Lambda \right\} , \tag{1.1}$$

is well known to be perturbatively nonrenormalizable. This has led to the widespread believe that a straightforward quantization of the metric degrees of freedom cannot lead to a mathematically consistent and predictive *fundamental* theory valid down to arbitrarily small spacetime distances. Einstein gravity was rather considered merely an *effective* theory whose range of applicability is limited to a phenomenological description of gravitational effects at distances much larger than the Planck length.

In particle physics one usually considers a theory fundamental if it is perturbatively renormalizable. The virtue of such models is that one can "hide" their infinities in only finitely many basic parameters (masses, gauge couplings, etc.) which are intrinsically undetermined within the theory and whose value must be taken from the experiment. All other couplings are then well-defined computable functions of those few parameters. In nonrenormalizable effective theories, on the other hand, the divergence structure is such that increasing orders of the loop expansion require an increasing number of new counter terms and, as a consequence, of undetermined free parameters. Typically, at high energies, all these unknown parameters enter on an equal footing so that the theory loses its predictive power.

However, there are examples of field theories which do "exist" as fundamental theories despite their perturbative nonrenormalizability [1, 2]. These models are "nonperturbatively renormalizable" along the lines of Wilson's modern formulation of renormalization theory [1]. They are constructed by performing the limit of infinite ultraviolet cutoff ("continuum limit") at a non-Gaussian renormalization group fixed point g_{*i} in the space $\{g_i\}$ of all (dimensionless, essential) couplings g_i which parametrize a general action functional. This construction has to be contrasted with the standard perturbative renormalization which, at least implicitly, is based upon the Gaussian fixed point at which all couplings vanish, $g_{*i} = 0$ [3, 4].

2. Asymptotic safety

In his "asymptotic safety" scenario Weinberg [5] has put forward the idea that, perhaps, a quantum field theory of gravity can be constructed nonperturbatively by invoking a non-Gaussian ultraviolet (UV) fixed point ($g_{*i} \neq 0$). The resulting theory would be "asymptotically safe" in the sense that at high energies unphysical singularities are likely to be absent.

The arena in which the idea is formulated is the so-called "theory space". By definition, it is the space of all action functionals $A[\cdot]$ which depend on a given set of fields and are invariant under certain symmetries. Hence the theory space $\{A[\cdot]\}$ is fixed once the field contents and the symmetries are fixed. The infinitely many generalized couplings g_i needed to parametrize a general action functional are local coordinates on theory space. In gravity one deals with functionals $A[g_{\mu\nu}, \cdots]$ which are required to depend on the metric in a diffeomorphism invariant way. (The dots represent matter fields and possibly background fields introduced for technical convenience.) Theory space carries a crucial geometric structure, namely a vector field which encodes the effect of a Kadanoff-Wilson-type block spin or "coarse graining" procedure, suitably reformulated in the continuum. The components β_i of this vector field are the beta-functions of the couplings g_i. They describe the dependence of $g_i \equiv g_i(k)$ on the coarse graining scale k:

$$k \, \partial_k g_i = \beta_i(g_1, g_2, \cdots) \tag{2.1}$$

By definition, k is taken to be a mass scale. Roughly speaking the running couplings $g_i(k)$ describe the dynamics of field averages, the averaging volume having a linear extension of the order $1/k$. The $g_i(k)$'s should be thought of as parametrizing a running action functional $\Gamma_k[g_{\mu\nu}, \cdots]$. By definition, the renormalization group (RG) trajectories, i.e. the solutions to the "exact renormalization group equation" (2.1) are the integral curves of the vector field $\vec{\beta} \equiv (\beta_i)$ defining the "RG flow".

The asymptotic safety scenario assumes that $\vec{\beta}$ has a zero at a point with coordinates g_{*i} not all of which are zero. Given such a non-Gaussian fixed point (NGFP) of the RG flow one defines its UV critical surface $\mathcal{S}_{UV}$ to consist of all points of theory space which are attracted into it in the limit $k \to \infty$. (Note that increasing k amounts to going in the direction *opposite* to the natural coarse graining flow.) The dimensionality $\dim(\mathcal{S}_{UV}) \equiv \Delta_{UV}$ is given by the number of attractive (for *increasing* cutoff k) directions in the space of couplings. The linearized flow near the fixed point is governed by the Jacobi matrix $\mathbf{B} = (B_{ij})$, $B_{ij} \equiv \partial_j \beta_i(g_*)$:

$$k \partial_k g_i(k) = \sum_j B_{ij} \left(g_j(k) - g_{*j} \right) . \tag{2.2}$$

The general solution to this equation reads

$$g_i(k) = g_{*i} + \sum_I C_I V_i^I \left(\frac{k_0}{k} \right)^{\theta_I} \tag{2.3}$$

where the V^I's are the right-eigenvectors of $\mathbf{B}$ with (complex) eigenvalues $-\theta_I$. Furthermore, k_0 is a fixed reference scale, and the C_I's are constants of integration. If $g_i(k)$ is to approach g_{*i} in the infinite cutoff limit $k \to \infty$ we must set $C_I = 0$ for all I with $\mathrm{Re}\,\theta_I < 0$. Hence the dimensionality Δ_{UV} equals the number of $\mathbf{B}$-eigenvalues with a negative real part, i.e. the number of θ_I's with a positive real part.

A specific quantum field theory is defined by a RG trajectory which exists globally, i.e. is well behaved all the way down from "$k = \infty$" in the UV to $k = 0$ in the IR. The key idea of asymptotic safety is to base the theory upon one of the trajectories running inside the hypersurface $\mathcal{S}_{UV}$ since these trajectories are manifestly well-behaved and free from fatal singularities (blowing up couplings, etc.) in the large$-k$ limit. Moreover, a theory based upon a trajectory inside $\mathcal{S}_{UV}$ can be predictive, the problem of an increasing number of counter terms and undetermined parameters which plagues effective theory does not arise.

In fact, in order to select a specific quantum theory we have to fix Δ_{UV} free parameters which are not predicted by the theory and must be taken from experiment. When we *lower* the cutoff, only Δ_{UV} parameters in the initial action are "relevant", and fixing these parameters amounts to picking a specific trajectory on $\mathcal{S}_{UV}$; the remaining "irrelevant" parameters are all attracted towards $\mathcal{S}_{UV}$ automatically. Therefore the theory has the more predictive power the smaller is the dimensionality of $\mathcal{S}_{UV}$, i.e. the fewer UV attractive eigendirections the non-Gaussian fixed point has. If $\Delta_{UV} < \infty$, the quantum field theory thus constructed is as predictive as a perturbatively renormalizable model with Δ_{UV} "renormalizable couplings", i.e. couplings relevant at the Gaussian fixed point.

It is plausible that $\mathcal{S}_{UV}$ is indeed finite dimensional. If the dimensionless g_i's arise as $g_i(k) = k^{-d_i} \bar{g}_i(k)$ by rescaling (with the cutoff k) the original couplings $\bar{g}_i$ with mass dimensions d_i, then $\beta_i = -d_i g_i + \cdots$ and $B_{ij} = -d_i \delta_{ij} + \cdots$ where the dots stand for the quantum corrections. Ignoring them, $\theta_i = d_i + \cdots$, and

Δ_{UV} equals the number of positive d_i's. Since adding derivatives or powers of fields to a monomial in the action always lowers d_i, there can be at most a finite number of positive d_i's and, therefore, of negative eigenvalues of $\mathbf{B}$. Thus, barring the presumably rather exotic possibility that the quantum corrections change the signs of infinitely many elements in $\mathbf{B}$, the dimensionality of $\mathcal{S}_{\mathrm{UV}}$ is finite [5].

We emphasize that in general the UV fixed point on which the above construction is based, if it exists, has no reason to be of the simple Einstein-Hilbert form (1.1). The initial point of the RG trajectory $\Gamma_{k\to\infty}$ is expected to contain many more invariants, both local (curvature polynomials) and nonlocal ones. For this reason the asymptotic safety scenario is *not* a quantization of General Relativity, and it cannot be compared in this respect to the loop quantum gravity approach, for instance. In a conventional field theory setting the functional $\Gamma_{k\to\infty}$ corresponds to the bare (or "classical") action S which usually can be chosen (almost) freely. It is one of the many attractive features of the asymptotic safety scenario that the bare action is fixed by the theory itself and actually can be *computed*, namely by searching for zeros of $\vec{\beta}$. In this respect it has, almost by construction, a degree of predictivity which cannot be reached by any scheme trying to quantize a given classical action.

3. RG flow of the effective average action

During the past few years, the asymptotic safety scenario in Quantum Einstein Gravity (QEG) has been mostly investigated in the framework of the effective average action [6]-[21], [4], a specific formulation of the Wilsonian RG which originally was developed for theories in flat space [22, 23, 24] and has been first applied to gravity in [6].

Quite generally, the effective average action Γ_k is a coarse grained free energy functional that describes the behavior of the theory at the mass scale k. It contains the quantum effects of all fluctuations of the dynamical variables with momenta larger than k, but not of those with momenta smaller than k. As k is decreased, an increasing number of degrees of freedom is integrated out. The method thus complies, at an intuitive level, with the coarse graining picture of the previous section. The successive averaging of the fluctuation variable is achieved by a k-dependent IR cutoff term $\Delta_k S$ which is added to the classical action in the standard Euclidean functional integral. This term gives a momentum dependent mass square $\mathcal{R}_k(p^2)$ to the field modes with momentum p. It is designed to vanish if $p^2 \gg k^2$, but suppresses the contributions of the modes with $p^2 < k^2$ to the path integral. When regarded as a function of k, Γ_k describes a curve in theory space that interpolates between the classical action $S = \Gamma_{k\to\infty}$ and the conventional effective action $\Gamma = \Gamma_{k=0}$. The change of Γ_k induced by an infinitesimal change of k is described by a functional differential equation, the exact RG equation. In a symbolic notation it reads

$$k \, \partial_k \Gamma_k = \frac{1}{2} \, \mathrm{STr} \left[\left(\Gamma_k^{(2)} + \mathcal{R}_k \right)^{-1} k \, \partial_k \mathcal{R}_k \right] . \tag{3.1}$$

For a detailed discussion of this equation we must refer to the literature [6]. Suffice it to say that, expanding $\Gamma_k[g_{\mu\nu}, \cdots]$ in terms of diffeomorphism invariant field monomials $I_i[g_{\mu\nu}, \cdots]$ with coefficients $\mathrm{g}_i(k)$, eq. (3.1) assumes the component form (2.1).

In general it is impossible to find exact solutions to eq. (3.1) and we are forced to rely upon approximations. A powerful nonperturbative approximation scheme is the truncation of theory space where the RG flow is projected onto a finite-dimensional subspace. In practice one makes an ansatz for Γ_k that comprises only a few couplings and inserts it into the RG equation. This leads to a, now finite, set of coupled differential equations of the form (2.1).

The simplest approximation one might try is the "Einstein-Hilbert trunca-tion" [6, 8] defined by the ansatz

$$\Gamma_k[g_{\mu\nu}] = (16\pi G_k)^{-1} \int d^d x \, \sqrt{g} \left\{ -R(g) + 2\bar{\lambda}_k \right\} \tag{3.2}$$

It applies to a d-dimensional Euclidean spacetime and involves only the cosmo-logical constant $\bar{\lambda}_k$ and the Newton constant G_k as running parameters. Inserting (3.2) into the RG equation (3.1) one obtains a set of two β-functions (β_λ, β_g) for the dimensionless cosmological constant $\lambda_k \equiv k^{-2} \bar{\lambda}_k$ and the dimensionless New-ton constant $g_k \equiv k^{d-2} G_k$, respectively. They describe a two-dimensional RG flow on the plane with coordinates $\mathrm{g}_1 \equiv \lambda$ and $\mathrm{g}_2 \equiv g$. At a fixed point (λ_*, g_*), both β-functions vanish simultaneously. In the Einstein-Hilbert truncation there exists both a trivial Gaussian fixed point (GFP) at $\lambda_* = g_* = 0$ and, quite remarkably, also a UV attractive NGFP at $(\lambda_*, g_*) \neq (0, 0)$.

In Fig. 1 we show part of the g-λ theory space and the corresponding RG flow for $d = 4$. The trajectories are obtained by numerically integrating the differential equations $k \, \partial_k \lambda = \beta_\lambda(\lambda, g)$ and $k \, \partial_k g = \beta_g(\lambda, g)$. The arrows point in the direction of increasing coarse graining, i.e. from the UV towards the IR. We observe that three types of trajectories emanate from the NGFP: those of Type Ia (Type IIIa) run towards negative (positive) cosmological constants, while the "separatrix", the unique trajectory (of Type IIa) crossing over from the NGFP to the GFP, has a vanishing cosmological constant in the IR. The flow is defined on the half-plane $\lambda < 1/2$ only; it cannot be continued beyond $\lambda = 1/2$ as the β-functions become singular there. In fact, the Type IIIa-trajectories cannot be integrated down to $k = 0$ within the Einstein-Hilbert approximation. They terminate at a non-zero k_{term} where they run into the $\lambda = 1/2$–singularity. Near k_{term} a more general truncation is needed in order to continue the flow.

In Weinberg's original paper [5] the asymptotic safety idea was tested in $d = 2 + \epsilon$ dimensions where $0 < \epsilon \ll 1$ was chosen so that the β-functions (actually β_g only) could be found by an ϵ-expansion. Before the advent of the

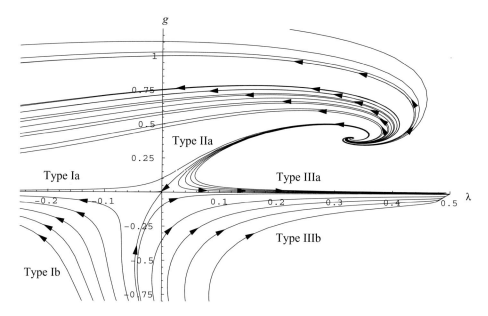

FIGURE 1. Part of theory space of the Einstein-Hilbert truncation
with its RG flow. The arrows point in the direction of decreasing
values of k. The flow is dominated by a non-Gaussian fixed point
in the first quadrant and a trivial one at the origin. (From [9].)

exact RG equations no practical tool was known which would have allowed a
nonperturbative calculation of the β-functions in the physically interesting case
of $d = 4$ spacetime dimensions. However, as we saw above, the effective average
action in the Einstein-Hilbert approximation does indeed predict the existence of
a NGFP in a nonperturbative setting. It was first analyzed in [13, 8, 9], and also
first investigations of its possible role in black hole physics [25] and cosmology
[26, 27] were performed already.

The detailed analyses of refs. [8, 9] demonstrated that the NGFP found has
all the properties necessary for asymptotic safety. In particular one has a pair of
complex conjugate critical exponents $\theta' \pm i\,\theta''$ with $\theta' > 0$, implying that the NGFP,
for $k \to \infty$, attracts all trajectories in the half-plane $g > 0$. (The lower half-plane
$g < 0$ is unphysical probably since it corresponds to a negative Newton constant.)
Because of the nonvanishing imaginary part $\theta'' \neq 0$, all trajectories spiral around
the NGFP before hitting it.

The question of crucial importance is whether the fixed point predicted by the
Einstein-Hilbert truncation actually approximates a fixed point in the exact theory,
or whether it is an artifact of the truncation. In refs. [8, 10, 9] evidence was found
which, in our opinion, strongly supports the hypothesis that there does indeed
exist a non-Gaussian fixed point in the exact 4-dimensional theory, with exactly

the properties required for the asymptotic safety scenario. In these investigations
the reliability of the Einstein-Hilbert truncation was tested both by analyzing the
cutoff scheme dependence within this truncation [8, 9] and by generalizing the
truncation ansatz itself [10]. The idea behind the first method is as follows.

The cutoff operator $\mathcal{R}_k(p^2)$ is specified by a matrix in field space and a
"shape function" $R^{(0)}(p^2/k^2)$ which describes the details of how the modes get
suppressed in the IR when p^2 drops below k^2. We checked the cutoff scheme
dependence of the various quantities of interest both by looking at their dependence
on the function $R^{(0)}$ and comparing two different matrix structures. Universal
quantities are particularly important in this respect because, by definition, they
are strictly cutoff scheme independent in the exact theory. Any truncation leads
to a residual scheme dependence of these quantities, however. Its magnitude is
a natural indicator for the quality of the truncation [28]. Typical examples of
universal quantities are the critical exponents θ_I. The existence or nonexistence
of a fixed point is also a universal, scheme independent feature, but its precise
location in parameter space is scheme dependent. Nevertheless it can be shown
that, in $d = 4$, the product $g_* \lambda_*$ must be universal [8] while g_* and λ_* separately
are not.

The detailed numerical analysis of the Einstein-Hilbert RG flow near the
NGFP [8, 9] shows that the universal quantities, in particular the product $g_* \lambda_*$,
are indeed scheme independent at a quite impressive level of accuracy. As the
many numerical "miracles" which lead to the almost perfect cancellation of the
$R^{(0)}$-dependence would have no reason to occur if there was not a fixed point in
the exact theory as an organizing principle, the results of this analysis can be
considered strong evidence in favor of a fixed point in the exact, un-truncated
theory.

The ultimate justification of any truncation is that when one adds further
terms to it its physical predictions do not change significantly any more. As a first
step towards testing the stability of the Einstein-Hilbert truncation against the
inclusion of other invariants [10] we took a (curvature)^2-term into account:

$$\Gamma_k[g_{\mu\nu}] = \int d^d x \sqrt{g} \left\{ (16\pi G_k)^{-1} \left[-R(g) + 2\bar{\lambda}_k \right] + \bar{\beta}_k R^2(g) \right\} \qquad (3.3)$$

Inserting (3.3) into the functional RG equation yields a set of β-functions $(\beta_\lambda, \beta_g,$
$\beta_\beta)$ for the dimensionless couplings λ_k, g_k and $\beta_k \equiv k^{4-d}\bar{\beta}_k$. They describe the
RG flow on the three-dimensional λ-g-β−space. Despite the extreme algebraic
complexity of the three β-functions it was possible to show [10, 11, 12] that they,
too, admit a NGFP $(\lambda_*, g_*, \beta_*)$ with exactly the properties needed for asymptotic
safety. In particular it turned out to be UV attractive in all three directions.
The value of β_* is extremely tiny, and close to the NGFP the projection of the
3-dimensional flow onto the λ-g−subspace is very well described by the Einstein-
Hilbert truncation which ignores the third direction from the outset. The λ_*-
and g_*-values and the critical exponents related to the flow in the λ-g−subspace,
as predicted by the 3-dimensional truncation, agree almost perfectly with those

from the Einstein-Hilbert approximation. Analyzing the scheme dependence of the universal quantities one finds again a highly remarkable $R^{(0)}$-independence − which is truly amazing if one visualizes the huge amount of nontrivial numerical compensations and cancellations among several dozens of $R^{(0)}$-dependent terms which is necessary to make $g_* \lambda_*$, say, approximately independent of the shape function $R^{(0)}$.

On the basis of these results we believe that the non-Gaussian fixed point occuring in the Einstein-Hilbert truncation is very unlikely to be an artifact of this truncation but rather may be considered the projection of a NGFP in the exact theory. The fixed point and all its qualitative properties are stable against variations of the cutoff and the inclusion of a further invariant in the truncation. It is particularly remarkable that within the scheme dependence the additional R^2-term has essentially no impact on the fixed point. These are certainly very nontrivial indications supporting the conjecture that 4-dimensional QEG indeed possesses a RG fixed point with the properties needed for its nonperturbative renormalizability.

This view is further supported by two conceptually independent investigations. In ref. [19] a proper time renormalization group equation rather than the flow equation of the average action has been used, and again a suitable NGFP was found. This framework is conceptually somewhat simpler than that of the effective average action; it amounts to an RG-improved 1-loop calculation with an IR cutoff. Furthermore, in refs. [29] the functional integral over the subsector of metrics admitting two Killing vectors has been performed *exactly*, and again a NGFP was found, this time in a setting and an approximation which is *very* different from that of the truncated Γ_k-flows. As for the inclusion of matter fields, both in the average action [14, 15, 16, 20] and the symmetry reduction approach [29] a suitable NGFP has been established for a broad class of matter systems.

4. Scale dependent metrics and the resolution function $\ell(k)$

In the following we take the existence of a suitable NGFP on the full theory space for granted and explore some of the properties of asymptotic safety, in particular we try to gain some understanding of what a "quantum spacetime" is like. Unless stated otherwise we consider pure Euclidean gravity in $d = 4$.

The running effective average action $\Gamma_k[g_{\mu\nu}]$ defines an infinite set of effective field theories, valid near the scale k which we may vary between $k = 0$ and $k = \infty$. Intuitively speaking, the solution $\langle g_{\mu\nu} \rangle_k$ of the scale dependent field equation

$$\frac{\delta \Gamma_k}{\delta g_{\mu\nu}(x)}[\langle g \rangle_k] = 0 \tag{4.1}$$

can be interpreted as the metric averaged over (Euclidean) spacetime volumes of a linear extension ℓ which typically is of the order of $1/k$. Knowing the scale dependence of Γ_k, i.e. the renormalization group trajectory $k \mapsto \Gamma_k$, we can derive the running effective Einstein equations (4.1) for any k and, after fixing appropriate boundary conditions and symmetry requirements, follow their solution $\langle g_{\mu\nu}\rangle_k$ from $k = \infty$ to $k = 0$.

The infinitely many equations of (4.1), one for each scale k, are valid *simultaneously*. They all refer *to the same* physical system, the "quantum spacetime". They describe its effective metric structure on different length scales. An observer using a "microscope" with a resolution $\approx k^{-1}$ will perceive the universe to be a Riemannian manifold with metric $\langle g_{\mu\nu}\rangle_k$. At every fixed k, $\langle g_{\mu\nu}\rangle_k$ is a smooth classical metric. But since the quantum spacetime is characterized by the infinity of metrics $\{\langle g_{\mu\nu}\rangle_k | k = 0, \cdots, \infty\}$ it can acquire very nonclassical and in particular fractal features. In fact, every proper distance calculated from $\langle g_{\mu\nu}\rangle_k$ is unavoidably scale dependent. This phenomenon is familiar from fractal geometry, a famous example being the coast line of England whose length depends on the size of the yardstick used to measure it [30].

Let us describe more precisely what it means to "average" over Euclidean spacetime volumes. The quantity we can freely tune is the IR cutoff scale k, and the "resolving power" of the microscope, henceforth denoted ℓ, is an a priori unknown function of k. (In flat space, $\ell \approx \pi/k$.) In order to understand the relationship between ℓ and k we must recall some more steps from the construction of $\Gamma_k[g_{\mu\nu}]$ in ref. [6].

The effective average action is obtained by introducing an IR cutoff into the path-integral over all metrics, gauge fixed by means of a background gauge fixing condition. Even without a cutoff the resulting effective action $\Gamma[g_{\mu\nu}; \bar{g}_{\mu\nu}]$ depends on two metrics, the expectation value of the quantum field, $g_{\mu\nu}$, and the background field $\bar{g}_{\mu\nu}$. This is a standard technique, and it is well known [31] that the functional $\Gamma[g_{\mu\nu}] \equiv \Gamma[g_{\mu\nu}; \bar{g}_{\mu\nu} = g_{\mu\nu}]$ obtained by equating the two metrics can be used to generate the 1PI Green's functions of the theory.

(We emphasize, however, that the average action method is manifestly *background independent* despite the temporary use of $\bar{g}_{\mu\nu}$ at an intermediate level. At no stage in the derivation of the β-functions it is necessary to assign a concrete metric to $\bar{g}_{\mu\nu}$, such as $\bar{g}_{\mu\nu} = \eta_{\mu\nu}$ in standard perturbation theory, say. The RG flow, i.e. the vector field $\vec{\beta}$, on the theory space of diffeomorphism invariant action functionals depending on $g_{\mu\nu}$ and $\bar{g}_{\mu\nu}$ is a highly universal object: it neither depends on any specific metric, nor on any specific action.)

The IR cutoff of the average action is implemented by first expressing the functional integral over all metric fluctuations in terms of eigenmodes of $\bar{D}^2$, the covariant Laplacian formed with the aid of the background metric $\bar{g}_{\mu\nu}$. Then the suppression term $\Delta_k S$ is introduced which damps the contribution of all $-\bar{D}^2$-modes with eigenvalues smaller than k^2. Coupling the dynamical fields to sources and Legendre-transforming leads to the scale dependent functional $\Gamma_k[g_{\mu\nu}; \bar{g}_{\mu\nu}]$, and the action with one argument again obtains by equating the two metrics:

$$\Gamma_k[g_{\mu\nu}] \equiv \Gamma_k[g_{\mu\nu}; \bar{g}_{\mu\nu} = g_{\mu\nu}]. \qquad (4.2)$$

It is this action which appears in the effective field equations (4.1).

A solution to those effective field equations represents a scale-dependent "one-point function" $\langle g_{\mu\nu}(x)\rangle_k$. Even though the metric operator is a highly singular object, as is any quantum field operator localized at a point, $\langle g_{\mu\nu}(x)\rangle_k$ is a smooth function of x in general, at most up to isolated singularities. In fact, within the Einstein-Hilbert truncation, the equation (4.1) has the same form as the classical Einstein equation, and all its well-known solutions, with G and $\bar{\lambda}$ replaced by the k-dependent G_k and $\bar{\lambda}_k$, respectively, provide examples of such one-point functions. The smoothness of $\langle g_{\mu\nu}(x)\rangle_k$, at every *fixed* value of $k \in (0,\infty)$, is due to the averaging over infinitely many configurations of the microscopic (i.e., quantum) metric. This average is performed with the path integral containing the cutoff term $\Delta_k S$. The occurrence of a smooth one-point function is familiar from standard field theory. A well-known text book example from quantum electrodynamics is the Uehling potential, the radiatively corrected field of an electric point charge. In quantum gravity, in a formalism without gauge fixing, one might encounter additional problems due to the fact that it is impossible to specify any particular point "x" in the quantum ensemble so that observables would always have to contain an integration over x. In the average action approach this problem does not arise since the renormalization group equation pertains to an explicitly gauge fixed path integral. As a result, for every given k, the labels "x" are in a one-to-one correspondence with the points of spacetime. It is a nontrivial issue, however, to make sure that when one compares solutions $\langle g_{\mu\nu}(x)\rangle_k$ for different values of k the coordinates x refer to the same point always. This can be done, for instance, by deriving a flow equation directly for the solution: $k\,\partial_k\langle g_{\mu\nu}(x)\rangle_k = \cdots$ [32]. A simple example of an equation of this kind (or rather its solution) is the relation (5.4) below. Once we have found a family of metrics $\langle g_{\mu\nu}(x)\rangle_k$ where "x" refers to the same point for any value of k we may perform only k-independent diffeomorphisms on this family if we want to maintain this property. A priori we could have changed the coordinates at each level k separately, but clearly this would destroy the scale-independent one-to-one correspondence between points and coordinates.

In the spirit of effective field theory, and by the very construction of the effective average action [24], $\langle g_{\mu\nu}\rangle_k$ should be thought of as the metric relevant in any single-scale physical process involving momenta of the order k, in the sense that fluctuations about the average are smallest if $\langle g_{\mu\nu}\rangle_k$ is used for this particular, physically determined value of k. The concrete identification of k depends on the physical situation or process under consideration. A typical example of a quantity which potentially can act as an IR cutoff, well-known from deep inelastic scattering, for instance, is the (4-momentum)^2, or virtuality, of a particle.

It is natural to ask how much of the spacetime structure is revealed by an experiment ("microscope") with a given characteristic scale k. Because of the identification of the two metrics in (4.2) we see that, in a sense, it is the eigenmodes of $\bar{D}^2 = D^2$, constructed from the argument of $\Gamma_k[g]$, which are cut off at k^2.

This last observation is essential for the following algorithm [23, 33] for the reconstruction of the averaging scale ℓ from the cutoff k. The input data is the set of metrics characterizing a quantum manifold, $\{\langle g_{\mu\nu} \rangle_k\}$. The idea is to deduce the relation $\ell = \ell(k)$ from the spectral properties of the scale dependent Laplacian $\Delta(k) \equiv D^2(\langle g_{\mu\nu} \rangle_k)$ built with the solution of the effective field equation. More precisely, for every fixed value of k, one solves the eigenvalue problem of $-\Delta(k)$ and studies the properties of the special eigenfunctions whose eigenvalue is k^2, or nearest to k^2 in the case of a discrete spectrum. We shall refer to an eigenmode of $-\Delta(k)$ whose eigenvalue is (approximately) the square of the cutoff k as a "cutoff mode" (COM) and denote the set of all COMs by $\mathsf{COM}(k)$.

If we ignore the k-dependence of $\Delta(k)$ for a moment (as it would be appropriate for matter theories in flat space) the COMs are, for a sharp cutoff, precisely the last modes integrated out when lowering the cutoff, since the suppression term in the path integral cuts out all modes of the metric fluctuation with eigenvalue smaller than k^2.

For a non-gauge theory in flat space the coarse graining or averaging of fields is a well defined procedure, based upon ordinary Fourier analysis, and one finds that in this case the length ℓ is essentially the wave length of the last modes integrated out, the COMs.

This observation motivates the following definition of ℓ in quantum gravity. We determine the COMs of $-\Delta(k)$, analyze how fast these eigenfunctions vary on spacetime, and read off a typical coordinate distance Δx^μ characterizing the scale on which they vary. For an oscillatory COM, for example, Δx^μ would correspond to an oscillation period. (In general there is a certain freedom in the precise identification of the Δx^μ belonging to a specific cutoff mode. This ambiguity can be resolved by refining the definition of Δx^μ on a case-by-case basis only.) Finally we use the metric $\langle g_{\mu\nu} \rangle_k$ itself in order to convert Δx^μ to a proper length. This proper length, by definition, is ℓ. Repeating the above steps for all values of k, we end up with a function $\ell = \ell(k)$. In general one will find that ℓ depends on the position on the manifold as well as on the direction of Δx^μ.

Applying the above algorithm on a non-dynamical flat spacetime one recovers the expected result $\ell(k) = \pi/k$. In ref. [33] a specific example of a QEG spacetime has been constructed, the quantum S^4, which is an ordinary 4-sphere at every fixed scale, with a k-dependent radius, though. In this case, too, the resolution function was found to be $\ell(k) = \pi/k$.

Thus the construction and interpretation of a QEG spacetime proceeds, in a nutshell, as follows. We start from a fixed RG trajectory $k \mapsto \Gamma_k$, derive its effective field equations at each k, and solve them. The resulting quantum mechanical counterpart of a classical spacetime is equipped with the infinity of Riemannian metrics $\{\langle g_{\mu\nu} \rangle_k | k = 0, \cdots, \infty\}$ where the parameter k is only a book keeping device a priori. It can be given a physical interpretation by relating it to the COM length scale characterizing the averaging procedure: One constructs the Laplacian $-D^2(\langle g_{\mu\nu} \rangle_k)$, diagonalizes it, looks how rapidly its k^2-eigenfunction varies, and "measures" the length of typical variations with the metric $\langle g_{\mu\nu}\rangle_k$ itself. In the ideal case one can solve the resulting $\ell = \ell(k)$ for $k = k(\ell)$ and reinterpret the metric $\langle g_{\mu\nu}\rangle_k$ as referring to a microscope with a known position and direction dependent resolving power ℓ. The price we have to pay for the background independence is that we cannot freely choose ℓ directly but rather k only.

5. Microscopic structure of the QEG spacetimes

One of the intriguing conclusions we reached in refs. [8, 10] was that the QEG spacetimes are fractals and that their effective dimensionality is scale dependent. It equals 4 at macroscopic distances ($\ell \gg \ell_{\text{Pl}}$) but, near $\ell \approx \ell_{\text{Pl}}$, it gets dynamically reduced to the value 2. For $\ell \ll \ell_{\text{Pl}}$ spacetime is, in a precise sense [8], a 2-dimensional fractal.

In ref. [26] the specific form of the graviton propagator on this fractal was applied in a cosmological context. It was argued that it gives rise to a Harrison-Zeldovich spectrum of primordial geometry fluctuations, perhaps responsible for the CMBR spectrum observed today. (In refs. [25, 26, 27], [34]-[39] various types of "RG improvements" were used to explore possible physical manifestations of the scale dependence of the gravitational parameters.)

A priori there exist several plausible definitions of a fractal dimensionality of spacetime. In our original argument [8] we used the one based upon the anomalous dimension η_N at the NGFP. We shall review this argument in the rest of this section. Then, in Section 6, we evaluate the spectral dimension for the QEG spacetimes [40] and demonstrate that it displays the same dimensional reduction $4 \to 2$ as the one based upon η_N. The spectral dimension has also been determined in Monte Carlo simulations of causal (i.e. Lorentzian) dynamical triangulations [41]-[44] and it will be interesting to compare the results.

For simplicity we use the Einstein-Hilbert truncation to start with, and we consider spacetimes with classical dimensionality $d = 4$. The corresponding RG trajectories are shown in Fig. 1. For $k \to \infty$, all of them approach the NGFP (λ_*, g_*) so that the dimensionful quantities run according to

$$G_k \approx g_*/k^2 \quad , \qquad\qquad \bar{\lambda}_k \approx \lambda_* \, k^2 \qquad\qquad (5.1)$$

The behavior (5.1) is realized in the asymptotic scaling regime $k \gg m_{\text{Pl}}$. Near $k = m_{\text{Pl}}$ the trajectories cross over towards the GFP. Since we are interested only in the limiting cases of very small and very large distances the following caricature of a RG trajectory will be sufficient. We assume that G_k and $\bar{\lambda}_k$ behave as in (5.1) for $k \gg m_{\text{Pl}}$, and that they assume constant values for $k \ll m_{\text{Pl}}$. The precise interpolation between the two regimes could be obtained numerically [9] but will not be needed here.

The argument of ref. [10] concerning the fractal nature of the QEG spacetimes is as follows. Within the Einstein-Hilbert truncation of theory space, the effective field equations (4.1) happen to coincide with the ordinary Einstein equation, but with G_k and $\bar{\lambda}_k$ replacing the classical constants. Without matter,

$$R_{\mu\nu}(\langle g \rangle_k) \; = \; \bar{\lambda}_k \, \langle g_{\mu\nu} \rangle_k \tag{5.2}$$

Since in absence of dimensionful constants of integration $\bar{\lambda}_k$ is the only quantity in this equation which sets a scale, every solution to (5.2) has a typical radius of curvature $r_c(k) \propto 1/\sqrt{\bar{\lambda}_k}$. (For instance, the S^4-solution has the radius $r_c = \sqrt{3/\bar{\lambda}_k}$.) If we want to explore the spacetime structure at a fixed length scale ℓ we should use the action $\Gamma_k[g_{\mu\nu}]$ at $k \approx \pi/\ell$ because with this functional a tree level analysis is sufficient to describe the essential physics at this scale, including the relevant quantum effects. Hence, when we observe the spacetime with a microscope of resolution ℓ, we will see an average radius of curvature given by $r_c(\ell) \equiv r_c(k = \pi/\ell)$. Once ℓ is smaller than the Planck length $\ell_{\mathrm{Pl}} \equiv m_{\mathrm{Pl}}^{-1}$ we are in the fixed point regime where $\bar{\lambda}_k \propto k^2$ so that $r_c(k) \propto 1/k$, or

$$r_c(\ell) \propto \ell \tag{5.3}$$

Thus, when we look at the structure of spacetime with a microscope of resolution $\ell \ll \ell_{\mathrm{Pl}}$, the average radius of curvature which we measure is proportional to the resolution itself. If we want to probe finer details and decrease ℓ we automatically decrease r_c and hence *increase* the average curvature. Spacetime seems to be more strongly curved at small distances than at larger ones. The scale-free relation (5.3) suggests that at distances below the Planck length the QEG spacetime is a special kind of fractal with a self-similar structure. It has no intrinsic scale because in the fractal regime, i.e. when the RG trajectory is still close to the NGFP, the parameters which usually set the scales of the gravitational interaction, G and $\bar{\lambda}$, are not yet "frozen out". This happens only later on, somewhere half way between the NGFP and the GFP, at a scale of the order of m_{Pl}. Below this scale, G_k and $\bar{\lambda}_k$ stop running and, as a result, $r_c(k)$ becomes independent of k so that $r_c(\ell) = \mathrm{const}$ for $\ell \gg \ell_{\mathrm{Pl}}$. In this regime $\langle g_{\mu\nu} \rangle_k$ is k-independent, indicating that the macroscopic spacetime is describable by a single smooth Riemannian manifold.

The above argument made essential use of the proportionality $\ell \propto 1/k$. In the fixed point regime it follows trivially from the fact that there exist no other relevant dimensionful parameters so that $1/k$ is the only length scale one can form. The algorithm for the determination of $\ell(k)$ described above yields the same answer.

It is easy to make the k-dependence of $\langle g_{\mu\nu} \rangle_k$ explicit. Picking an arbitrary reference scale k_0 we rewrite (5.2) as $[\bar{\lambda}_{k_0}/\bar{\lambda}_k] \, R^\mu{}_\nu(\langle g \rangle_k) = \bar{\lambda}_{k_0} \, \delta^\mu_\nu$. Since $R^\mu{}_\nu(c\,g) = c^{-1} R^\mu{}_\nu(g)$ for any constant $c > 0$, the average metric and its inverse scale as

$$\langle g_{\mu\nu}(x) \rangle_k \; = \; [\bar{\lambda}_{k_0}/\bar{\lambda}_k] \, \langle g_{\mu\nu}(x) \rangle_{k_0} \tag{5.4}$$

$$\langle g^{\mu\nu}(x) \rangle_k \; = \; [\bar{\lambda}_k/\bar{\lambda}_{k_0}] \, \langle g^{\mu\nu}(x) \rangle_{k_0} \tag{5.5}$$

These relations are valid provided the family of solutions considered exists for all scales between k_0 and k, and $\bar{\lambda}_k$ has the same sign always.

As we discussed in ref. [8] the QEG spacetime has an effective dimensionality which is k-dependent and hence noninteger in general. The discussion was based upon the anomalous dimension η_N of the operator $\int \sqrt{g}\,R$. It is defined as $\eta_N \equiv -k\,\partial_k \ln Z_{Nk}$ where $Z_{Nk} \propto 1/G_k$ is the wavefunction renormalization of the metric [6]. In a sense which we shall make more precise in a moment, the effective dimensionality of spacetime equals $4 + \eta_N$. The RG trajectories of the Einstein-Hilbert truncation (within its domain of validity) have $\eta_N \approx 0$ for $k \to 0$[1] and $\eta_N \approx -2$ for $k \to \infty$, the smooth change by two units occuring near $k \approx m_{\mathrm{Pl}}$. As a consequence, the effective dimensionality is 4 for $\ell \gg \ell_{\mathrm{Pl}}$ and 2 for $\ell \ll \ell_{\mathrm{Pl}}$.

The UV fixed point has an anomalous dimension $\eta \equiv \eta_N(\lambda_*, g_*) = -2$. We can use this information in order to determine the momentum dependence of the dressed graviton propagator for momenta $p^2 \gg m_{\mathrm{Pl}}^2$. Expanding the Γ_k of (3.2) about flat space and omitting the standard tensor structures we find the inverse propagator $\widetilde{\mathcal{G}}_k(p)^{-1} \propto Z_N(k)\,p^2$. The conventional dressed propagator $\widetilde{\mathcal{G}}(p)$, the one contained in $\Gamma \equiv \Gamma_{k=0}$, obtains from the exact $\widetilde{\mathcal{G}}_k$ by taking the limit $k \to 0$. For $p^2 > k^2 \gg m_{\mathrm{Pl}}^2$ the actual cutoff scale is the physical momentum p^2 itself[2] so that the k-evolution of $\widetilde{\mathcal{G}}_k(p)$ stops at the threshold $k = \sqrt{p^2}$. Therefore

$$\widetilde{\mathcal{G}}(p)^{-1} \propto Z_N\left(k = \sqrt{p^2}\right) p^2 \propto (p^2)^{1-\frac{\eta}{2}} \tag{5.6}$$

because $Z_N(k) \propto k^{-\eta}$ when $\eta \equiv -\partial_t \ln Z_N$ is (approximately) constant. In d dimensions, and for $\eta \neq 2 - d$, the Fourier transform of $\widetilde{\mathcal{G}}(p) \propto 1/(p^2)^{1-\eta/2}$ yields the following propagator in position space:

$$\mathcal{G}(x;y) \propto \frac{1}{|x-y|^{d-2+\eta}}\,. \tag{5.7}$$

This form of the propagator is well known from the theory of critical phenomena, for instance. (In the latter case it applies to large distances.) Eq. (5.7) is not valid directly at the NGFP. For $d = 4$ and $\eta = -2$ the dressed propagator is $\widetilde{\mathcal{G}}(p) = 1/p^4$ which has the following representation in position space:

$$\mathcal{G}(x;y) = -\frac{1}{8\pi^2} \ln\left(\mu\,|x-y|\right)\,. \tag{5.8}$$

Here μ is an arbitrary constant with the dimension of a mass. Obviously (5.8) has the same form as a $1/p^2$-propagator in 2 dimensions.

Slightly away from the NGFP, before other physical scales intervene, the propagator is of the familiar type (5.7) which shows that the quantity η_N has the standard interpretation of an anomalous dimension in the sense that fluctuation effects modify the decay properties of $\mathcal{G}$ so as to correspond to a spacetime of effective dimensionality $4 + \eta_N$.

---

[1] In the case of type IIIa trajectories [9, 38] the macroscopic k-value is still far above k_{term}, i.e. in the "GR regime" described in [38].
[2] See Section 1 of ref. [36] for a detailed discussion of "decoupling" phenomena of this kind.

Thus the properties of the RG trajectories imply the following "dimensional reduction": Spacetime, probed by a "graviton" with $p^2 \ll m_{\mathrm{Pl}}^2$ is 4-dimensional, but it appears to be 2-dimensional for a graviton with $p^2 \gg m_{\mathrm{Pl}}^2$ [8].

It is interesting to note that in d classical dimensions, where the macroscopic spacetime is d-dimensional, the anomalous dimension at the fixed point is $\eta = 2-d$. Therefore, for any d, the dimensionality of the fractal as implied by η_N is $d+\eta = 2$ [8, 10].

6. The spectral dimension

Next we turn to the spectral dimension $\mathcal{D}_{\mathrm{s}}$ of the QEG spacetimes. This particular definition of a fractal dimension is borrowed from the theory of diffusion processes on fractals [45] and is easily adapted to the quantum gravity context [46, 44]. In particular it has been used in the Monte Carlo studies mentioned above.

Let us study the diffusion of a scalar test particle on a d-dimensional classical Euclidean manifold with a fixed smooth metric $g_{\mu\nu}(x)$. The corresponding heat-kernel $K_g(x, x'; T)$ giving the probability for the particle to diffuse from x' to x during the fictitious diffusion time T satisfies the heat equation $\partial_T K_g(x, x'; T) = \Delta_g K_g(x, x'; T)$ where $\Delta_g \equiv D^2$ denotes the scalar Laplacian: $\Delta_g \phi \equiv g^{-1/2} \partial_\mu (g^{1/2} \cdot g^{\mu\nu} \partial_\nu \phi)$. The heat-kernel is a matrix element of the operator $\exp(T \Delta_g)$. In the random walk picture its trace per unit volume, $P_g(T) \equiv V^{-1} \int d^d x \sqrt{g(x)} \cdot K_g(x, x; T) \equiv V^{-1} \operatorname{Tr} \exp(T \Delta_g)$, has the interpretation of an average return probability. (Here $V \equiv \int d^d x \sqrt{g}$ denotes the total volume.) It is well known that P_g possesses an asymptotic expansion (for $T \to 0$) of the form $P_g(T) = (4\pi T)^{-d/2} \sum_{n=0}^\infty A_n T^n$. For an infinite flat space, for instance, it reads $P_g(T) = (4\pi T)^{-d/2}$ for all T. Thus, knowing the function P_g, one can recover the dimensionality of the target manifold as the T-independent logarithmic derivative $d = -2\, d \ln P_g(T)/d \ln T$. This formula can also be used for curved spaces and spaces with finite volume V provided T is not taken too large [44].

In QEG where we functionally integrate over all metrics it is natural to replace $P_g(T)$ by its expectation value. Symbolically, $P(T) \equiv \langle P_\gamma(T) \rangle$ where $\gamma_{\mu\nu}$ denotes the microscopic metric (integration variable) and the expectation value is with respect to the ordinary path integral (without IR cutoff) containing the fixed point action. Given $P(T)$, we define the spectral dimension of the quantum spacetime in analogy with the classical formula:

$$\mathcal{D}_{\mathrm{s}} = -2 \frac{d \ln P(T)}{d \ln T} \tag{6.1}$$

Let us now evaluate (6.1) using the average action method. The fictitious diffusion process takes place on a "manifold" which, at every fixed scale, is described by a smooth Riemannian metric $\langle g_{\mu\nu} \rangle_k$. While the situation appears to be classical at fixed k, nonclassical features emerge in the regime with nontrivial RG running since there the metric depends on the scale at which the spacetime structure is probed.

The nonclassical features are encoded in the properties of the diffusion operator. Denoting the covariant Laplacians corresponding to the metrics $\langle g_{\mu\nu} \rangle_k$ and $\langle g_{\mu\nu} \rangle_{k_0}$ by $\Delta(k)$ and $\Delta(k_0)$, respectively, eqs. (5.4) and (5.5) imply that they are related by

$$\Delta(k) = [\bar{\lambda}_k / \bar{\lambda}_{k_0}] \Delta(k_0) \tag{6.2}$$

When $k, k_0 \gg m_{\text{Pl}}$ we have, for example,

$$\Delta(k) = (k/k_0)^2 \Delta(k_0) \tag{6.3}$$

Recalling that the average action Γ_k defines an effective field theory at the scale k we have that $\langle \mathcal{O}(\gamma_{\mu\nu}) \rangle \approx \mathcal{O}(\langle g_{\mu\nu} \rangle_k)$ if the operator $\mathcal{O}$ involves typical covariant momenta of the order k and $\langle g_{\mu\nu} \rangle_k$ solves eq. (4.1). In the following we exploit this relationship for the RHS of the diffusion equation, $\mathcal{O} = \Delta_\gamma K_\gamma(x, x'; T)$. It is crucial here to correctly identify the relevant scale k.

If the diffusion process involves only a small interval of scales near k over which $\bar{\lambda}_k$ does not change much the corresponding heat equation must contain the $\Delta(k)$ for this specific, fixed value of k:

$$\partial_T K(x, x'; T) = \Delta(k) K(x, x'; T) \tag{6.4}$$

Denoting the eigenvalues of $-\Delta(k_0)$ by $\mathcal{E}_n$ and the corresponding eigenfunctions by ϕ_n, this equation is solved by

$$K(x, x'; T) = \sum_n \phi_n(x) \, \phi_n(x') \exp\left(- F(k^2) \, \mathcal{E}_n \, T \right) \tag{6.5}$$

Here we introduced the convenient notation $F(k^2) \equiv \bar{\lambda}_k / \bar{\lambda}_{k_0}$. Knowing this propagation kernel we can time-evolve any initial probability distribution $p(x; 0)$ according to $p(x; T) = \int d^4x' \sqrt{g_0(x')} \, K(x, x'; T) \, p(x'; 0)$ with g_0 the determinant of $\langle g_{\mu\nu} \rangle_{k_0}$. If the initial distribution has an eigenfunction expansion of the form $p(x; 0) = \sum_n C_n \phi_n(x)$ we obtain

$$p(x; T) = \sum_n C_n \, \phi_n(x) \exp\left(- F(k^2) \, \mathcal{E}_n \, T \right) \tag{6.6}$$

If the C_n's are significantly different from zero only for a single eigenvalue $\mathcal{E}_N$, we are dealing with a single-scale problem. In the usual spirit of effective field theories we would then identify $k^2 = \mathcal{E}_N$ as the relevant scale at which the running couplings are to be evaluated. However, in general the C_n's are different from zero over a wide range of eigenvalues. In this case we face a multiscale problem where different modes ϕ_n probe the spacetime on different length scales.

If $\Delta(k_0)$ corresponds to flat space, say, the eigenfunctions $\phi_n \equiv \phi_p$ are plane waves with momentum p^μ, and they resolve structures on a length scale ℓ of order $\pi/|p|$. Hence, in terms of the eigenvalue $\mathcal{E}_n \equiv \mathcal{E}_p = p^2$ the resolution is $\ell \approx \pi/\sqrt{\mathcal{E}_n}$. This suggests that when the manifold is probed by a mode with eigenvalue $\mathcal{E}_n$ it "sees" the metric $\langle g_{\mu\nu} \rangle_k$ for the scale $k = \sqrt{\mathcal{E}_n}$. Actually the identification $k = \sqrt{\mathcal{E}_n}$ is correct also for curved space since, in the construction of Γ_k, the parameter k is introduced precisely as a cutoff in the spectrum of the covariant Laplacian.

Therefore we conclude that under the spectral sum of (6.6) we must use the scale $k^2 = \mathcal{E}_n$ which depends explicitly on the resolving power of the corresponding mode. Likewise, in eq. (6.5), $F(k^2)$ is to be interpreted as $F(\mathcal{E}_n)$. Thus we obtain the traced propagation kernel

$$
\begin{aligned}
P(T) &= V^{-1} \sum_n \exp\left[- F(\mathcal{E}_n)\,\mathcal{E}_n\,T \right] \\
&= V^{-1} \operatorname{Tr} \exp\left[F\left(- \Delta(k_0) \right) \Delta(k_0)\,T \right]
\end{aligned}
\tag{6.7}
$$

It is convenient to choose k_0 as a macroscopic scale in a regime where there are no strong RG effects any more.

Furthermore, let us assume for a moment that at k_0 the cosmological constant is tiny, $\bar{\lambda}_{k_0} \approx 0$, so that $\langle g_{\mu\nu} \rangle_{k_0}$ is an approximately flat metric. In this case the trace in eq. (6.7) is easily evaluated in a plane wave basis:

$$
P(T) = \int \frac{d^4 p}{(2\pi)^4} \exp\left[-p^2\, F(p^2)\, T \right]
\tag{6.8}
$$

The T-dependence of (6.8) determines the fractal dimensionality of spacetime via (6.1). In the limits $T \to \infty$ and $T \to 0$ where the random walks probe very large and small distances, respectively, we obtain the dimensionalities corresponding to the largest and smallest length scales possible. The limits $T \to \infty$ and $T \to 0$ of $P(T)$ are determined by the behavior of $F(p^2) \equiv \bar{\lambda}(k = \sqrt{p^2})/\bar{\lambda}_{k_0}$ for $p^2 \to 0$ and $p^2 \to \infty$, respectively.

For a RG trajectory where the renormalization effects stop below some threshold we have $F(p^2 \to 0) = 1$. In this case (6.8) yields $P(T) \propto 1/T^2$, and we conclude that the macroscopic spectral dimension is $\mathcal{D}_s = 4$.

In the fixed point regime we have $\bar{\lambda}_k \propto k^2$, and therefore $F(p^2) \propto p^2$. As a result, the exponent in (6.8) is proportional to p^4 now. This implies the $T \to 0$−behavior $P(T) \propto 1/T$. It corresponds to the spectral dimension $\mathcal{D}_s = 2$.

This result holds for all RG trajectories since only the fixed point properties were used. In particular it is independent of $\bar{\lambda}_{k_0}$ on macroscopic scales. Indeed, the above assumption that $\langle g_{\mu\nu} \rangle_{k_0}$ is flat was not necessary for obtaining $\mathcal{D}_s = 2$. This follows from the fact that even for a curved metric the spectral sum (6.7) can be represented by an Euler-Mac Laurin series which always implies (6.8) as the leading term for $T \to 0$.

Thus we may conclude that on very large and very small length scales the spectral dimensions of the QEG spacetimes are

$$
\begin{aligned}
\mathcal{D}_s(T \to \infty) &= 4 \\
\mathcal{D}_s(T \to 0) &= 2
\end{aligned}
\tag{6.9}
$$

The dimensionality of the fractal at sub-Planckian distances is found to be 2 again, as in the first argument based upon η_N. Remarkably, the equality of $4 + \eta$ and $\mathcal{D}_s$ is a special feature of 4 classical dimensions. Generalizing for d classical dimensions, the fixed point running of Newton's constant becomes $G_k \propto k^{2-d}$ with a dimension-dependent exponent, while $\bar{\lambda}_k \propto k^2$ continues to have a quadratic k-dependence. As a result, the $\widetilde{\mathcal{G}}(k)$ of eq. (5.6) is proportional to $1/p^d$ in general so that, for any d, the 2-dimensional looking graviton propagator (5.8) is obtained. (This is equivalent to saying that $\eta = 2 - d$, or $d + \eta = 2$, for arbitrary d.)

On the other hand, the impact of the RG effects on the diffusion process is to replace the operator Δ by Δ^2, for any d, since the cosmological constant always runs quadratically. Hence, in the fixed point regime, eq. (6.8) becomes $P(T) \propto \int d^d p \exp\left[-p^4 T\right] \propto T^{-d/4}$. This T-dependence implies the spectral dimension

$$\mathcal{D}_s(d) \;=\; d/2 \tag{6.10}$$

This value coincides with $d + \eta$ if, and only if, $d = 4$. It is an intriguing speculation that this could have something to do with the observed macroscopic dimensionality of spacetime.

For the sake of clarity and to be as explicit as possible we described the computation of $\mathcal{D}_s$ within the Einstein-Hilbert truncation. However, it is easy to see [40] that the only nontrivial ingredient of this computation, the scaling behavior $\Delta(k) \propto k^2$, is in fact an exact consequence of asymptotic safety. If the fixed point exists, simple dimensional analysis implies $\Delta(k) \propto k^2$ at the un-truncated level, and this in turn gives rise to (6.10). If QEG is asymptotically safe, $\mathcal{D}_s = 2$ at sub-Planckian distances is an *exact* nonperturbative result for all of its spacetimes.

It is interesting to compare the result (6.9) to the spectral dimensions which were recently obtained by Monte Carlo simulations of the causal dynamical triangulation model of quantum gravity [44]:

$$\mathcal{D}_s(T \to \infty) \;=\; 4.02 \pm 0.1$$
$$\mathcal{D}_s(T \to 0) \;=\; 1.80 \pm 0.25 \tag{6.11}$$

These figures, too, suggest that the long-distance and short-distance spectral dimension should be 4 and 2, respectively. The dimensional reduction from 4 to 2 dimensions is a highly nontrivial dynamical phenomenon which seems to occur in both QEG and the discrete triangulation model. We find it quite remarkable that the discrete and the continuum approach lead to essentially identical conclusions in this respect. This could be a first hint indicating that the discrete model and QEG in the average action formulation describe the same physics.

7. Summary

In the first part of this article we reviewed the asymptotic safety scenario of quantum gravity, and the evidence supporting it coming from the average action approach. We explained why it is indeed rather likely that 4-dimensional Quantum Einstein Gravity can be defined ("renormalized") nonperturbatively along the lines of asymptotic safety. The conclusion is that it seems quite possible to construct a quantum field theory of the spacetime metric which is not only an effective, but rather a fundamental one and which is mathematically consistent and predictive on the smallest possible length scales even. If so, it is not necessary to leave the realm of quantum field theory in order to construct a satisfactory quantum gravity. This is at variance with the basic credo of string theory, for instance, which is also claimed to provide a consistent gravity theory. Here a very high price has to be paid for curing the problems of perturbative gravity, however: one has to live with infinitely many (unobserved) matter fields.

In the second part of this review we described the spacetime structure in non-perturbative, asymptotically safe gravity. The general picture of the QEG space-times which emerged is as follows. At sub-Planckian distances spacetime is a fractal of dimensionality $\mathcal{D}_s = 4 + \eta = 2$. It can be thought of as a self-similar hierarchy of superimposed Riemannian manifolds of any curvature. As one considers larger length scales where the RG running of the gravitational parameters comes to a halt, the "ripples" in the spacetime gradually disappear and the structure of a classical 4-dimensional manifold is recovered.

References

[1] K.G. Wilson, J. Kogut, Phys. Rept. 12 (1974) 75; K.G. Wilson, Rev. Mod. Phys. 47 (1975) 773.

[2] G. Parisi, Nucl. Phys. B 100 (1975) 368, Nucl. Phys. B 254 (1985) 58; K. Gawedzki, A. Kupiainen, Nucl. Phys. B 262 (1985) 33, Phys. Rev. Lett. 54 (1985) 2191, Phys. Rev. Lett. 55 (1985) 363; B. Rosenstein, B.J. Warr, S.H. Park, Phys. Rept. 205 (1991) 59; C. de Calan, P.A. Faria da Veiga, J. Magnen, R. Sénéor, Phys. Rev. Lett. 66 (1991) 3233.

[3] J. Polchinski, Nucl. Phys. B 231 (1984) 269.

[4] For a review see: C. Bagnuls and C. Bervillier, Phys. Rept. 348 (2001) 91;
T.R. Morris, Prog. Theor. Phys. Suppl. 131 (1998) 395;
J. Polonyi, Central Eur. J. Phys. 1 (2004) 1.

[5] S. Weinberg in General Relativity, an Einstein Centenary Survey, S.W. Hawking and W. Israel (Eds.), Cambridge University Press (1979); S. Weinberg, hep-th/9702027.

[6] M. Reuter, Phys. Rev. D 57 (1998) 971 and hep-th/9605030.

[7] D. Dou and R. Percacci, Class. Quant. Grav. 15 (1998) 3449.

[8] O. Lauscher and M. Reuter, Phys. Rev. D 65 (2002) 025013 and hep-th/0108040.

[9] M. Reuter and F. Saueressig, Phys. Rev. D 65 (2002) 065016 and hep-th/0110054.

[10] O. Lauscher and M. Reuter, Phys. Rev. D 66 (2002) 025026 and hep-th/0205062.

[11] O. Lauscher and M. Reuter, Class. Quant. Grav. 19 (2002) 483 and hep-th/0110021.

[12] O. Lauscher and M. Reuter, Int. J. Mod. Phys. A 17 (2002) 993 and hep-th/0112089.

[13] W. Souma, Prog. Theor. Phys. 102 (1999) 181.

[14] R. Percacci and D. Perini, Phys. Rev. D 67 (2003) 081503.

[15] R. Percacci and D. Perini, Phys. Rev. D 68 (2003) 044018.

[16] D. Perini, Nucl. Phys. Proc. Suppl. 127 C (2004) 185.

[17] M. Reuter and F. Saueressig, Phys. Rev. D 66 (2002) 125001 and hep-th/0206145; Fortschr. Phys. 52 (2004) 650 and hep-th/0311056.

[18] D. Litim, Phys. Rev. Lett. 92 (2004) 201301.

[19] A. Bonanno, M. Reuter, JHEP 02 (2005) 035 and hep-th/0410191.

[20] R. Percacci and D. Perini, hep-th/0401071.

[21] R. Percacci, hep-th/0409199.

[22] C. Wetterich, Phys. Lett. B 301 (1993) 90.

[23] M. Reuter and C. Wetterich, Nucl. Phys. B 417 (1994) 181, Nucl. Phys. B 427 (1994) 291, Nucl. Phys. B 391 (1993) 147, Nucl. Phys. B 408 (1993) 91; M. Reuter, Phys. Rev. D 53 (1996) 4430, Mod. Phys. Lett. A 12 (1997) 2777.

[24] For a review see: J. Berges, N. Tetradis and C. Wetterich, Phys. Rept. 363 (2002) 223; C. Wetterich, Int. J. Mod. Phys. A 16 (2001) 1951.

[25] A. Bonanno and M. Reuter, Phys. Rev. D 62 (2000) 043008 and hep-th/0002196; Phys. Rev. D 60 (1999) 084011 and gr-qc/9811026.

[26] A. Bonanno and M. Reuter, Phys. Rev. D 65 (2002) 043508 and hep-th/0106133; M. Reuter and F. Saueressig, JCAP 09 (2005) 012 and hep-th/0507167.

[27] A. Bonanno and M. Reuter, Phys. Lett. B 527 (2002) 9 and astro-ph/0106468; Int. J. Mod. Phys. D 13 (2004) 107 and astro-ph/0210472.

[28] J.-I. Sumi, W. Souma, K.-I. Aoki, H. Terao, K. Morikawa, hep-th/0002231.

[29] P. Forgács and M. Niedermaier, hep-th/0207028; M. Niedermaier, JHEP 12 (2002) 066; Nucl. Phys. B 673 (2003) 131.

[30] B. Mandelbrot, The Fractal Geometry of Nature, Freeman, New York (1977).

[31] L.F. Abbott, Nucl. Phys. B 185 (1981) 189; B.S. DeWitt, Phys. Rev. 162 (1967) 1195; M.T. Grisaru, P. van Nieuwenhuizen and C.C. Wu, Phys. Rev. D 12 (1975) 3203; D.M. Capper, J.J. Dulwich and M. Ramon Medrano, Nucl. Phys. B 254 (1985) 737; S.L. Adler, Rev. Mod. Phys. 54 (1982) 729.

[32] M. Reuter and J.-M. Schwindt, work in progress.

[33] M. Reuter and J.-M. Schwindt, JHEP 01 (2006) 070 and hep-th/0511021.

[34] E. Bentivegna, A. Bonanno and M. Reuter, JCAP 01 (2004) 001, and astro-ph/0303150.

[35] A. Bonanno, G. Esposito and C. Rubano, Gen. Rel. Grav. 35 (2003) 1899; Class. Quant. Grav. 21 (2004) 5005.

[36] M. Reuter and H. Weyer, Phys. Rev. D 69 (2004) 104022 and hep-th/0311196.

[37] M. Reuter and H. Weyer, Phys. Rev. D 70 (2004) 124028 and hep-th/0410117.

[38] M. Reuter and H. Weyer, JCAP 12 (2004) 001 and hep-th/0410119.

[39] J. Moffat, JCAP 05 (2005) 003 and astro-ph/0412195; J.R. Brownstein and J. Moffat, astro-ph/0506370 and astro-ph/0507222.

[40] O. Lauscher and M. Reuter, JHEP 10 (2005) 050 and hep-th/0508202.

[41] A. Dasgupta and R. Loll, Nucl. Phys. B 606 (2001) 357; J. Ambjørn, J. Jurkiewicz and R. Loll, Nucl. Phys. B 610 (2001) 347; Phys. Rev. Lett. 85 (2000) 924; R. Loll, Nucl. Phys. Proc. Suppl. 94 (2001) 96; J. Ambjørn, gr-qc/0201028.

[42] J. Ambjørn, J. Jurkiewicz and R. Loll, Phys. Rev. Lett. 93 (2004) 131301.

[43] J. Ambjørn, J. Jurkiewicz and R. Loll, Phys. Lett. B 607 (2005) 205.

[44] J. Ambjørn, J. Jurkiewicz and R. Loll, preprints hep-th/0505113 and hep-th/0505154.

[45] D. ben-Avraham and S. Havlin, *Diffusion and reactions in fractals and disordered systems*, Cambridge University Press, Cambridge (2004).

[46] H. Kawai, M. Ninomiya, Nucl. Phys. B 336 (1990) 115; R. Floreanini and R. Percacci, Nucl. Phys. B 436 (1995) 141; I. Antoniadis, P.O. Mazur and E. Mottola, Phys. Lett. B 444 (1998) 284.

Oliver Lauscher
Institute of Theoretical Physics
University of Leipzig
Augustusplatz 10-11
D 04109 Leipzig
Germany
e-mail: lauscher@itp.uni-leipzig.de

Martin Reuter
Institute of Physics
University of Mainz
Staudingerweg 7
D 55099 Mainz
Germany
e-mail: reuter@thep.physik.uni-mainz.de

Quantum Gravity
B. Fauser, J. Tolksdorf and E. Zeidler, Eds., 315–326
© 2006 Birkhäuser Verlag Basel/Switzerland

Noncommutative QFT and Renormalization

Harald Grosse and Raimar Wulkenhaar

Abstract. Since the two pillars of modern physics: Quantum Field Theory and general relativity are incompatible, one tries to take fluctuating geometries into account through deforming space-time. The resulting noncommutative Quantum Field Theory shows the IR/UV mixing. We modify the action for a scalar model in 4 dimensions and show, that a renormalizable field theory results. For the proof we fist transform to a matrix model and use the Wilson-Polchinski approach to renormalization. An efficient power-counting theorem allows to eliminate all higher genus contributions. By taking finite differences we reduce the infinite number of possible two point and four point functions to only two relevant/marginal operators, thus completing the proof. At a special point of the parameter space the model becomes self-dual, the beta function vanishes and the model connects to integrable systems.

Mathematics Subject Classification (2000). Primary 81T15; Secondary 81T75.

Keywords. Noncommutative quantum field theory, renormalization, infrared/ultraviolet mixing, θ-deformation, matrix models.

1. Introduction

Four-dimensional quantum field theory suffers from infrared and ultraviolet divergences as well as from the divergence of the renormalized perturbation expansion. Despite the impressive agreement between theory and experiments and despite many attempts, these problems are not settled and remain a big challenge for theoretical physics. Furthermore, attempts to formulate a quantum theory of gravity have not yet been fully successful. It is astonishing that the two pillars of modern physics, quantum field theory and general relativity, seem incompatible. This convinced physicists to look for more general descriptions: After the formulation of supersymmetry (initiated by Bruno Zumino and Julius Wess) and supergravity, string theory was developed, and anomaly cancellation forced the introduction of six additional dimensions. On the other hand, loop gravity was formulated, and led to spin networks and space-time foams. Both approaches are not fully satisfactory. A third impulse came from noncommutative geometry developed by Alain Connes, providing a natural interpretation of the Higgs effect at the classical level. This finally led to noncommutative quantum field theory, which is the subject of this contribution. It allows to incorporate fluctuations of space into quantum field theory. There are of course relations among these three developments. In particular, the field theory limit of string theory leads to certain noncommutative field theory models, and some models defined over fuzzy spaces are related to spin networks.

The argument that space-time should be modified at very short distances goes back to Schrödinger and Heisenberg. Noncommutative coordinates appeared already in the work of Peierls for the magnetic field problem, and are obtained after projecting onto a particular Landau level. Pauli communicated this to Oppenheimer, whose student Snyder [27] wrote down the first deformed space-time algebra preserving Lorentz symmetry. After the development of noncommutative geometry by Connes [8], it was first applied in physics to the integer quantum Hall effect. Gauge models on the two-dimensional noncommutative tori were formulated, and the relevant projective modules over this space were classified.

Through interactions with John Madore the first author realized that such Fuzzy geometries allow to obtain natural cutoffs for quantum field theory [13]. This line of work was further developed together with Peter Prešnajder and Ctirad Klimčík [12]. At almost the same time, Filk [11] developed his Feynman rules for the canonically deformed four-dimensional field theory, and Doplicher, Fredenhagen and Roberts [9] published their work on deformed spaces. The subject experienced a major boost after one realized that string theory leads to noncommutative field theory under certain conditions [25, 26], and the subject developed very rapidly; see e.g. [10, 19, 30].

2. Noncommutative Quantum Field Theory

The formulation of Noncommutative Quantum Field Theory (NCFT) follows a dictionary worked out by mathematicians. Starting from some manifold $\mathcal{M}$ one obtains the commutative algebra of smooth functions over $\mathcal{M}$, which is then quantized along with additional structure. Space itself then looks locally like a phase space in quantum mechanics. Fields are elements of the algebra resp. a finitely generated projective module, and integration is replaced by a suitable trace operation.

Following these lines, one obtains field theory on quantized (or deformed) spaces, and Feynman rules for a perturbative expansion can be worked out. However some unexpected features such as IR/UV mixing arise upon quantization, which are described below. In 2000 Minwalla, van Raamsdonk and Seiberg realized [21] that perturbation theory for field theories defined on the Moyal plane faces a serious problem. The planar contributions show the standard singularities which can be handled by a renormalization procedure. The nonplanar one loop contributions are finite for generic momenta, however they become singular at exceptional momenta. The usual UV divergences are then reflected in new singularities in the infrared, which is called IR/UV mixing. This spoils the usual renormalization procedure: Inserting many such loops to a higher order diagram generates singularities of any inverse power. Without imposing a special structure such as supersymmetry, the renormalizability seems lost; see also [6, 7].

However, progress was made recently, when we were able to give a solution of this problem for the special case of a scalar four-dimensional theory defined on the Moyal-deformed space $\mathbb{R}^4_\theta$ [16]. The IR/UV mixing contributions were taken into account through a modification of the free Lagrangian by adding an oscillator term with parameter Ω, which modifies the spectrum of the free Hamiltonian. The harmonic oscillator term was obtained as a result of the renormalization proof. The model fulfills then the Langmann-Szabo duality [18] relating short distance and long distance behavior. Our proof followed ideas of Polchinski. There are indications that a constructive procedure might be possible and give a nontrivial ϕ^4 model, which is currently under investigation [23]. At $\Omega = 1$ the model becomes self-dual, and we are presently studying this model in greater details.

Nonperturbative aspects of NCFT have also been studied in recent years. The most significant and surprising result is that the IR/UV mixing can lead to a new phase denoted as "striped phase" [17], where translational symmetry is spontaneously broken. The existence of such a phase has indeed been confirmed in numerical studies [4, 20]. To understand better the properties of this phase and the phase transitions, further work and better analytical techniques are required, combining results from perturbative renormalization with nonperturbative techniques. Here a particular feature of scalar NCFT is very suggestive: the field can be described as a hermitian matrix, and the quantization is defined nonperturbatively by integrating over all such matrices. This provides a natural starting point for nonperturbative studies. In particular, it suggests and allows to apply ideas and techniques from random matrix theory.

Remarkably, gauge theories on quantized spaces can also be formulated in a similar way [1, 2, 5, 28]. The action can be written as multi-matrix models, where the gauge fields are encoded in terms of matrices which can be interpreted as "covariant coordinates". The field strength can be written as commutator, which induces the usual kinetic terms in the commutative limit. Again, this allows a natural nonperturbative quantization in terms of matrix integrals.

Numerical studies for gauge theories have also been published including the 4-dimensional case [3], which again show a very intriguing picture of nontrivial phases and spontaneous symmetry breaking. These studies also strongly suggest the nonperturbative stability and renormalizability of NC gauge theory, adding to the need of further theoretical work.

3. Renormalization of ϕ^4-theory on the $4D$ Moyal plane

We briefly sketch the methods used by ourselves [16] in the proof of renormalizability for scalar field theory defined on the 4-dimensional quantum plane $\mathbb{R}^4_\theta$, with commutation relations $[x_\mu, x_\nu] = i\theta_{\mu\nu}$. The IR/UV mixing was taken into account through a modification of the free Lagrangian, by adding an oscillator term which modifies the spectrum of the free Hamiltonian:

$$ S = \int d^4x \left(\frac{1}{2} \partial_\mu \phi \star \partial^\mu \phi + \frac{\Omega^2}{2} (\tilde{x}_\mu \phi) \star (\tilde{x}^\mu \phi) + \frac{\mu^2}{2} \phi \star \phi + \frac{\lambda}{4!} \phi \star \phi \star \phi \star \phi \right)(x) \, . \quad (1) $$

Here, $\tilde{x}_\mu = 2(\theta^{-1})_{\mu\nu} x^\nu$ and $\star$ is the Moyal star product

$$ (a \star b)(x) := \int d^4y \, \frac{d^4k}{(2\pi)^4} a(x + \tfrac{1}{2}\theta \cdot k) b(x+y) \, e^{iky} \, , \qquad \theta_{\mu\nu} = -\theta_{\nu\mu} \in \mathbb{R} \, . \quad (2) $$

The harmonic oscillator term in (1) was found as a result of the renormalization proof. The model is covariant under the Langmann-Szabo duality relating short distance and long distance behavior. At $\Omega = 1$ the model becomes self-dual, and connected to integrable models. This leads to the hope that a constructive procedure around this particular case allows the construction of a nontrivial interacting ϕ^4 model, which would be an extremely interesting and remarkable achievement.

The renormalization proof proceeds by using a matrix base, which leads to a dynamical matrix model of the type:

$$ S[\phi] = (2\pi\theta)^2 \sum_{m,n,k,l \in \mathbb{N}^2} \left(\frac{1}{2} \phi_{mn} \Delta_{mn;kl} \phi_{kl} + \frac{\lambda}{4!} \phi_{mn} \phi_{nk} \phi_{kl} \phi_{lm} \right) , \quad (3) $$

where

$$ \Delta_{\substack{m^1 \ n^1 \\ m^2 \ n^2}; \substack{k^1 \ l^1 \\ k^2 \ l^2}} = \left(\mu^2 + \frac{2+2\Omega^2}{\theta}(m^1 + n^1 + m^2 + n^2 + 2) \right) \delta_{n^1 k^1} \delta_{m^1 l^1} \delta_{n^2 k^2} \delta_{m^2 l^2} $$
$$ - \frac{2-2\Omega^2}{\theta} \left(\sqrt{k^1 l^1} \, \delta_{n^1+1,k^1} \delta_{m^1+1,l^1} + \sqrt{m^1 n^1} \, \delta_{n^1-1,k^1} \delta_{m^1-1,l^1} \right) \delta_{n^2 k^2} \delta_{m^2 l^2} $$
$$ - \frac{2-2\Omega^2}{\theta} \left(\sqrt{k^2 l^2} \, \delta_{n^2+1,k^2} \delta_{m^2+1,l^2} + \sqrt{m^2 n^2} \, \delta_{n^2-1,k^2} \delta_{m^2-1,l^2} \right) \delta_{n^1 k^1} \delta_{m^1 l^1} \, . $$
$$ (4) $$

The interaction part becomes a trace of product of matrices, and no oscillations occur in this basis. The propagator obtained from the free part is quite complicated, in 4 dimensions it is:

$$G_{\substack{m^1 \ n^1 \ ; \ k^1 \ l^1 \\ m^2 \ n^2 \ ; \ k^2 \ l^2}}$$

$$= \frac{\theta}{2(1+\Omega)^2} \sum_{v^1 = \frac{|m^1 - l^1|}{2}}^{\frac{m^1+l^1}{2}} \sum_{v^2 = \frac{|m^2 - l^2|}{2}}^{\frac{m^2+l^2}{2}} B\left(1 + \frac{\mu^2\theta}{8\Omega} + \frac{1}{2}(m^1 + k^1 + m^2 + k^2) - v^1 - v^2, 1 + 2v^1 + 2v^2\right)$$

$$\times \ _2F_1\left(\begin{array}{c} 1 + 2v^1 + 2v^2, \ \frac{\mu^2\theta}{8\Omega} - \frac{1}{2}(m^1 + k^1 + m^2 + k^2) + v^1 + v^2 \\ 2 + \frac{\mu^2\theta}{8\Omega} + \frac{1}{2}(m^1 + k^1 + m^2 + k^2) + v^1 + v^2 \end{array}\middle| \frac{(1-\Omega)^2}{(1+\Omega)^2}\right)\left(\frac{1-\Omega}{1+\Omega}\right)^{2v^1 + 2v^2}$$

$$\times \prod_{i=1}^{2} \delta_{m^i + k^i, n^i + l^i} \sqrt{\binom{n^i}{v^i + \frac{n^i - k^i}{2}}\binom{k^i}{v^i + \frac{k^i - n^i}{2}}\binom{m^i}{v^i + \frac{m^i - l^i}{2}}\binom{l^i}{v^i + \frac{l^i - m^i}{2}}}. \tag{5}$$

These propagators (in 2 and 4 dimensions) show asymmetric decay properties:

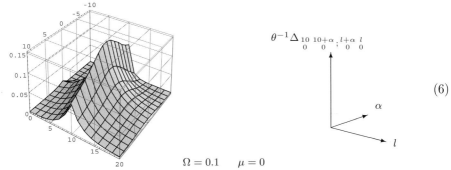

$$\theta^{-1}\Delta_{\substack{10 \ 10+\alpha \ ; \ l+\alpha \ l \\ 0 \ 0 \ \ ; \ 0 \ 0}} \tag{6}$$

$$\Omega = 0.1 \qquad \mu = 0$$

They decay exponentially on particular directions (in l-direction in the picture), but have power law decay in others (in α-direction in the picture). These decay properties are crucial for the perturbative renormalizability of the models.

Our proof in [15,16] then followed the ideas of Polchinski [22]. The quantum field theory corresponding to the action (3) is defined — as usual — by the partition function

$$Z[J] = \int \left(\prod_{m,n} d\phi_{mn}\right) \exp\left(-S[\phi] - \sum_{m,n} \phi_{mn} J_{nm}\right). \tag{7}$$

The strategy due to Wilson consists in integrating in the first step only those field modes ϕ_{mn} which have a matrix index bigger than some scale $\theta\Lambda^2$. The result is an effective action for the remaining field modes which depends on Λ. One can now adopt a smooth transition between integrated and not integrated field modes so that the Λ-dependence of the effective action is given by a certain differential equation, the Polchinski equation.

Now, renormalization amounts to prove that the Polchinski equation admits a regular solution for the effective action which depends on only a finite number of initial data. This requirement is hard to satisfy because the space of effective actions is infinite dimensional and as such develops an infinite dimensional space of singularities when starting from generic initial data.

The Polchinski equation can be iteratively solved in perturbation theory where it can be graphically written as

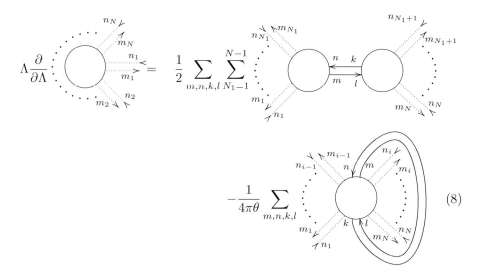

$$(8)$$

The graphs are graded by the number of vertices and the number of external legs. Then, to the Λ-variation of a graph on the lhs there only contribute graphs with a smaller number of vertices and a bigger number of legs. A general graph is thus obtained by iteratively adding a propagator to smaller building blocks, starting with the initial ϕ^4-vertex, and integrating over Λ. Here, these propagators are differentiated cut-off propagators $Q_{mn;kl}(\Lambda)$ which vanish (for an appropriate choice of the cut-off function) unless the maximal index is in the interval $[\theta\Lambda^2, 2\theta\Lambda^2]$. As the field carry two matrix indices and the propagator four of them, the graphs are ribbon graphs familiar from matrix models.

We have then proven that the cut-off propagator $Q(\Lambda)$ is bounded by $\frac{C}{\theta\Lambda^2}$. This was achieved numerically in [16] and later confirmed analytically in [23]. A nonvanishing frequency parameter Ω is required for such a decay behavior. As the volume of each two-component index $m \in \mathbb{N}^2$ is bounded by $C'\theta^2\Lambda^4$ in graphs of the above type, the power counting degree of divergence is (at first sight) $\omega = 4S - 2I$, where I is the number of propagators and S the number of summation indices.

It is now important to take into account that if three indices of a propagator $Q_{mn;kl}(\Lambda)$ are given, the fourth one is determined by $m+k = n+l$, see (5). Then, for simple planar graphs one finds that $\omega = 4 - N$ where N is the number of external legs. But this conclusion is too early, there is a difficulty in presence of completely inner vertices, which require additional index summations. The graph

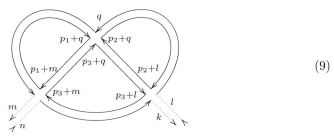

$$(9)$$

entails four independent summation indices p_1, p_2, p_3 and q, whereas for the powercounting degree $2 = 4 - N = 4S - 5 \cdot 2$ we should only have $S = 3$ of them. It turns out that due to the quasi-locality of the propagator (the exponential decay in l-direction in (6)), the sum over q for fixed m can be estimated without the need of the volume factor.

Remarkably, the quasi-locality of the propagator not only ensures the correct powercounting degree for planar graphs, it also renders all nonplanar graphs superficially convergent. For instance, in the nonplanar graphs

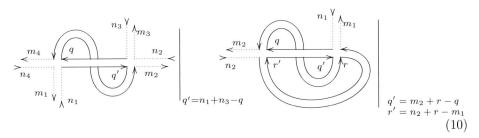

$$(10)$$

the summation over q and q, r, respectively, is of the same type as over q in (9) so that the graphs in (10) can be estimated without any volume factor.

After all, we have obtained the powercounting degree of divergence

$$\omega = 4 - N - 4(2g + B - 1) \qquad (11)$$

for a general ribbon graph, where g is the genus of the Riemann surface on which the graph is drawn and B the number of holes in the Riemann surface. Both are directly determined by the graph. It should be stressed, however, that although the number ω (11) follows from counting the required volume factors, its proof in our scheme is not so obvious: The procedure consists in adding a new cut-off propagator to a given graph, and in doing so the topology (B, g) has many possibilities to arise from the topologies of the smaller parts for which we have estimates by induction. The proof that in every situation of adding a new propagator one obtains (11) goes alone over 20 pages in [15]. Moreover, the boundary conditions for the integration have to be correctly chosen to confirm (11), see below.

The powercounting behavior (11) is good news because it implies that (in contrast to the situation without the oscillator potential) all nonplanar graphs are superficially convergent. However, this does not mean that all problems are solved: The remaining planar two- and four-leg graphs which are divergent carry matrix indices, and (11) suggests that these are divergent independent of the matrix indices. An infinite number of adjusted initial data would be necessary in order to remove these divergences.

Fortunately, a more careful analysis shows that the powercounting behavior is improved by the index jump along the trajectories of the graph. For example, the index jump for the graph (9) is defined as $J = \|k-n\|_1 + \|q-l\|_1 + \|m-q\|_1$. Then, the amplitude is suppressed by a factor of order $\left(\dfrac{\max(m, n \ldots)}{\theta \Lambda^2} \right)^{\frac{J}{2}}$ compared with the naive estimation. Thus, only planar four-leg graphs with $J = 0$ and planar two-leg graphs with $J = 0$ or $J = 2$ are divergent (the total jumps are even). For these cases, we have invented a discrete Taylor expansion about the graphs with vanishing indices. Only the leading terms of the expansion, i.e. the reference graphs with vanishing indices, are divergent whereas the difference between original graph and reference graph is convergent. Accordingly, in our scheme only the reference graphs must be integrated in a way that involves initial conditions. For example, if the contribution to the rhs of the Polchinski equation (8) is given by the graph

$$\Lambda \frac{\partial}{\partial \Lambda} A^{(2)\text{planar,1PI}}_{mn;nk;kl;lm}[\Lambda] = \sum_{p \in \mathbb{N}^2} \left(\vcenter{\hbox{}} \right)(\Lambda), \qquad (12)$$

the Λ-integration is performed as follows:

$$A^{(2)\text{planar,1PI}}_{mn;nk;kl;lm}[\Lambda]$$

$$= -\int_\Lambda^\infty \frac{d\Lambda'}{\Lambda'} \sum_{p \in \mathbb{N}^2} \left[\vcenter{\hbox{}} - \vcenter{\hbox{}} \right][\Lambda']$$

$$+ \vcenter{\hbox{}} \left[\left(\int_{\Lambda_R}^\Lambda \frac{d\Lambda'}{\Lambda'} \sum_{p \in \mathbb{N}^2} \vcenter{\hbox{}} \right)[\Lambda'] + A^{(2,1,0)\text{1PI}}_{00;00;00;00}[\Lambda_R] \right].$$

$$(13)$$

Only one initial condition, $A^{(2,1,0)\text{1PI}}_{00;00;00;00}[\Lambda_R]$, is required for an infinite number of planar four-leg graphs (distinguished by the matrix indices). We need one further initial condition for the two-leg graphs with $J = 2$ and two more initial condition for the two-leg graphs with $J = 0$ (for the leading quadratic and the subleading logarithmic divergence). This is one condition more than in a commutative ϕ^4-theory, and this additional condition justifies a posteriori our starting point of adding one new term to the action (1), the oscillator term Ω.

This being established, it was straightforward to derive beta functions for the coupling constant flow. To one-loop order we have found [14]

$$\beta_\lambda = \frac{\lambda_{\mathrm{phys}}^2}{48\pi^2}\frac{(1-\Omega_{\mathrm{phys}}^2)}{(1+\Omega_{\mathrm{phys}}^2)^3}\ , \qquad \beta_\Omega = \frac{\lambda_{\mathrm{phys}}\Omega_{\mathrm{phys}}}{96\pi^2}\frac{(1-\Omega_{\mathrm{phys}}^2)}{(1+\Omega_{\mathrm{phys}}^2)^3}\ , \quad (14)$$

$$\beta_\mu = -\frac{\lambda_{\mathrm{phys}}\left(4\mathcal{N}\ln(2) + \frac{(8+\theta\mu_{\mathrm{phys}}^2)\Omega_{\mathrm{phys}}^2}{(1+\Omega_{\mathrm{phys}}^2)^2}\right)}{48\pi^2\theta\mu_{\mathrm{phys}}^2(1+\Omega_{\mathrm{phys}}^2)}\ , \qquad \gamma = \frac{\lambda_{\mathrm{phys}}}{96\pi^2}\frac{\Omega_{\mathrm{phys}}^2}{(1+\Omega_{\mathrm{phys}}^2)^3}\ . \quad (15)$$

Together with the differential equation for the β-functions,

$$\lim_{\mathcal{N}\to\infty}\left(\mathcal{N}\frac{\partial}{\partial\mathcal{N}} + N\gamma + \mu^2\beta_\mu\frac{\partial}{\partial\mu_0^2} + \beta_\lambda\frac{\partial}{\partial\lambda} + \beta_\Omega\frac{\partial}{\partial\Omega}\right)\Gamma_{m_1n_1;\ldots;m_Nn_N}[\mu,\lambda,\Omega,\mathcal{N}] = 0\ ,$$
$$(16)$$

(14) shows that the ratio of the coupling constants $\frac{\lambda}{\Omega^2}$ remains bounded along the renormalization group flow up to first order. Starting from given small values for Ω_R, λ_R at $\mathcal{N}_R$, the frequency grows in a small region around $\ln\frac{\mathcal{N}}{\mathcal{N}_R} = \frac{48\pi^2}{\lambda_R}$ to $\Omega \approx 1$. The coupling constant approaches $\lambda_\infty = \frac{\lambda_R}{\Omega_R^2}$, which can be made small for sufficiently small λ_R. This leaves the chance of a nonperturbative construction [24] of the model.

In particular, the β-function vanishes at the self-dual point $\Omega = 1$, indicating special properties of the model.

4. Matrix-model techniques

Our recent interests turned towards dynamical matrix models, which are closely connected to integrable models. We briefly explain this method. Consider e.g. the scalar field theory defined by (3). Since ϕ is a hermitian matrix, it can be diagonalized as $\phi = U^{-1}diag(\phi_i)U$ where ϕ_i are the real eigenvalues. Hence the field theory can be reformulated in terms of the eigenvalues ϕ_i and the unitary matrix U. The main idea is now the following: consider the probability measure for the (suitably rescaled) eigenvalues ϕ_i induced by the path integral by integrating out U:

$$Z = \int \mathcal{D}\phi\ \exp(-S(\phi))) = \int d\phi_i\Delta^2(\phi_i)\int dU\ \exp(-S(U^{-1}(\phi_i)U))$$

$$= \int d\phi_i\ \exp(-\tilde{\mathcal{F}}(\phi) - (2\pi\theta)^{d/2}\sum_i V(\phi_i) + \sum_{i\neq j}\log|\phi_i - \phi_j|), \quad (17)$$

where the analytic function

$$e^{-\tilde{\mathcal{F}}(\phi)} := \int dU\ \exp(-S_{kin}(U^{-1}(\phi)U)) \quad (18)$$

is introduced, which depends only on the eigenvalues of ϕ. The crucial point is that the logarithmic terms in the effective action above implies a repulsion of the eigenvalues ϕ_i, which therefore arrange themselves according to some distribution similar as in the standard matrix models of the form $\tilde{S} = \int d\phi \exp(\text{Tr}\tilde{V}(\phi))$. This is related to the fact that nonplanar diagrams are suppressed. The presence of the unknown function $\tilde{\mathcal{F}}(\phi)$ in (17) cannot alter this conclusion qualitatively, since it is analytic. The function $\tilde{\mathcal{F}}(\phi)$ can be determined approximately by considering the weak coupling regime. For example, the effective action of the eigenvalue sector for the ϕ^4 model in the noncommutative regime $\frac{1}{\theta} \ll \Lambda^2$ becomes essentially

$$\tilde{S}(\phi) = f_0(m) + \frac{2N}{\alpha_0^2(m)}\text{Tr}\phi^2 + g\phi^4, \tag{19}$$

where $\alpha_0^2(m)$ depends on the degree of divergence of a basic diagram [29].

This effective action (19) can now be studied using standard results from random matrix theory. For example, this allows to study the renormalization of the effective potential using matrix model techniques. The basic mechanism is the following: In the free case, the eigenvalue sector follows Wigner's semicircle law, where the size of the eigenvalue distribution depends on m via $\alpha_0(m)$. Turning on the coupling g alters that eigenvalue distribution. The effective or renormalized mass can be found by matching that distribution with the "closest" free distribution. To have a finite renormalized mass then requires a negative mass counterterm as usual.

This approach is particularly suitable to study the thermodynamical properties of the field theory. For the ϕ^4 model, the above effective action (19) implies a phase transition at strong coupling, to a phase which was identified with the striped or matrix phase in [29]. Based on the known universality properties of matrix models, these results on phase transitions are expected to be realistic, and should not depend on the details of the unknown function $\tilde{\mathcal{F}}(\phi)$. The method is applicable to 4 dimensions, where a critical line is found which terminates at a nontrivial point, with finite critical coupling. This can be seen as evidence for a new nontrivial fixed-point in the 4-dimensional NC ϕ^4 model. This is in accordance with results from the RG analysis of [14], which also point to the existence of nontrivial ϕ^4 model in 4 dimensions.

References

[1] Jan Ambjorn, Y. M. Makeenko, J. Nishimura, and R. J. Szabo. Nonperturbative dynamics of noncommutative gauge theory. *Phys. Lett.*, B480:399–408, 2000.

[2] Wolfgang Behr, Frank Meyer, and Harold Steinacker. Gauge theory on fuzzy $S^2 \times S^2$ and regularization on noncommutative $\mathbb{R}^4$. 2005.

[3] W. Bietenholz et al. Numerical results for u(1) gauge theory on 2d and 4d noncommutative spaces. *Fortsch. Phys.*, 53:418–425, 2005.

[4] Wolfgang Bietenholz, Frank Hofheinz, and Jun Nishimura. Phase diagram and dispersion relation of the non- commutative lambda phi**4 model in d = 3. *JHEP*, 06:042, 2004.

[5] Ursula Carow-Watamura and Satoshi Watamura. Noncommutative geometry and gauge theory on fuzzy sphere. *Commun. Math. Phys.*, 212:395–413, 2000.

[6] Iouri Chepelev and Radu Roiban. Renormalization of quantum field theories on noncommutative r**d. i: Scalars. *JHEP*, 05:037, 2000.

[7] Iouri Chepelev and Radu Roiban. Convergence theorem for non-commutative feynman graphs and renormalization. *JHEP*, 03:001, 2001.

[8] Alain Connes. Noncommutative differential geometry. *Inst. Hautes Etudes Sci. Publ. Math.*, 62:257, 1986.

[9] Sergio Doplicher, Klaus Fredenhagen, and John E. Roberts. The quantum structure of space-time at the planck scale and quantum fields. *Commun. Math. Phys.*, 172:187–220, 1995.

[10] Michael R. Douglas and Nikita A. Nekrasov. Noncommutative field theory. *Rev. Mod. Phys.*, 73:977–1029, 2001.

[11] T. Filk. Divergencies in a field theory on quantum space. *Phys. Lett.*, B376:53–58, 1996.

[12] H. Grosse, C. Klimcik, and P. Presnajder. Towards finite quantum field theory in noncommutative geometry. *Int. J. Theor. Phys.*, 35:231–244, 1996.

[13] H. Grosse and J. Madore. A noncommutative version of the schwinger model. *Phys. Lett.*, B283:218–222, 1992.

[14] Harald Grosse and Raimar Wulkenhaar. The beta-function in duality-covariant non-commutative ϕ^4 theory. *Eur. Phys. J.*, C35:277–282, 2004.

[15] Harald Grosse and Raimar Wulkenhaar. Power-counting theorem for non-local matrix models and renormalisation. *Commun. Math. Phys.*, 254:91–127, 2005.

[16] Harald Grosse and Raimar Wulkenhaar. Renormalisation of ϕ^4 theory on noncommutative $\mathbb{R}^4$ in the matrix base. *Commun. Math. Phys.*, 256:305–374, 2005.

[17] Steven S. Gubser and Shivaji L. Sondhi. Phase structure of non-commutative scalar field theories. *Nucl. Phys.*, B605:395–424, 2001.

[18] Edwin Langmann and Richard J. Szabo. Duality in scalar field theory on noncommutative phase spaces. *Phys. Lett.*, B533:168–177, 2002.

[19] John Madore, Stefan Schraml, Peter Schupp, and Julius Wess. Gauge theory on noncommutative spaces. *Eur. Phys. J.*, C16:161–167, 2000.

[20] Xavier Martin. A matrix phase for the ϕ^4 scalar field on the fuzzy sphere. *JHEP*, 04:077, 2004.

[21] Shiraz Minwalla, Mark Van Raamsdonk, and Nathan Seiberg. Noncommutative perturbative dynamics. *JHEP*, 02:020, 2000.

[22] Joseph Polchinski. Renormalization and effective lagrangians. *Nucl. Phys.*, B231:269–295, 1984.

[23] Vincent Rivasseau, Fabien Vignes-Tourneret, and Raimar Wulkenhaar. Renormalization of noncommutative phi**4-theory by multi-scale analysis. 2005.

[24] Rivasseau, Vincent. From perturbative to constructive renormalization, Princeton series in physics, Princeton Univ. Press, Princeton, 1991. http://www.slac.stanford.edu/spires/find/hep/www?key=2573300

[25] Volker Schomerus. D-branes and deformation quantization. *JHEP*, 06:030, 1999.

[26] Nathan Seiberg and Edward Witten. String theory and noncommutative geometry. *JHEP*, 09:032, 1999.

[27] Hartland S. Snyder. Quantized space-time. *Phys. Rev.*, 71:38–41, 1947.

[28] Harold Steinacker. Quantized gauge theory on the fuzzy sphere as random matrix model. *Nucl. Phys.*, B679:66–98, 2004.

[29] Harold Steinacker. A non-perturbative approach to non-commutative scalar field theory. *JHEP*, 03:075, 2005.

[30] Richard J. Szabo. Quantum field theory on noncommutative spaces. *Phys. Rept.*, 378:207–299, 2003.

Harald Grosse
Universität Wien, Institut für Theoretische Physik
Boltzmanngasse 5
A-1090 Wien, Austria
e-mail: harald.grosse@univie.ac.at

Raimar Wulkenhaar
Universität Münster, Mathematisches Institut
Einsteinstraße 62
D-48149 Münster
Germany
e-mail: raimar@math.uni-muenster.de

Index